MONOGRAPHS AND RESEARCH NOTES IN MATHEMATICS

Stochastic Cauchy Problems in Infinite Dimensions

Generalized and Regularized Solutions

Irina V. Melnikova

CRC Press is an imprint of the
Taylor & Francis Group, an **informa** business
A CHAPMAN & HALL BOOK

MONOGRAPHS AND RESEARCH NOTES IN MATHEMATICS

Series Editors

John A. Burns
Thomas J. Tucker
Miklos Bona
Michael Ruzhansky

Published Titles

Application of Fuzzy Logic to Social Choice Theory, John N. Mordeson, Davender S. Malik and Terry D. Clark

Blow-up Patterns for Higher-Order: Nonlinear Parabolic, Hyperbolic Dispersion and Schrödinger Equations, Victor A. Galaktionov, Enzo L. Mitidieri, and Stanislav Pohozaev

Complex Analysis: Conformal Inequalities and the Bieberbach Conjecture, Prem K. Kythe

Computational Aspects of Polynomial Identities: Volume l, Kemer's Theorems, 2nd Edition Alexei Kanel-Belov, Yakov Karasik, and Louis Halle Rowen

Cremona Groups and Icosahedron, Ivan Cheltsov and Constantin Shramov

Diagram Genus, Generators, and Applications, Alexander Stoimenow

Difference Equations: Theory, Applications and Advanced Topics, Third Edition, Ronald E. Mickens

Dictionary of Inequalities, Second Edition, Peter Bullen

Iterative Optimization in Inverse Problems, Charles L. Byrne

Line Integral Methods for Conservative Problems, Luigi Brugnano and Felice Iavernaro

Lineability: The Search for Linearity in Mathematics, Richard M. Aron, Luis Bernal González, Daniel M. Pellegrino, and Juan B. Seoane Sepúlveda

Modeling and Inverse Problems in the Presence of Uncertainty, H. T. Banks, Shuhua Hu, and W. Clayton Thompson

Monomial Algebras, Second Edition, Rafael H. Villarreal

Nonlinear Functional Analysis in Banach Spaces and Banach Algebras: Fixed Point Theory Under Weak Topology for Nonlinear Operators and Block Operator Matrices with Applications, Aref Jeribi and Bilel Krichen

Partial Differential Equations with Variable Exponents: Variational Methods and Qualitative Analysis, Vicenţiu D. Rădulescu and Dušan D. Repovš

A Practical Guide to Geometric Regulation for Distributed Parameter Systems, Eugenio Aulisa and David Gilliam

Reconstruction from Integral Data, Victor Palamodov

Signal Processing: A Mathematical Approach, Second Edition, Charles L. Byrne

Sinusoids: Theory and Technological Applications, Prem K. Kythe

Special Integrals of Gradshteyn and Ryzhik: the Proofs – Volume l, Victor H. Moll

Published Titles Continued

Special Integrals of Gradshteyn and Ryzhik: the Proofs – Volume II, Victor H. Moll

Stochastic Cauchy Problems in Infinite Dimensions: Generalized and Regularized Solutions, Irina V. Melnikova

Submanifolds and Holonomy, Second Edition, Jürgen Berndt, Sergio Console, and Carlos Enrique Olmos

Forthcoming Titles

Actions and Invariants of Algebraic Groups, Second Edition, Walter Ferrer Santos and Alvaro Rittatore

Analytical Methods for Kolmogorov Equations, Second Edition, Luca Lorenzi

Geometric Modeling and Mesh Generation from Scanned Images, Yongjie Zhang

Groups, Designs, and Linear Algebra, Donald L. Kreher

Handbook of the Tutte Polynomial, Joanna Anthony Ellis-Monaghan and Iain Moffat

Microlocal Analysis on Rˆn and on NonCompact Manifolds, Sandro Coriasco

Practical Guide to Geometric Regulation for Distributed Parameter Systems, Eugenio Aulisa and David S. Gilliam

Symmetry and Quantum Mechanics, Scott Corry

CRC Press
Taylor & Francis Group
6000 Broken Sound Parkway NW, Suite 300
Boca Raton, FL 33487-2742

© 2016 by Taylor & Francis Group, LLC
CRC Press is an imprint of Taylor & Francis Group, an Informa business

No claim to original U.S. Government works

Printed on acid-free paper
Version Date: 20151207

International Standard Book Number-13: 978-1-4822-1050-7 (Hardback)

This book contains information obtained from authentic and highly regarded sources. Reasonable efforts have been made to publish reliable data and information, but the author and publisher cannot assume responsibility for the validity of all materials or the consequences of their use. The authors and publishers have attempted to trace the copyright holders of all material reproduced in this publication and apologize to copyright holders if permission to publish in this form has not been obtained. If any copyright material has not been acknowledged please write and let us know so we may rectify in any future reprint.

Except as permitted under U.S. Copyright Law, no part of this book may be reprinted, reproduced, transmitted, or utilized in any form by any electronic, mechanical, or other means, now known or hereafter invented, including photocopying, microfilming, and recording, or in any information storage or retrieval system, without written permission from the publishers.

For permission to photocopy or use material electronically from this work, please access www.copyright.com (http://www.copyright.com/) or contact the Copyright Clearance Center, Inc. (CCC), 222 Rosewood Drive, Danvers, MA 01923, 978-750-8400. CCC is a not-for-profit organization that provides licenses and registration for a variety of users. For organizations that have been granted a photocopy license by the CCC, a separate system of payment has been arranged.

Trademark Notice: Product or corporate names may be trademarks or registered trademarks, and are used only for identification and explanation without intent to infringe.

Visit the Taylor & Francis Web site at
http://www.taylorandfrancis.com

and the CRC Press Web site at
http://www.crcpress.com

To my family
Boris, Alexandra, and Nikolai

Contents

Preface	**ix**
Introduction	**xi**
Symbol Description	**xvii**

I Well-Posed and Ill-Posed Abstract Cauchy Problems: The Concept of Regularization   1

1 Semi-group methods for construction of exact, approximated, and regularized solutions   3

     1.1   The Cauchy problem and strongly continuous semi-groups of solution operators   5

     1.2   The Cauchy problem with generators of regularized semi-groups: integrated, convoluted, and R-semi-groups   14

     1.3 R-semi-groups and regularizing operators in the construction of approximated solutions to ill-posed problems   36

2 Distribution methods for construction of generalized solutions to ill-posed Cauchy problems   43

     2.1   Solutions in spaces of abstract distributions   44

     2.2   Solutions in spaces of abstract ultra-distributions   54

     2.3   Solutions to the Cauchy problem for differential systems in Gelfand–Shilov spaces   59

3 Examples. Supplements   75

     3.1   Examples of regularized semi-groups and their generators   . .   75

     3.2   Examples of solutions to Petrovsky correct, conditionally correct, and incorrect systems   84

     3.3   Definitions and properties of spaces of test functions   93

     3.4   Generalized Fourier and Laplace transforms. Structure theorems   104

viii   *Contents*

II Infinite-Dimensional Stochastic Cauchy Problems

111

4 Weak, regularized, and mild solutions to Itô integrated stochastic Cauchy problems in Hilbert spaces

113

    4.1   Hilbert space-valued variables, processes, and stochastic integrals. Main properties and results 114

    4.2   Solutions to Cauchy problems for equations with additive noise and generators of regularized semi-groups 138

    4.3   Solutions to Cauchy problems for semi-linear equations with multiplicative noise 166

    4.4   Extension of the Feynman–Kac theorem to the case of relations between stochastic equations and PDEs in Hilbert spaces . . 179

5 Infinite-dimensional stochastic Cauchy problems with white noise processes in spaces of distributions

197

    5.1   Generalized solutions to linear stochastic Cauchy problems with generators of regularized semi-groups 198

    5.2   Quasi-linear stochastic Cauchy problem in abstract Colombeau spaces 218

6 Infinite-dimensional extension of white noise calculus with application to stochastic problems

229

    6.1   Spaces of Hilbert space-valued generalized random variables: $(\mathcal{S})_{-\rho}(\mathbb{H})$. Basic examples 230

    6.2   Analysis of $(\mathcal{S})_{-\rho}(\mathbb{H})$-valued processes 241

    6.3   S-transform and Wick product. Hitsuda–Skorohod integral. Main properties. Connection with Itô integral 247

    6.4   Generalized solutions to stochastic Cauchy problems in spaces of abstract stochastic distributions 253

Bibliography   **275**

Index   **283**

Preface

In recent decades there has been growing realization that elements of chance play an essential role in many processes around us, including processes in physics, biology, and finance. Mathematical models that give an accurate description of these processes lead to stochastic equations in finite- and infinite-dimensional spaces. So far most of the literature on stochastic equations has been focused on the finite-dimensional case.

This book is devoted to stochastic differential equations for random processes with values in Hilbert spaces. The main object is the stochastic Cauchy problem

$$X'(t) = AX(t) + F(t, X) + B(t, X)\mathbb{W}(t), \quad t \in [0, T], \quad X(0) = \zeta, \quad (P.1)$$

where A is the generator of a semi-group of operators in a Hilbert space H, $\mathbb{W}$ is a white noise with values in another Hilbert space $\mathbb{H}$, B is an operator from $\mathbb{H}$ to H, and F is a non-linear term.

Due to the well-known irregularity of white noise, the Cauchy problem (P.1) is usually replaced with the related integral equation by constructing the stochastic integral with respect to a Wiener process W, the "primitive" of $\mathbb{W}$. Problems of this type with a "good" operator A that generates a C_0-semi-group have been extensively studied in the literature.

In this book, we consider a much wider class of operators A, namely, the operators that do not necessarily generate C_0-semi-groups, but generate regularized semi-groups. Typical examples include generators of integrated, convoluted, and R-semi-groups. Moreover, along with the "classical" approach to stochastic problems, which consists of solving the corresponding integral equations, we consider the Cauchy problem in its initial form (P.1) with white noise processes in spaces of distributions and obtain generalized solutions.

The motivation for writing the book was two-fold. First, to give an account of modern semi-group and distribution methods in their interrelations with the methods of infinite-dimensional stochastic analysis, accessible to nonspecialists. Second, to show how the idea of regularization, which we treat as the regularization in a broad sense, runs through all these methods. We hope that this idea will be useful for numerical realization and applications of the theory.

The stated objectives are implemented in two parts of the book. In Part I we give a self-contained introduction to modern semi-group and abstract distribution methods for solving the homogeneous (deterministic) Cauchy problem. We discuss basic properties of regularized semi-groups and illustrate them

ix

Preface

with numerous examples, paying special attention to differential systems in Gelfand–Shilov spaces. In Part II the semi-group and distribution methods are used for solving stochastic problems along with the methods of infinite-dimensional stochastic analysis. This part also includes novel material that extends the white noise analysis to Hilbert spaces and allows us to obtain new types of solutions to stochastic problems.

I began my career in mathematics as a graduate student of Prof. Valentin K. Ivanov, one of the founders of the theory of ill-posed problems. I am grateful to him for his help and encouragement over the years.

I would like to thank Dr. Alexei Filinkov for long-term cooperation. During my visits to the University of Adelaide we wrote our first joint book for CRC Press, *The Abstract Cauchy Problem: Three Approaches*, and planned to write another one on stochastic problems. Unfortunately, Alexei had to withdraw from this project.

It is my pleasure to thank my colleagues and friends Profs. Edward J. Allen, Jean Francois Colombeau, Angelo Favini, Andrzej Kaminski, Michael Oberguggenberger, Stevan Pilipović, and Dora Seleši for many useful discussions of problems related to the topic of the book.

Last but not least, I am grateful to the members of my group at Ural Federal University: Drs. Uliana Alekseeva, Maxim Alshanskiy, Valentina Parfenenkova, and post-graduates Vadim Bovkun and Olga Starkova. I owe much to them in preparing this book. However, the final responsibility for the content of this book lies solely with me.

Ekaterinburg
Ural Federal University   Irina V. Melnikova

Introduction

Models of various dynamic processes with random perturbations lead to stochastic equations in infinite-dimensional spaces (see, e.g., [1, 2, 3, 13, 32]). Most important for applications is the first-order Cauchy problem (P.1), where A is the generator of a semi-group in a Hilbert space H and $\mathbb{W}$ is a white noise process with values in another Hilbert space $\mathbb{H}$.

In the finite-dimensional case the white noise process was initially defined as a process with identically distributed and independent (at different times) random values that have zero expectations and infinite variations. Defined in such a way, the white noise does not satisfy the classical conditions which are imposed on the inhomogeneity in the theory of differential-operator equations [56]. Therefore stochastic differential equations are replaced by the related integral equations with stochastic integrals wrt (with respect to) a Brownian motion.

In the infinite-dimensional case, one has to consider stochastic integrals wrt a Wiener process $W = \{W(t), t \geq 0\}$, which is a "primitive" of $\mathbb{W}$. The integral form of problem (P.1) is as follows:

$$X(t) = \zeta + \int_0^t AX(s)\,ds + \int_0^t F(s, X)\,ds + \int_0^t B(s, X)\,dW(s), \quad t \in [0, T],$$

commonly written

$$dX(t) = AX(t)\,dt + F(t, X)\,dt + B(t, X)\,dW(t), \quad t \in [0, T], \quad X(0) = \zeta. \quad \text{(I.1)}$$

Problem (I.1) has been studied in the case when A is the generator of a C_0-semi-group, i.e., when the corresponding homogeneous Cauchy problem is well-posed (see, e.g., [20, 22, 34, 58, 81]).

In this book we consider (I.1) with generators of a much wider class of regularized semi-groups, important in applications. Moreover, along with (I.1), we consider the Cauchy problem in its initial form (P.1) with the white noise defined in spaces of distributions and obtain generalized solutions.

The book consists of two parts, presenting necessary deterministic techniques and results on stochastic equations, respectively.

In Part I we study the abstract homogeneous Cauchy problem

$$u'(t) = Au(t), \quad t \in [0, \tau), \quad \tau \leq \infty, \quad u(0) = f, \quad \text{(I.2)}$$

in connection with properties of the semi-group generated by A in a Banach

xi

Introduction

space $\mathcal{X}$. Special attention is paid to the Cauchy problem for the systems of differential equations

$$\frac{\partial u(x;t)}{\partial t} = A\left(i\frac{\partial}{\partial x}\right)u(x;t), \quad t \in [0,T], \quad x \in \mathbb{R}^n, \quad u(x;0) = f(x). \quad (\text{I.3})$$

The study of the well-posedness of Cauchy problems revealed that their solution operators $U(t)$, $t \geq 0$, have the semi-group property

$$U(t)U(\tau) = U(t+\tau), \quad t,\tau \geq 0.$$

A large amount of work in semi-group theory was devoted to connections between the uniform well-posedness of (I.2), the behavior of the resolvent of A, and the existence of a C_0-semi-group generated by A (see, e.g., [39, 43, 47, 56, 96]). Based on these results, we formulated two necessary and sufficient conditions for the uniform well-posedness of (I.2) with a closed densely defined A:

- A is the generator of a C_0-semi-group $\{U(t),\, t \geq 0\}$;
- the resolvent of A is defined in a half-plane $Re\lambda > \omega$ and satisfies the conditions

$$\|\mathcal{R}^k(\lambda)\| \leq \frac{C}{(Re\lambda - \omega)^k}, \quad Re\lambda > \omega, \quad k \in \mathbb{N}_0. \quad (\text{I.4})$$

This result is referred to as the MFPHY (Miyadera–Feller–Phillips–Hille–Yosida) theorem.

Nowadays various applications lead to problems, both deterministic and stochastic, with operators A that do not satisfy the conditions. Three approaches to such problems (I.2) are presented in [79].

The first one is based on semi-group techniques. Modern semi-group methods are aimed at constructing families of bounded linear operators such as integrated, convoluted, and R-semi-groups, which are more general than C_0-semi-groups. These families $\{S(t),\, t \in [0,\tau)\}$ may be global, i.e., $\tau = \infty$, or local, i.e., $\tau < \infty$. They allow us to obtain solutions of (I.2) for initial data f from certain subclasses of $dom\, A \subset \mathcal{X}$. These solutions are stable wrt small (in the sense of a stronger norm than the original one in $\mathcal{X}$) changes of f.

The second approach is based on the theory of abstract distributions. In contrast to the semi-group approach, it yields a generalized solution for any $f \in \mathcal{X}$. The term "abstract" means that the distributions are not functionals, but $\mathcal{X}$-valued operators defined on spaces of test functions.

The third approach to solving problems that are not uniformly well-posed uses the methods developed in the theory of regularization of ill-posed problems. The regularization consists of approximating the solution corresponding to initial data $f \in \mathcal{X}$ given with an error ($\|f_\delta - f\| \leq \delta$) by solutions u_ε of well-posed problems depending on a regularizing parameter ε. Under proper coordination of the parameters $\varepsilon = \varepsilon(\delta)$ the following convergence of the regularized solutions to the exact solution holds:

$$u_{\varepsilon(\delta)}(t) = \mathbf{R}_{\varepsilon(\delta)}(t)f_\delta \to u(t) \quad \text{as} \quad \delta \to 0 \quad \text{for each} \quad t \in [0,T].$$

Introduction xiii

Here $\mathbf{R}_{\varepsilon(\delta)}(t)$ denotes the corresponding regularizing operator.

In Part I we study different types of well-posedness of problems (I.2) and (I.3) and construct solutions by semi-group, distribution, and regularization methods. These methods are extensively used later in Part II for solving stochastic problems. The difference in the selection of the material compared to [79] is that here we study non-degenerate problems, paying special attention to problems with differential operators and regularization of ill-posed problems. We show that regularizing operators are closely related to $R = R(\varepsilon)-$ semi-groups and that all solutions constructed for ill-posed problems (I.2) and (I.3) by semi-group and distribution methods are regularized in a broad sense.

Now we give a summary of Part I by chapter.

Chapter 1 is devoted to the modern semi-group methods of solving abstract Cauchy problems (I.2) with operators A that generate strongly continuous semi-groups of solution operators, particularly C_0-semi-groups, as well as regularized semi-groups, including integrated, convoluted, and R-semi-groups.

Chapter 2 is devoted to generalized (wrt t) solutions to (I.2) in spaces of abstract distributions and to generalized (wrt x) solutions to (I.3) in Gelfand–Shilov spaces. The choice of a space of distributions depends on the type of the semi-group generated by A. In addition, using the connection between the semi-group techniques and generalized Fourier transform techniques developed in Gelfand–Shilov spaces, we obtain the generalized (wrt t and x) solutions to (I.3):

$$u(x;t) = (U(t)f)(x) = G_t(x) * f(x), \quad t \in [0,T], \quad x \in \mathbb{R}^n,$$

where the Green function $G_t(x) = \mathcal{F}^{-1}\left[e^{t\mathbf{A}(\cdot)}\right](x)$ and corresponding distribution spaces are determined by the growth rate of $e^{t\mathbf{A}(\cdot)}$, the family of solution operators for the Fourier transformed Cauchy problem.[1]

Chapter 3 presents supplementary material that helps us understand the techniques and methods of the book. In Sections 3.1 and 3.2 we give different examples of operators that generate regularized semi-groups. Special attention is paid to examples of differential operators generating Petrovsky correct, conditionally correct, and incorrect systems in the Gelfand–Shilov classification [36, 77]. Important results of the distribution theory, including definitions of different spaces of test functions and their properties, are presented in Section 3.3. A summary of integral transforms and structure theorems for distributions is presented in Section 3.4. The results are given in a unified manner for the convenience of the reader.

The aim of Part II is construction of solutions for the integrated stochastic problem (I.1) with generators of regularized semi-groups and Wiener processes in Hilbert spaces as well as solutions for the "differential" problem (P.1) with white noise in spaces of abstract distributions. As in Part I, we present the results in a form accessible to nonspecialists in these branches who are inter-

---

[1] If we write a generalized function as a function of a variable, we mean that it is a distribution applied to test functions depending on this variable.

xiv   *Introduction*

ested in methods of constructing solutions to infinite-dimensional stochastic problems arising in applications.

Now we give a summary of Part II by chapter.

Chapter 4 is devoted to the construction of weak and mild solutions for the Itô integrated problem (I.1) in Hilbert spaces. Necessary facts of the theory of random processes, related properties of linear operators, and stochastic integrals in Hilbert spaces are presented in Section 4.1, which has an auxiliary character.

In Section 4.2 we consider the basic type of (I.1), namely, the linear stochastic Cauchy problem with additive noise and A generating a regularized semi-group. Here conditions of existence and uniqueness of weak and weak regularized solutions are obtained and solutions are constructed.

In Section 4.3 mild solutions to the semi-linear problem are explored and their connections with weak solutions are shown.

In Section 4.4 we study relationships between solutions of stochastic problems and solutions of deterministic PDEs with derivatives in Hilbert spaces. We show that certain probability characteristics of solutions to stochastic problems satisfy these PDEs. The results extend the Feynman–Kac theorem to the case of Hilbert spaces [21, 82, 83].

In Chapter 5 we study the stochastic Cauchy problem (P.1) as it is, i.e., in differential form, where $\mathbb{W}$ is a generalized $\mathbb{H}$-valued white noise process considered as the derivative of a Wiener process in spaces of distributions and A is the generator of a regularized semi-group in H. Such problems are ill-posed and there are two reasons for that. The first one, as in the case of (I.1), is related to A, generally, generating only a regularized semi-group. For such an A the solution operators of the corresponding problem (I.2), playing a crucial role in solving stochastic problems, are unbounded in H. The other reason is inherently related to the stochastic term and manifests itself in the irregularity of the white noise. In Chapter 4 we overcome these obstacles by studying the problem in the integrated form with stochastic integrals wrt Wiener processes and constructing regularized solutions. It turns out that introduction of generalized solutions allows us to do this in a different way (see, e.g., [7, 67, 68, 72, 73, 74, 80]). Thus, we study the problem (P.1) in spaces of distributions.

In Section 5.1 we study solutions to the linear Cauchy problem in the generalized statement

$$\langle \varphi, X' \rangle = A\langle \varphi, X \rangle + \langle \varphi, \delta \rangle \zeta + \langle \varphi, B\mathbb{W} \rangle, \quad \langle \varphi, \mathbb{W}' \rangle := -\langle \varphi', W \rangle,$$

on test functions φ chosen in dependence on properties of the semi-group generated by A. For the case of A generating an integrated semi-group, we construct a generalized solution in spaces of abstract distributions. For the case of a convoluted semi-group, we do this in spaces of ultra-distributions. For the case of the generator of an R-semi-group in H, we construct (H_{-k})-valued generalized solutions, where the abstract Sobolev spaces H_{-k} are introduced in such a way that the operator R^{-1} is bounded there. Special attention is paid

Introduction

to the stochastic Cauchy problem with differential operators $A\left(i\partial/\partial x\right)$ generating R-semi-groups. For corresponding systems, we obtain generalized (wrt t and $x \in \mathbb{R}$) solutions using the technique of Gelfand–Shilov spaces [36, 37] presented in Section 3.3. In conclusion, the relationship between generalized and weak solutions is established.

In Section 5.2 we construct generalized solutions to semi-linear stochastic Cauchy problems. We managed to overcome the additional difficulties associated with the problem of multiplication of distributions. This is due to application of the Colombeau technique [18, 19, 91, 92, 93, 94] which we extend to the case of Hilbert space-valued distributions [69, 70].

It should be noted that the considered weak and weak regularized solutions to (I.1) as well as generalized solutions to (P.1) are predictable. In the framework of white noise analysis one can consider and solve certain equations without the condition of predictability. Such equations are called anticipating. This gives a perspective of introducing "dependence on the future" into mathematical models; for example, in financial mathematics when modeling markets which admit insider information.

In Chapter 6 we obtain generalized (wrt the random variable ω) solutions to the problem (P.1) with a singular white noise $\mathbb{W}$ taking values in spaces of abstract generalized random variables. We begin with the definition of these spaces.

The spaces of generalized random variables with values in $\mathbb{R}^n$ appeared in the last decades of the 20th-century [40, 41, 42] and were developed by many authors (see, e.g., [4, 7, 31, 44, 50, 55, 60, 64, 75, 90, 99, 100, 101, 107]).

Since white noise can be thought of as the derivative of Brownian motion, whose sample paths are continuous but nowhere differentiable, the sample paths of the white noise can be considered as elements of the Schwartz space of distributions $\mathcal{S}'$. Therefore the white noise probability space $(\Omega, \mathcal{F}, P)$ is constructed by taking $\Omega = \mathcal{S}'$ and introducing Gaussian normalized measure $P = \mu$ on the σ-algebra $\mathcal{F} = \mathcal{B}(\mathcal{S}')$ of Borel subsets of $\mathcal{S}'$. The existence of this measure is due to the famous Bochner–Minlos–Sazonov theorem. The white noise calculus offers a framework where all random variables are considered as functionals defined on $\mathcal{S}'$.

In order to embrace all the needed functionals, a few generalizations of the Schwartz distribution theory to the case of functions defined on $\mathcal{S}'$ were introduced within the white noise calculus. With the help of the theory of rigged Hilbert spaces [38] numerous stochastic analogs of the Gelfand triple $\mathcal{S} \subset L^2(\mathbb{R}) \subset \mathcal{S}'$ were constructed. One of them is

$$(\mathcal{S})_\rho \subset (L^2) \subset (\mathcal{S})_{-\rho}, \quad 0 \le \rho \le 1,$$

where (L^2) is the space of all random variables with finite second moments defined on $\mathcal{S}'$, $(\mathcal{S})_\rho$ is referred to as the Kondratiev space of test random variables, and $(\mathcal{S})_{-\rho}$ is called the Kondratiev space of generalized random variables. The values of the white noise process belong to $(\mathcal{S})_{-\rho}$. Moreover, the white noise process becomes infinitely-differentiable as a function of t

xvi   *Introduction*

with values in this space. The construction allows us to introduce the noise term directly into a differential equation and to state and solve stochastic differential equations with no restrictions connected with predictability.

In Section 6.1 we define the spaces of $\mathbb{H}$-valued generalized random variables $(\mathcal{S})_{-\rho}(\mathbb{H})$ as the spaces of linear continuous operators acting from $(\mathcal{S})_\rho$ to $\mathbb{H}$. The space $(\mathcal{S})_{-\rho}(\mathbb{H})$ becomes an extension of $(L^2)(\mathbb{H}) := L^2(\mathcal{S}',\mu;\mathbb{H})$. We show that the values of the $\mathbb{H}$-valued Q-Wiener and cylindrical Wiener processes, as well as the values of a Q-white noise and a singular white noise, lie in $(\mathcal{S})_{-\rho}(\mathbb{H})$.

In Section 6.2 we develop analysis of $(\mathcal{S})_{-\rho}(\mathbb{H})$-valued functions of $t \in \mathbb{R}$ introducing differentiation and integration. We show that Q-white noise and singular white noise are the derivatives of the $\mathbb{H}$-valued Q-Wiener and cylindrical Wiener process, respectively.

In Section 6.3 we introduce the concepts of the Wick product, Hitsuda–Skorohod integral, and S-transform and study their properties.

In Section 6.4 we obtain and study solutions to the linear problem with additive and multiplicative singular white noise in the spaces $(\mathcal{S})_{-\rho}(\mathbb{H})$. Some examples of the problems are given. In conclusion, the relationship between generalized wrt ω solutions and weak solutions of the corresponding integral problem is established.

Here again, as in Part I, all the solutions to the stochastic Cauchy problems obtained in Part II are regularized in a broad sense. The weak solutions are regularized by elements from *dom A^**. Different generalized solutions are regularized by means of test functions from spaces defined in the theories of abstract distributions, of Gelfand–Shilov generalized functions, and of abstract stochastic distributions.

We conclude with some observations.

First, we consider equations (deterministic and stochastic) for $t \in [0,\tau)$, or for $t \in [0,T]$, $T < \tau \leq \infty$. The choice of the interval depends on the type of semi-group generated by A: if A generates a global semi-group $\{S(t), t \geq 0\}$, we can consider the problems on $[0,\infty)$ or on $[0,T]$, as the specific model requires; if A generates just a local semi-group $\{S(t), t \in [0,\tau)\}$, the problem can be considered on $[0,\tau)$ or on $[0,T]$ with $T < \tau$.

Second, when we use the term "construction of a solution," we do not mean their actual structure; we usually prove the existence and uniqueness of solutions having a certain form. This is what is required in applications, where some relations of solutions to stochastic problems with their various probability characteristics are needed.

Chapters consist of sections which are reflected in the table of contents. Some sections contain items, but they are not reflected in the table of contents. Everywhere in the book we use triple numbering for definitions, formulas, theorems, and so on: the first figure is the number of the chapter, the second is the number of the section, and the third is the ordinal number of the definition, the formula, or the theorem, respectively. The end of the proof of a theorem is marked by the symbol □.

Symbol Description

A	operator of the abstract Cauchy problem
$\mathcal{A}$	notion of classes of semi-groups with the Abel-summability property
$A\left(i\frac{\partial}{\partial x}\right), \mathbf{A}\left(i\frac{\partial}{\partial x}\right)$	matrix differential operator
$A_{jk}\left(i\frac{\partial}{\partial x}\right)$	linear differential operator of finite order
$\mathcal{B}(\Omega)$	Borel σ-algebra on all subsets of Ω
$\boldsymbol{Cov}(u)$	covariance operator of a random variable u
$\mathcal{D}$	L. Schwartz's test function space
$\mathcal{D}'$	space of linear continuous functionals on $\mathcal{D}$ (space of distributions)
$\mathcal{D}'(\mathcal{X})$	space of linear continuous operators from $\mathcal{D}$ to $\mathcal{X}$ (space of abstract distributions)
$\mathcal{D}^{\{M_q\}}, \mathcal{D}^{\{M_q\},B}, \mathcal{D}_a^{\{M_q\},B}, \mathcal{D}_{\{M_q\}}, \ldots$	ultra-differentiable test function spaces (subspaces of $\mathcal{D}$)
$\det Q$	determinant of a matrix Q
$dom\,A$	domain of an operator A
$[dom\,A]$	Banach space $\{dom\,A, \ \|x\|_A = \|x\| + \|Ax\|\}$
$\boldsymbol{E}(u)$	expectation of a random variable u
$f * g$	convolution of f and g
$f \otimes g$	tensor product of f and g
$\mathcal{F}_t$	filter of σ-algebras
$\mathcal{F}u, \mathcal{F}[u], \widetilde{u}$	Fourier transform of u
$\mathcal{F}^{-1}u, \mathcal{F}^{-1}[u]$	inverse Fourier transform of u
$G_t(x), G_R(t,x)$	Green function, regularized Green function
$h_k(\cdot)$	Hermit polynomials
$\mathbf{h}_\alpha(\cdot)$	stochastic Hermit polynomials
$H_n := H \times \cdots \times H$	direct product of spaces
$Im\lambda, \Im\lambda$	imaginary part of $\lambda \in \mathbb{C}$
K	function defining a K-convoluted semi-group of operators
$L^2(G)$	space of functions square summable on G
$\mathcal{L}u, \mathcal{L}[u], \widetilde{u}$	Laplace transform of u
$\mathcal{L}^{-1}u, \mathcal{L}^{-1}[u]$	inverse Laplace transform of u
$\mathcal{L}(\mathcal{X})$	space of linear bounded operators on $\mathcal{X}$
$\mathcal{L}(\mathcal{X}, \mathcal{Y})$	space of linear bounded operators from $\mathcal{X}$ to $\mathcal{Y}$
$\mathcal{L}_{\mathrm{HS}}(H, H_1)$	space of Hilbert–Schmidt operators from H to H_1
$\mathbb{N}, \mathbb{R}, \mathbb{R}_+, \mathbb{C}$	sets of the natural, real, positive, complex numbers
$\mathbb{N}_0$	$\mathbb{N} \cup \{0\}$

xviii   *Symbol Description*

$\mathcal{N}(m, Q)$	Gaussian distribution law with mean m and covariance Q
$\mathcal{O}(f)$	characteristics of the growth of a function f
R	operator smoothing out initial data in the abstract Cauchy problem and defining an R-semi-group
$\mathbf{R}_\alpha(t)$	regularizing operator (operator regularizing the Cauchy problem solution at the time moment t)
$\mathcal{R}(\lambda),\ \lambda \in \rho(A)$	resolvent of A (the resolvent of an operator B has the identifying symbol, for example, $\mathcal{R}_B(\lambda),\ \lambda \in \rho(B)$)
$ran\,A$	range of A
$Re\lambda,\ \Re\lambda$	real part of $\lambda \in \mathbb{C}$
$\mathcal{S}$	space of rapidly decreasing test functions
$\mathcal{S}'$	space of linear continuous functionals on $\mathcal{S}$ (space of distributions of slow growth)
$\mathcal{S}_\alpha,\ \mathcal{S}^\beta,\ \mathcal{S}^\beta_\alpha,$ $\mathcal{S}^\beta_{\alpha,A},\ \mathcal{S}^{\beta,B}_\alpha, \ldots$	subspaces of $\mathcal{S}$
$\mathcal{S}'_\omega$	space of exponentially bounded distributions
$(\mathcal{S})_{-\rho}(\mathbb{H})$	space of all linear continuous operators on $\mathcal{S}_\rho$ to $\mathbb{H}$
$S\,\xi,\ S\,\Phi$	S-transform of a generalized random variable ξ or Φ
$Sp\,(A)$	spectrum of A
$\operatorname{supp}\varphi$	support of φ
t	time variable $t \geq 0$
$Tr\,A$	trace of an operator A
$u;\ u(x),\ x \in \mathbb{R}^n;$ $u(\cdot);\ u(\cdot;t);\ \ldots$	functions (we usually denote an argument by the point if the function has several arguments and it should be pointed out which one changes)
$\{U(t),\ t \in [0,\infty)\},$ $\{S(t),\ t \in [0,\tau)\},\ \ldots$	families of operators
x	space variable $x \in \mathbb{R}^n$
$\langle x, f \rangle$	value of the functional f on the element x
$\langle x, y \rangle,\ (x, y)$	scalar product of x and y
X, Y	solutions to stochastic equations
$\mathcal{X},\ \mathcal{Y},\ \mathcal{Z},\ \ldots$	Banach spaces
$\|\cdot\|_\mathcal{X}$	norm in a space $\mathcal{X}$
wrt	with respect to
$W_Q,\ W$	Q-Wiener, cylindrical Wiener processes, respectively
$\mathbb{W}_Q, \mathbb{W}$	Q-white noise, white noise, respectively
$W^{\Omega,b}_{M,a}$	space of entire functions, whose growth is defined by functions M and Ω and by parameters a and b
$\beta(\cdot)$	Brownian motion
δ	Dirac delta-function
$\Lambda(s), s \in \mathbb{C}^n$	characteristic function of a differential system
$\Lambda_\omega,\ \Lambda^M_{\alpha,\gamma,\omega}$	regions in the complex plane, whose boundaries are defined by parameters $\alpha,\ \gamma,\ \omega$ and by function M

$\xi_k(\cdot)$	Hermit functions
$\Phi,\ \Psi,\ \ldots$	linear topological spaces
$\widetilde{\Phi},\ \widetilde{\Psi},\ \ldots$	spaces of Fourier transforms of spaces $\Phi,\ \Psi,\ \ldots$, respectively
$\Phi',\ \Psi',\ \ldots$	spaces dual to $\Phi,,\ \Psi,\ \ldots$, respectively
$\rho(A)$	resolvent set of A
$\Theta \diamond \mathbb{W},\quad \Phi \diamond \mathbb{W}$	Wick product of $\Theta, \Phi \in (\mathcal{S})_{-\rho}(\mathcal{L}_{\mathsf{HS}}(\mathbb{H}, H))$ and $\mathbb{W}$
Ω	probability space
ω	random variable $\omega \in \Omega$
$(\Omega, \mathcal{F}, P),\ (\Omega, \mathcal{F}_t, P),$ $(\mathcal{S}', \mathcal{B}(\mathcal{S}'), \mu)$	probability triples

Part I

Well-Posed and Ill-Posed Abstract Cauchy Problems: The Concept of Regularization

Chapter 1

Semi-group methods for construction of exact, approximated, and regularized solutions

In this chapter we explore the Cauchy problem written in the abstract form with a closed linear operator A in a Banach space $\mathcal{X}$:

$$u'(t) = Au(t), \quad t \in [0, \tau), \quad \tau \leq \infty, \quad u(0) = f. \tag{1.0.1}$$

We use appropriate semi-group methods in dependence on properties of solution operators $\{U(t), t \in [0, \tau)\}$ generated by A. Here the time variable t takes values on $[0, \tau)$, $\tau \leq \infty$. This allows us to apply the techniques of global or local semi-groups for constructing solutions on any segment $[0, T]$ with $T < \tau$.

The MFPHY theorem gives necessary and sufficient conditions for well-posedness of (1.0.1) in terms of behavior of solution operators and the resolvent of A:

Theorem 1.0.1 (MFPHY theorem) *Let A be a closed linear operator densely defined on $\mathcal{X}$. Then the following statements are equivalent:*

(i) the problem (1.0.1) is uniformly well-posed on $\operatorname{dom} A$ for $t \geq 0$;

(ii) A generates a C_0-semi-group of solution operators $\{U(t), t \geq 0\}$;

(iii) the resolvent of A satisfies the MFPHY condition (I.4).

Due to the exponential boundedness of C_0-semi-groups $\|U(t)\| \leq Ce^{\omega t}$, $t \geq 0$, the Laplace transform can be applied to the operators $U(t)$ to obtain the resolvent of its generator:

$$\mathcal{R}(\lambda) = \mathcal{L}[U](\lambda) = \int_0^\infty e^{-\lambda t} U(t)\, dt, \quad Re\lambda > \omega. \tag{1.0.2}$$

The equality (1.0.2) illustrates the following profound connection between a semi-group and the resolvent of its generator. An exponentially bounded family of operators $\{U(t), t \geq 0\}$ satisfies the semi-group relation

$$U(t + s) = U(t)U(s), \quad t, s \geq 0,$$

if and only if its Laplace transform satisfies the resolvent identity

$$(\mu - \lambda)\mathcal{R}(\lambda)\mathcal{R}(\mu) = \mathcal{R}(\lambda) - \mathcal{R}(\mu), \quad Re\lambda, \ Re\mu > \omega.$$

1. Semi-group methods for construction of solutions

Moreover, the criterion for the Cauchy problem to be well-posed in the form of estimates (I.4) is a result of (1.0.2).

The techniques of (local and generalized) Laplace transform can be used in exploring the well-posedness of (1.0.1) and corresponding properties of the resolvent in a more general case, namely, when the solutions of the Cauchy problem are not exponentially bounded or exist only locally and generate regularized semi-groups. According to the type of regularization of the abstract Cauchy problem via a regularized semi-group, we distinguish the following classes of ill-posed problems determined by the geometric properties of the set of regular points of A together with the behavior of its resolvent:

(R1) $\Lambda_\omega = \{\lambda \in \mathbb{C} : \ Re\lambda > \omega, \ \omega \in \mathbb{R}\} \subseteq \rho(A)$ and

$$\left\| \frac{d^k}{d\lambda^k} \left(\frac{\mathcal{R}(\lambda)}{\lambda^n} \right) \right\| \leq \frac{Ck!}{(Re\lambda - \omega)^{k+1}}, \quad \lambda \in \Lambda_\omega, \quad k \in \mathbb{N}_0;$$

(R2) $\Lambda^{\ln}_{n,\nu,\omega} = \{\lambda \in \mathbb{C} : \ Re\lambda > n\nu \ln|\lambda| + \omega\} \subseteq \rho(A)$ and

$$\|\mathcal{R}(\lambda)\| \leq C|\lambda|^n, \quad \lambda \in \Lambda^{\ln}_{n,\nu,\omega};$$

(R3) $\Lambda^M_{\alpha,\gamma,\omega} = \{\lambda \in \mathbb{C} : \ Re\lambda > \alpha M(\gamma|\lambda|) + \omega\} \subseteq \rho(A)$ and

$$\|\mathcal{R}(\lambda)\| \leq Ce^{\omega M(\gamma|\lambda|)}, \quad \lambda \in \Lambda^M_{\alpha,\gamma,\omega},$$

where M is a certain positive nondecreasing function;

(R4) the regular points of A fill no interval of the type $\lambda > \omega$ in the right-hand half-plane.

Note that (R1) coincides with MFPHY condition as $n = 0$. Contraction of the set $\rho(A)$ and the change of its resolvent behavior when we pass from class (R1) to (R4) reflect strengthening of the ill-posedness of (1.0.1), which is directly connected with the character of peculiarities of solution operators of the problem.

In Section 1.1 we study strongly continuous with respect to $t \geq 0$ semi-groups (C_0-semi-groups) and strongly continuous with respect to $t > 0$ semi-groups (semi-groups of classes C_1 and $\mathcal{A}$) and consider types of well-posedness of the Cauchy problem related to these semi-groups. In Section 1.2 we explore these questions for regularized semi-groups, namely, integrated, convoluted, and R-semi-groups. In Section 1.3 we study connections of R-semi-groups with regularizing operators.

1.1 The Cauchy problem and strongly continuous semi-groups of solution operators

We begin studying strongly continuous semi-groups with the C_0-semi-groups and consider the uniform well-posedness of the abstract Cauchy problem related to their generators.

1.1.1 The Cauchy problem with generators of C_0-semi-groups

As we mentioned above, uniformly well-posed problems are of great importance in exploring the Cauchy problem. Fundamental results in the field obtained with the help of the semi-group theory and Laplace transform techniques are reflected in the MFPHY theorem. It connects uniform well-posedness of the abstract Cauchy problem with a C_0-semi-group generated by A as well as with the behavior of the resolvent of A.

Let us give the definition of uniform well-posedness.

Definition 1.1.1 *The Cauchy problem (1.0.1) is called uniformly well-posed on $D \subseteq \operatorname{dom} A$ if for any $f \in D$*

(a) *there exists a unique solution $u \in C([0, \infty), \operatorname{dom} A) \cap C^1([0, \infty), \mathcal{X})$;*

(b) *the solution is stable with respect to change of the initial data and the stability is uniform with respect to $t \in [0, T]$, $T < \infty$:*

$$\sup_{t \in [0,T]} \|u(t)\| \leq C_T \|f\|.$$

It follows from the definition that the solution of a well-posed problem is defined on $[0, \infty)$. Therefore, $U(t)$, $t \geq 0$, defined as solution operators of such a Cauchy problem, $U(t)f := u(t)$, $f \in D$, also depend on the parameter $t \in [0, \infty)$. If (1.0.1) is uniformly well-posed on $\operatorname{dom} A$ and the operator A is densely defined, then the solution operators have the semi-group property. This is indeed the case since for each $f \in \operatorname{dom} A$ and $t, h \geq 0$ the elements $U(t + h)f$ and $U(t)U(h)f$ are the solutions of (1.0.1) with the initial data $U(h)f$ at the moment $t = 0$. Since the solution is unique, they coincide:

$$U(t + h)f = U(t)U(h)f, \quad t, h \geq 0, \quad f \in \operatorname{dom} A. \tag{1.1.3}$$

In a similar manner using the uniqueness property one can show that the operators $U(t)$ are linear for each $t \geq 0$. Hence, under condition $\overline{\operatorname{dom} A} = \mathcal{X}$ the equality (1.1.3) can be extended to the whole space $\mathcal{X}$ by continuity. The stability of the Cauchy problem solutions implies boundness of the operators obtained. Moreover, it follows from the uniform well-posedness of the problem that the solution operators actually form a semi-group of a special type, namely, a C_0-semi-group.

6   *1. Semi-group methods for construction of solutions*

Definition 1.1.2 *A family of bounded linear operators $\{U(t),\, t \geq 0\}$ acting in a Banach space $\mathcal{X}$ and satisfying the conditions*

(U1) $U(t+h) = U(t)U(h), \quad t, h \geq 0$

(U2) $U(0) = I$

(U3) $U(t)f$ *is continuous with respect to $t \geq 0$ for any $f \in \mathcal{X}$*

is called a semi-group of class C_0 (shortened to C_0-semi-group).
    An operator defined by

$$U'(0)f := \lim_{h \to 0} h^{-1}(U(h) - I)f, \tag{1.1.4}$$

with

$$dom\, U'(0) = \{f \in \mathcal{X} : \lim_{h \to 0} h^{-1}(U(h) - I)f \;\; exists \,\}, \tag{1.1.5}$$

is called an (infinitesimal) generator of the family.

The properties of C_0-semi-groups and their generators have been thoroughly studied, and one can find numerous publications on the subject (see, e.g., [9, 10, 29, 30, 56, 76, 79, 113]). Let us pick out fundamental properties of these semi-groups. As we will see further, these properties are the basis for setting the properties of more general semi-groups.

    1.   A C_0-semi-group is non-degenerate, i.e., $U(t)f = 0$ for each $t > 0$ implies $f = 0$.

This is indeed the case since if $U(t)f = 0$ for each $t > 0$ for a certain $f \in \mathcal{X}$, then the properties (U2) and (U3) imply

$$f = U(0)f = \lim_{t \to 0} U(t)f = 0.$$

    2.   The operators of a C_0-semi-group commute with the generator on its domain. Operators forming a C_0-semi-group with the generator A are the solution operators[1] for (1.0.1).

The properties follow from (U1) and (1.1.4):

$$
\begin{aligned}
U(t)U'(0)f &= U(t)\lim_{h \to 0} h^{-1}(U(h) - I)f \\
&= \lim_{h \to 0} h^{-1}(U(h) - I)U(t)f = U'(0)U(t)f, \qquad f \in dom\, U'(0), \\
\frac{dU(t)}{dt}f &= \lim_{h \to 0} h^{-1}(U(t+h) - U(t))f = \lim_{h \to 0} h^{-1}(U(h)U(t) - U(t))f \\
&= \lim_{h \to 0} h^{-1}(U(h) - I))U(t)f = AU(t)f, \qquad f \in dom\, A.
\end{aligned}
$$

---

[1]Sometimes the semi-group operators are called evolution operators; nevertheless, some more general families than the solution operators may be called evolution operators as well. For example, integrated, convoluted, and other semi-groups considered further are families of evolution operators and not the solution ones.

1.1. The Cauchy problem and strongly continuous semi-groups

3. The generator of a C_0-semi-group is a closed operator.

Let $f_n \in dom\, U'(0)$, $f_n \to f$ and $U'(0)f_n \to g$. Then the continuity of semi-group operators with respect to t together with property 2 implies

$$h^{-1}\left(U(h) - I\right)f_n = h^{-1}\int_0^h \frac{dU(s)}{ds}f_n\, ds = h^{-1}\int_0^h U(s)U'(0)f_n\, ds.$$

Passing to the limit when $n \to \infty$ gives

$$h^{-1}\left(U(h) - I\right)f = h^{-1}\int_0^h U(s)g\, ds.$$

The right-hand side of this equality has a limit equal to g as $h \to 0$. Hence $f \in dom\, U'(0)$ and $g = U'(0)f$.

4. The generator of a C_0-semi-group is densely defined.

In order to prove it, consider the set

$$\mathcal{X}_{a,b} := \left\{u_{a,b} = \int_a^b U(s)f\, ds,\ \ f \in \mathcal{X},\ 0 \le a < b\right\},$$

and show that $\mathcal{X}_{a,b}$ is a subset of $dom\, U'(0)$ and is dense in $\mathcal{X}$.

For any $u_{a,b} \in \mathcal{X}_{a,b}$ we have

$$
\begin{aligned}
h^{-1}(U(h) - I)u_{a,b} \ &=\ h^{-1}\int_a^b (U(h+s) - U(s))f\, ds \\
&=\ h^{-1}\left(\int_{a-h}^{b-h} U(t)f\, dt - \int_a^b U(s)f\, ds\right) \\
&\xrightarrow[h \to 0]{}\ (U(b) - U(a))\, f.
\end{aligned}
$$

Therefore $u_{a,b} \in dom\, U'(0)$.

Now let us show that $\mathcal{X}_{a,b}$ is dense in $\mathcal{X}$. Assume the contrary, namely, $\overline{\mathcal{X}_{a,b}} \ne \mathcal{X}$. Then the corollary of the Hahn–Banach theorem implies existence of a nonzero functional $F \in \mathcal{X}^*$ equal to zero on the subspace $\overline{\mathcal{X}_{a,b}}$, i.e.,

$$F(u_{a,b}) = \int_a^b F(U(s)f)\, ds = 0.$$

Hence, $F(U(0)f) = F(f) = 0$ for any $f \in \mathcal{X}$. The obtained contradiction proves the equality $\overline{\mathcal{X}_{a,b}} = \mathcal{X}$. Hence $\overline{dom\, U'(0)} = \mathcal{X}$.

5. A C_0-semi-group is an exponentially bounded family of operators:

$$\exists\, C > 0,\ \omega \in \mathbb{R}\ :\quad \|U(t)\| \le Ce^{\omega t},\quad t \ge 0.$$

8   *1. Semi-group methods for construction of solutions*

To show this, write each $t \geq 0$ as $t = n + s$ with $n \in \mathbb{N}_0$ and $0 \leq s < 1$. Then the semi-group property (U1) gives the equality $U(t) = U^n(1)U(s)$. Uniform boundedness of $\|U(s)\|$ with respect to $0 \leq s < 1$ follows from the uniform boundedness principle and we have

$$\|U(t)\| \leq C\|U(1)\|^n = Ce^{n \ln \|U(1)\|} \leq Ce^{t \ln \|U(1)\|} = Ce^{\omega t},$$

where $C = \sup_{0 \leq s < 1} \|U(s)\| < \infty$ and $\omega = \ln \|U(1)\|$.

6. The Laplace transform of a C_0-semi-group of operators exists for each λ from the half-plane $Re\lambda > \omega$ and coincides with the resolvent of the generator:

$$\widetilde{U}(\lambda) = (\lambda I - U'(0))^{-1} = \mathcal{R}_{U'(0)}(\lambda), \quad Re\lambda > \omega. \tag{1.1.6}$$

Strong continuity and exponential boundedness of a C_0-semi-group allow us to apply the Laplace transform to it and yield boundedness of the Laplace transform:

$$\widetilde{U}(\lambda)f = \int_0^\infty e^{-\lambda t} U(t)f \, dt, \quad f \in \mathcal{X}, \qquad \|\widetilde{U}(\lambda)\| \leq \frac{C}{Re\lambda - \omega}, \quad Re\lambda > \omega.$$

Apply $U'(0)$ to the integral for $f \in dom\, U'(0)$. Since $U'(0)$ is a closed operator, it can be inserted under the integral sign. Further, applying property 2 and integrating by parts, we obtain

$$U'(0) \int_0^\infty e^{-\lambda t} U(t)f \, dt = \int_0^\infty e^{-\lambda t} U'(0)U(t)f \, dt$$

$$= \int_0^\infty e^{-\lambda t} U'(t)f \, dt = -f + \lambda \int_0^\infty e^{-\lambda t} U(t)f \, dt, \quad f \in dom\, U'(0).$$

This equality can be extended to the whole space by continuity:

$$(\lambda I - U'(0)) \int_0^\infty e^{-\lambda t} U(t)f \, dt = f, \quad f \in \mathcal{X}.$$

On the other hand, we have

$$\int_0^\infty e^{-\lambda t} U(t)(\lambda I - U'(0))f \, dt = f, \quad f \in dom\, U'(0).$$

These two equalities imply (1.1.6).

It turns out that a property inverse to the previous one is valid for the operators of a C_0-semi-group.

7. The following theorem shows complete interconnection between the semi-group property (U1) of an exponentially bounded family of operators and the resolvent identity for its Laplace transform.

1.1. The Cauchy problem and strongly continuous semi-groups

Theorem 1.1.1 [8, 79]. *Let $\{U(t), t \geq 0\}$ be an exponentially bounded, strongly continuous family of linear operators. Then the operator function*

$$\widetilde{U}(\lambda) = \int_0^\infty e^{-\lambda t} U(t)\, dt, \qquad Re\lambda > \omega,$$

satisfies the resolvent identity

$$\widetilde{U}(\lambda)\widetilde{U}(\mu) = \frac{\widetilde{U}(\lambda) - \widetilde{U}(\mu)}{\mu - \lambda}, \qquad Re\lambda > Re\mu > \omega,$$

if and only if the operators of the family satisfy the semi-group relation (U1).

8. There exists a definition of a C_0-semi-group and its generator which is equivalent to Definition 1.1.2.

The resolvent identity implies the following equality for any C_0-semi-group $\{U(t), t \geq 0\}$:

$$\lambda I - (\widetilde{U}(\lambda))^{-1} = \mu I - (\widetilde{U}(\mu))^{-1}, \qquad Re\lambda, Re\mu > \omega.$$

This means that there exists an operator

$$A := \lambda I - (\widetilde{U}(\lambda))^{-1}, \qquad dom\, A = ran\, \widetilde{U}(\lambda), \quad Re\lambda > \omega, \tag{1.1.7}$$

and it coincides with $U'(0)$. Now on the basis of the proved properties of generators of C_0-semi-groups we can introduce one more (equivalent) definition of a C_0-semi-group and its generator.

Definition 1.1.3 *Let A be a closed linear operator in a Banach space $\mathcal{X}$. A strongly continuous and exponentially bounded family of linear operators $\{U(t), t \geq 0\}$ acting in $\mathcal{X}$ and satisfying the equalities*

$$A \int_0^t U(s) f\, ds = U(t)f - f, \qquad t \geq 0, \quad f \in \mathcal{X},$$

$$U(t)Af = AU(t)f, \qquad t \geq 0, \quad f \in dom\, A,$$

is called a C_0-semi-group generated by A.

Extensions of this definition will be taken as definitions of different regularized semi-groups and their generators in the next section.

9. Let A be the generator of a C_0-semi-group. Then the estimates

$$\left\| R^{(k)}(\lambda) \right\| \leq \frac{Ck!}{(Re\lambda - \omega)^{k+1}}, \qquad Re\lambda > \omega, \quad k \in \mathbb{N}, \tag{1.1.8}$$

follow from the equality (1.1.6). Due to the resolvent identity the estimates can be written in the equivalent form (I.4).

10 1. Semi-group methods for construction of solutions

Remark 1.1.1 The above proofs of properties 1–6 and the estimates (1.1.8) for C_0-semi-groups constitute the proofs of implications (i)$\Longleftrightarrow$(ii)$\Longrightarrow$(iii) in the MFPHY theorem. As for (iii)$\Longrightarrow$(ii), we refer the reader to [79], where the main methods of constructing solution operators to the Cauchy problem with A satisfying (1.1.8) are collected. Constructions based on the Yosida approximations and the Widder–Post inversion formula are presented there.

In conclusion, before we pass to more general semi-groups of classes C_1 and $\mathcal{A}$, we note one more important property of C_0-semi-groups, which remains true for the semi-groups of class C_1 and is the basis for the definition of semi-groups of class $\mathcal{A}$.

10. For the resolvent of the generator of a C_0-semi-group the following equality holds:

$$\lim_{\lambda \to \infty} \lambda \mathcal{R}(\lambda) f = f, \qquad f \in \mathcal{X}, \quad \lambda \in \rho(A) \cap \mathbb{R}. \tag{1.1.9}$$

Let A be the generator of a C_0-semi-group and $f \in \operatorname{dom} A$. It can be represented in the form $f = \mathcal{R}(\lambda_0) y$, $y \in \mathcal{X}$. By the resolvent identity we have

$$\lambda \mathcal{R}(\lambda) \mathcal{R}(\lambda_0) y = \lambda \frac{\mathcal{R}(\lambda) y}{\lambda_0 - \lambda} - \lambda \frac{\mathcal{R}(\lambda_0) y}{\lambda_0 - \lambda}, \qquad \lambda, \lambda_0 \in \rho(A).$$

Hence, taking into account (I.4) as $\lambda \to \infty$, we obtain (1.1.9) on $\operatorname{dom} A$. Due to the boundedness of $\|\lambda \mathcal{R}(\lambda)\|_{\mathcal{L}(\mathcal{X})}$ as $\lambda \to \infty$ and the density of $\operatorname{dom} A$, the equality can be continued to the whole space $\mathcal{X}$.

1.1.2 Semi-groups of classes C_1 and $\mathcal{A}$ and of growth order α. The Cauchy problem with generators of these semi-groups

For the C_0-semi-groups considered in the previous subsection the strong continuity at the point $t = 0$ is the fundamentally important property. It follows from the properties of these semi-groups proved above that the strong continuity at zero implies continuity at any point $t \in (0, \infty)$. It also follows that the semi-group operators are the solution operators for the uniformly well-posed Cauchy problem (1.0.1). However, Cauchy problems with solutions which are not continuous at zero include important differential problems and often arise in applications along with the well-posed ones.

In the present subsection we consider the semi-groups related to such problems. Being strongly continuous as $t > 0$, they lose the property of strong continuity at $t = 0$ and are defined with the help of weaker conditions than C_0-semi-groups.

For the semi-groups of class C_1 this is the condition of C-summability:

$$\lim_{\eta \to 0} C(\eta) f = f, \quad f \in \mathcal{X}, \quad \text{where} \quad C(\eta) f := \frac{1}{\eta} \int_0^\eta U(s) f \, ds. \tag{1.1.10}$$

1.1. The Cauchy problem and strongly continuous semi-groups

For the semi-groups of class $\mathcal{A}$ this is the condition of $\mathcal{A}$-summability:

$$\lim_{\lambda \to \infty} \lambda \widetilde{U}(\lambda) f = f, \quad f \in \mathcal{X}, \quad \text{where} \quad \widetilde{U}(\lambda) f = \int_0^\infty e^{-\lambda t} U(t) f \, dt. \quad (1.1.11)$$

For semi-groups of growth order α this is the condition on the rate of growth with respect to t as $t \to 0$.

Definition 1.1.4 *A family of bounded linear operators $\{U(t), t > 0\}$ in a Banach space $\mathcal{X}$ satisfying the conditions*

(U1′) $U(t + h) = U(t)U(h), \quad t, h \geq 0$

(U3′) *the operator function $U(\cdot)$ is strongly continuous with respect to $t > 0$*

is called a strongly continuous semi-group.

    The operator $U'(0)$ defined by (1.1.4)–(1.1.5) is called an infinitesimal operator and the operator $A := \overline{U'(0)}$, if it exists, is called the generator of the family.

The semi-groups of classes C_1 and $(0, \mathcal{A})$ contain some subclasses.

Definition 1.1.5 *A strongly continuous semi-group $\{U(t), t > 0\}$ is called a semi-group of class $(0, C_1)$ (a semi-group of class $(1, C_1)$) if the condition (1.1.10) holds and*

$$\int_0^1 \|U(t)f\| \, dt < \infty, \quad f \in \mathcal{X} \qquad \left(\int_0^1 \|U(t)\| \, dt < \infty \right). \quad (1.1.12)$$

A semi-group is called a semi-group of class $(0, \mathcal{A})$ (class $(1, \mathcal{A})$) if the conditions (1.1.11) and (1.1.12) hold.

    The semi-groups of classes $(0, C_1)$ and $(1, C_1)$ form the class C_1 and the semi-groups of classes $(0, \mathcal{A})$ and $(1, \mathcal{A})$ form the class $\mathcal{A}$.

As we will demonstrate below, generally, the infinitesimal operator is not closed and the Laplace transform of a strongly continuous semi-group is not the resolvent of the infinitesimal operator.

    Let us consider the properties of the semi-groups introduced and compare them with the properties of C_0-semi-groups.

    1. Exponential estimates. Behavior of the resolvent.

When studying the properties of C_0-semi-groups we showed the important relation (1.1.9) which serves as a basis for the definition of a semi-group of class $\mathcal{A}$. Furthermore, the relation (1.1.10), being the basis of the definition of the semi-groups of class C_1, obviously holds true for a C_0-semi-group. We

12   *1. Semi-group methods for construction of solutions*

will show that (1.1.9) holds under a weaker than (U3) condition of continuity (1.1.10). Thus the following embeddings hold:

$$\{\text{semi-groups of class } C_0\} \subset \{\text{semi-groups of class } C_1\} \subset$$
$$\subset \{\text{semi-groups of class } \mathcal{A}\}. \quad (1.1.13)$$

Proposition 1.1.1 *Let $\{U(t), t > 0\}$ be a semi-group of class C_1. Then the condition (1.1.11) holds.*

Proof. Let us denote $\omega_0 := \lim_{t\to\infty} t^{-1} \ln \|U(t)\|_{\mathcal{L}(\mathcal{X})}$. It is called a type of the semi-group. It is shown in [43] that $-\infty \le \omega_0 < \infty$ and for any $\eta > 0$

$$\|U(t)\| \le M(\omega, \eta)e^{\omega t}, \qquad t \ge \eta, \quad \omega > \omega_0. \quad (1.1.14)$$

Hence the Laplace transform $\widetilde{U}(\lambda)$ is defined for $Re\lambda > \omega$. Integrating by parts the right-hand side of the equality (1.1.11), we obtain

$$\lambda \widetilde{U}(\lambda)f - f = \lambda^2 \int_0^\infty t\, e^{-\lambda t} \left(C(t)f - f\right) dt.$$

The estimate

$$\left\| \lambda^2 \int_0^\eta e^{-\lambda t} t \left(C(t)f - f\right) dt \right\| \le M \sup_{0<t<\eta} \|C(t)f - f\|, \qquad f \in \mathcal{X},$$

and the property of exponential boundedness (1.1.14) imply

$$\lambda^2 \int_\eta^\infty e^{-\lambda t} \left(\int_0^t U(s)f - sf \right) ds \xrightarrow[\lambda \to \infty]{} 0, \qquad f \in \mathcal{X},$$

which proves the property (1.1.11). $\qquad\qquad\qquad\qquad\qquad\qquad\square$

Thus the Laplace transform for semi-groups of classes C_1 and $\mathcal{A}$ exists and similarly to the case of class C_0, the limit relation (1.1.11) holds for it.

2. Properties of generators and infinitesimal operators.

We continue to compare the properties of the semi-groups introduced above with the properties of C_0-semi-groups. We first make clear whether the important equality of the Laplace transform of a semi-group holds for the resolvent of its generator. Then we compare the properties of the generators.

It is shown in [43] (Theorem 11.5.1) that for an (exponentially bounded) strongly continuous semi-group $\{U(t), t > 0\}$ and for those $f \in \mathcal{X}$ that provide convergence in (1.1.10) and (1.1.12), the equality

$$(\lambda I - U'(0))\widetilde{U}(\lambda)f = f, \qquad Re\lambda > \omega, \quad (1.1.15)$$

1.1. The Cauchy problem and strongly continuous semi-groups

holds and for $f \in dom\, U'(0)$ the equality

$$\widetilde{U}(\lambda)(\lambda I - U'(0))f = f, \qquad Re\lambda > \omega, \qquad (1.1.16)$$

holds true. It follows that the Laplace transform of a C_1 class semi-group is the resolvent of the infinitesimal operator. Since the infinitesimal operator of such a semi-group is closed, it coincides with its generator. As for the infinitesimal operator of a semi-group of class $\mathcal{A}$, it is not necessarily closed (see an example in [97]).

Let us show that the infinitesimal operator $U'(0)$ of a semi-group of class C_1 or of class $\mathcal{A}$ is densely defined and that for the domain of $U'(0)$ the following embedding holds $\mathcal{X}_1 \subset dom\, U'(0)$, where

$$\mathcal{X}_1 := \left\{ u_{a,b} = \int_a^b U(s)f\, ds, \quad f \in \mathcal{X}_0,\ 0 < a < b \right\} \subset \mathcal{X}_0 := \bigcup_{t>0} U(t)(\mathcal{X}),$$

as well as the equalities $\overline{\mathcal{X}_1} = \mathcal{X}_0,\ \overline{\mathcal{X}_0} = \mathcal{X}$.

The embedding and density of $\mathcal{X}_1$ in $\mathcal{X}_0$ can be proved by the same scheme as property 4 of C_0-semi-groups, changing the set $\mathcal{X}_{a,b}$ by $\mathcal{X}_1$ and $\mathcal{X}$ by $\mathcal{X}_0$. The equality $\overline{\mathcal{X}_0} = \mathcal{X}$ follows from the Hahn–Banach theorem: if $\overline{\mathcal{X}_0} \neq \mathcal{X}$, then there exist $f_0 \notin \mathcal{X}_0$ and $F \in \mathcal{X}^*$ such that $F(\mathcal{X}_0) = 0$ and $F(f_0) \neq 0$. However,

$$0 = \lambda \int_0^\infty e^{-\lambda t} F(U(t)f_0)\, dt \xrightarrow[h \to 0]{} F(f_0),$$

which contradicts $F(f_0) \neq 0$.

It follows from the density of $dom\, U'(0)$ in $\mathcal{X}$ and the equalities (1.1.15)–(1.1.16) that the Laplace transform of a semi-group of class $\mathcal{A}$ is the resolvent of its (densely defined) generator $\overline{U'(0)}$:

$$\mathcal{R}_{\overline{U'(0)}}(\lambda) = \int_0^\infty e^{-\lambda t} U(t)\, dt, \qquad Re\lambda > \omega. \qquad (1.1.17)$$

It follows from (1.1.17) and the proved properties of semi-groups of classes C_1 and $\mathcal{A}$ that, similar to the case of a C_0-semi-group, the infinitesimal generator can be (equivalently) defined by the equality (1.1.7). However, we cannot retain the equivalence of the definition to the one obtained similarly to Definition 1.1.3.

3. Connection with the Cauchy problem (1.0.1).

We formulate a result on the connection between a semi-group and the Cauchy problem for the case of a semi-group of class $\mathcal{A}$. The latter is the widest one among all classes of strongly continuous semi-groups studied up to now.

Theorem 1.1.2 [43] *Let* $\{U(t),\, t > 0\}$ *be a semi-group of class* $\mathcal{A}$. *Then for any* $n \geq 1$

$$\frac{d^n}{dt^n} U(t)f = (U'(0))^n\, U(t)f = U(t)\left(\overline{U'(0)}\right)^n f, \qquad f \in dom\left(\overline{U'(0)}\right)^n.$$

14   1. Semi-group methods for construction of solutions

If $f \in dom\left(\overline{U'(0)}\right)^2$, then $\lim_{t \to 0} U(t)f = f$.

At the end of the subsection we define one more class of semi-groups, which not only loses the property of strong continuity in zero, but even admits singularities with power rate of growth.

Definition 1.1.6 *Let $\alpha \geq 0$. A strongly continuous wrt $t > 0$ semi-group of operators $\{U(t),\, t > 0\}$ on a Banach space $\mathcal{X}$ satisfying the conditions:*

($U_\alpha 1$) *the set $\mathcal{X}_0 = \bigcup_{t>0} U(t)(\mathcal{X})$ is dense in $\mathcal{X}$*

($U_\alpha 2$) *the semi-group is non-degenerate*

($U_\alpha 3$) *the function $\|t^\alpha U(t)\|_{\mathcal{L}(\mathcal{X})}$ is bounded as $t \to +0$*

is called a semi-group of growth order α.

Comparing the semi-groups of growth order α with the semi-groups considered before, we must mention the following.

1. The property of a semi-group of growth order α to be non-degenerate and the property of the set $\mathcal{X}_0$ to be dense, as a consequence of conditions at point zero, hold true for all the classes of semi-groups considered above, including that introduced in Definition 1.1.6.

2. The Laplace transform of a semi-group of growth order α, generally, is not well defined, due to the power singularity at zero.

3. The classes of semi-groups C_1 and $\mathcal{A}$ and the class of semi-groups of growth order α are not ordered by inclusion. Nevertheless, since for semi-groups of classes $(1, C_1)$ and $(1, \mathcal{A})$ the function $\|U(t)\|$, $t > 0$, does not have non-integrable singularities at $t = 0$, we can continue the embeddings (1.1.13) as follows:

$$\{\text{semi-groups of class } (1, C_1)\} \subset \{\text{semi-groups of class } (1, \mathcal{A})\} \subset$$
$$\subset \{\text{semi-groups of order } \alpha\}. \quad (1.1.18)$$

1.2 The Cauchy problem with generators of regularized semi-groups: integrated, convoluted, and R-semi-groups

In Section 1.1 we considered the Cauchy problem (1.0.1) with a generator of a strongly continuous semi-group of solution operators $\{U(t),\, t \geq 0\}$ and

1.2. The Cauchy problem with generators of regularized semi-groups   15

showed the properties of solutions to the problem. They have the form $u(t) = U(t)\xi$, $t \geq 0$, for such semi-groups. The problem is well-posed in the case of C_0-semi-groups and can have singularities at zero in the case of other strongly continuous semi-groups. The aim of the present section is to investigate the Cauchy problem (1.0.1) with generators of regularized semi-groups, which generally are not semi-groups of solution operators, just more general families of bounded operators.

1.2.1   The Cauchy problem with generators of exponentially bounded integrated semi-groups

We begin with the Cauchy problem (1.0.1) with A satisfying the condition (R1) and show the type of well-posedness corresponding to such A, the "semi-group" family corresponding to this type of well-posedness, and a way of regularizing of this problem.

The estimates (R1) are a generalization of MFPHY conditions. They coincide with estimates (I.4) for $n = 0$. As a consequence, the "semi-group" that the operator A generates should be a generalization of a C_0-semi-group. The families of operators introduced by Arendt and called integrated semi-groups serve as such a generalization [8].

Definition 1.2.1 *Let* $n \in \mathbb{N}$. *A family of bounded linear operators* $\{S_n(t), t \geq 0\}$ *in a Banach space* $\mathcal{X}$ *that satisfies the conditions*

$(S_n 1)$ $\dfrac{1}{(n-1)!} \displaystyle\int_0^s \left((s-r)^{n-1} S_n(t+r) - (t+s-r)^{n-1} S_n(r) \right) dr$

$$= S_n(t)S_n(s), \quad s,t \geq 0, \quad S_n(0) = 0$$

$(S_n 2)$ $S_n(\cdot)$ *is a strongly continuous wrt* $t \geq 0$ *operator function*

$(S_n 3)$ *there exist* $C > 0$ *and* $\omega \in \mathbb{R}$ *such that* $\|S_n(t)\| \leq C e^{\omega t}$, $t \geq 0$

is called an exponentially bounded n*-times integrated semi-group.*

It is easy to see that the characteristic property $(S_n 1)$ for the family just introduced is the n-times integrated semi-group property (U1). This fact clarifies the name "integrated semi-group." Namely, if there exists a C_0-semi-group $\{U(t), t \geq 0\}$, then $S_n(\cdot)$ is nothing but the n-tuple integral of $U(\cdot)$:

$$S_n(t) = \int_0^t \int_0^{t_1} \cdots \int_0^{t_{n-1}} U(t_n)\, dt_n \ldots dt_2\, dt_1, \quad t \geq 0.$$

This explains why the C_0-semi-groups are often called 0 times integrated semi-groups for reasons of similarity, although we cannot formally substitute $n = 0$ in the definition of n-times integrated semi-groups.

Due to exponential boundedness of integrated semi-groups, as a basis for

16   *1. Semi-group methods for construction of solutions*

the definition of a generator for the family introduced, one can take the definition connected with the Laplace transform. In order to give such a definition we formulate the following generalization of Theorem 1.1.1.

Proposition 1.2.1 [8, 79] *Let $\{S_n(t),\ t \geq 0\}$ be a strongly continuous, exponentially bounded family of bounded linear operators and*

$$r(\lambda) := \int_0^\infty \lambda^n e^{-\lambda t} S_n(t) dt, \qquad Re\lambda > \omega. \tag{1.2.1}$$

Then $r(\cdot)$ satisfies the resolvent identity if and only if $S_n(t),\ t \geq 0$, satisfies the relation (S_n1).[2]

Proceeding from this relation for the family, we give the following definition of its generator.

Definition 1.2.2 *Let $\{S_n(t),\ t \geq 0\}$ be a non-degenerate n-times integrated exponentially bounded semi-group and let operator $r(\lambda)$, $Re\lambda > \omega$, be defined by (1.2.1). The operator $A := \lambda I - r(\lambda)^{-1}$ is called the generator of the semigroup.*

The generator introduced is obviously closed. In contrast to the case of a C_0-semi-group, the generator defined in this fashion does not coincide with an infinitesimal generator from Definition 1.1.2.

The following criterion can be obtained on the base of Proposition 1.2.1.

Theorem 1.2.1 *Let $n \in \mathbb{N}_0$, $\omega \geq 0$. A linear operator A is the generator of a non-degenerate $(n + 1)$-times integrated exponentially bounded semi-group $\{S_{n+1}(t),\ t \geq 0\}$ satisfying the condition*

$$\limsup_{h \to 0} h^{-1} \|S_{n+1}(t + h) - S_{n+1}(t)\| \leq Ce^{\omega t}, \qquad t \geq 0,$$

if and only if the resolvent of A satisfies (R1).

It is interesting that for $n = 0$ the theorem implies that existence of a 1-times integrated semi-group is equivalent to the MFPHY conditions. At first sight this result is weaker than that of the MFPHY theorem. The point is that the generator of an integrated semi-group, unlike that of a C_0-semi-group, may be a non-densely defined operator. Later we investigate the properties of n-times integrated semi-groups and prove that a stronger result guaranteeing for $n = 0$ the equivalence of assertions (ii) and (iii) of the MFPHY theorem is valid if $\overline{dom\ A} = \mathcal{X}$.

Now in the next few theorems we summarize the main properties of integrated exponentially bounded semi-groups. (Proofs appear in [79].)

---

[2] A note about notations: in this section we denote an n-times integrated semi-group by S_n, a K-convoluted one by S_K, and sometimes an R-semi-group by S_R. Further on, particularly in Part II, if there is no danger of confusion, we use the notation for all these semi-groups without a subscript, just S.

1.2. The Cauchy problem with generators of regularized semi-groups 17

Theorem 1.2.2 *Let A be the generator of a non-degenerate n-times integrated exponentially bounded semi-group $\{S_n(t), t \geq 0\}$, $n \in \mathbb{N}_0$. Then*

1) *if $f \in dom\ A$, then $S_n(t)f \in dom\ A$ and*

$$AS_n(t)f = S_n(t)Af, \quad \int_0^t S_n(s)Af\ ds = S_n(t)f - \frac{t^n}{n!}f, \quad t \geq 0; \quad (1.2.2)$$

2) *if $f \in \overline{dom\ A}$, then $\int_0^t S_n(s)f\ ds \in dom\ A$ and*

$$A\int_0^t S_n(s)f\ ds = S_n(t)f - \frac{t^n}{n!}f, \quad t \geq 0; \quad (1.2.3)$$

3) *if $f \in dom\ A^p$, $p = 1, 2, \ldots, n$, then*

$$S_n^{(p)}(t)f = S_n(t)A^p f + \sum_{k=0}^{p-1} \frac{t^{n-p+k}}{(n-p+k)!}A^k f, \quad t \geq 0; \quad (1.2.4)$$

4) *if $f \in dom\ A^{n+1}$, then*

$$\frac{d}{dt}S_n^{(n)}(t)f = AS_n^{(n)}(t)f = S_n^{(n)}Af, \quad t \geq 0. \quad (1.2.5)$$

Theorem 1.2.3 *Let A be a linear densely defined on $\mathcal{X}$ operator. Then A satisfies (R1) if and only if A generates a non-degenerate n-times integrated semi-group $\{S_n(t), t \geq 0\}$ such that*

$$\|S_n(t)\| \leq Ce^{\omega t}, \quad t \geq 0.$$

Based on the properties of n-times integrated semi-groups, we study the connection of these families to the Cauchy problem.

Note that the equality (1.2.4) with $p = 1$ due to commutativity of a semi-group with its generator on $dom\ A$ takes the form

$$S_n'(t)f = AS_n(t)f + \frac{t^{n-1}}{(n-1)!}f, \quad t \geq 0, \quad f \in dom\ A.$$

This equality shows that $v(t) := S_n(t)f$ for $t \geq 0$ and $f \in dom\ A$ is a solution of the Cauchy problem

$$v'(t) = Av(t) + \frac{t^{n-1}}{(n-1)!}f, \quad t \geq 0, \quad f \in dom\ A, \quad v(0) = 0. \quad (1.2.6)$$

Thus we obtain that the operators of an n-times integrated semi-group are solving operators for the Cauchy problem (1.2.6). The problem can be considered as a regularizing problem for (1.0.1) and n-tuple integration can be considered as a way of regularization of the original ill-posed problem.

18   *1. Semi-group methods for construction of solutions*

Property 4 of Theorem 1.2.2 shows the connection between the family of operators and the original Cauchy problem (1.0.1). Let us introduce a well-posedness of the Cauchy problem (1.0.1) corresponding to the case considered. It is naturally weaker compared with the uniform well-posedness introduced in the previous section.

Definition 1.2.3 *The Cauchy problem (1.0.1) is called uniformly (n, ω)-well-posed, if for each $f \in dom\, A^{n+1}$ and for each $T > 0$*

(a) *there exists the unique solution of the problem*

$$u \in C\left([0, T], dom\, A\right) \cap C^1\left([0, T], \mathcal{X}\right);$$

(b) *there exist $C > 0$, $\omega \in \mathbb{R}$ such that $\|u(t)\| \leq Ce^{\omega t}\|f\|_n$, $t \geq 0$, where $\|f\|_n := \|f\| + \|Af\| + \ldots + \|A^n f\|$.*

The connection of a uniformly (n, ω)-well-posed Cauchy problem with integrated semi-groups is illustrated by the following theorem.

Theorem 1.2.4 *Let A be a linear operator densely defined on $\mathcal{X}$ with a non-empty resolvent set. Then the following assertions are equivalent:*

(i) the Cauchy problem (1.0.1) is uniformly (n, ω)-well-posed;

(ii) A is the generator of a non-degenerate n-times integrated exponentially bounded semi-group of operators $\{S_n(t),\ t \geq 0\}$.

Proof. $(i) \Longrightarrow (ii)$. Let $f \in dom\, A^{n+1}$. Then there exists the unique solution u of (1.0.1) such that $\|u(t)\| \leq Ce^{\omega t}\|f\|_n$, $t \geq 0$. Hence, for each $\mu \in \rho(A)$, the function $w(t) = \mathcal{R}(\mu)u(t)$, $t \geq 0$, is a solution of the Cauchy problem with the initial value $\mathcal{R}(\mu)f$. The estimate $\|w(t)\| \leq C_1 e^{\omega t}\|f\|_{n-1}$, $t \geq 0$, is valid for it. Denote

$$v_1(t) = \int_0^t u(s)\, ds,$$

then

$$\begin{aligned} v_1(t) &= \mathcal{R}(\mu)(\mu I - A)\int_0^t u(s)\, ds = \mu\mathcal{R}(\mu)\int_0^t u(s)\, ds - \mathcal{R}(\mu)\int_0^t Au(s)\, ds \\ &= \mu\int_0^t w(s)\, ds - \mathcal{R}(\mu)\left(u(t) - u(0)\right) = \mu\int_0^t w(s)\, ds + \mathcal{R}(\mu)f - w(t), \end{aligned}$$

and $\|v_1(t)\| \leq C_2 e^{\omega t}\|f\|_{n-1}$.

By induction we obtain that the n-times integrated solution of (1.0.1),

$$v_n(t) = \int_0^t \frac{1}{(n-1)!}(t-s)^{n-1}u(s)\, ds, \qquad t \geq 0,$$

is an exponentially bounded function: $\|v_n(t)\| \leq C_n e^{\omega t}\|f\|$.

1.2. The Cauchy problem with generators of regularized semi-groups   19

We define a family of linear bounded operators $S_n(t) : dom\, A^{n+1} \to \mathcal{X}$ for $t \geq 0$, putting $S_n(t)f := v_n(t)$, $f \in dom\, A^{n+1}$.

We show that conditions $\overline{dom\, A} = \mathcal{X}$ and $\rho(A) \neq \emptyset$ imply $\overline{dom\, A^{n+1}} = \mathcal{X}$. Since the resolvent set of A is not empty, any $f \in dom\, A$ can be represented as $f = \mathcal{R}(\mu)g$, $g \in \mathcal{X}$, $\mu \in \rho(A)$. Due to the density of $dom\, A$ in $\mathcal{X}$, there exists a sequence $f_n \in dom\, A$ such that $f_n \to g$. Then

$$h_n = \mathcal{R}(\mu)f_n \in dom\, A^2 \quad \text{and} \quad h_n \to \mathcal{R}(\mu)g = f,$$

which implies $\overline{dom\, A^2} = \overline{dom\, A} = \mathcal{X}$. By repeating the procedure $(n-1)$ times, we obtain the required equality $\overline{dom\, A^{n+1}} = \overline{dom\, A} = \mathcal{X}$.

Now the operators $S_n(t)$, $t \geq 0$, can be continuously extended to $\mathcal{X}$. The operator function $S_n(\cdot)$ obtained on $\mathcal{X}$ is exponentially bounded as $t \geq 0$, is continuous wrt t for every $f \in dom\, A^{n+1}$, and therefore is strongly continuous. It is easy to show that the operators

$$r(\mu) = \int_0^\infty \mu^n e^{-\mu t} S_n(t)\, dt, \quad Re\mu > \omega,$$

coincide with the resolvent $\mathcal{R}(\mu)$ of A. Thus $\{S_n(t),\, t \geq 0\}$ is the n-times integrated semi-group generated by A.

$(ii) \Longrightarrow (i)$. For each $f \in dom\, A^{n+1}$ in accordance with (1.2.4) we have

$$S_n^{(n)}(t)f = S_n(t)A^n f + \sum_{k=0}^{n-1} \frac{t^k}{k!} A^k f, \qquad t \geq 0.$$

Denote

$$u(t) := S_n^{(n)}(t)f, \quad f \in dom\, A^{n+1}, \quad t \geq 0.$$

Then $u(0) = f$ and $\|u(t)\| \leq Ce^{\omega t}\|f\|_n$. It follows from the property (1.2.5) that $u(t) \in dom\, A$ and

$$u'(t) = Au(t), \qquad t \geq 0.$$

Now we prove that the solution obtained is unique. Let v be one more solution of (1.0.1) with $v(0) = f \in dom\, A^{n+1}$. Then $\mathcal{R}^n(\mu)v$, $\mu \in \rho(A)$, is a solution of (1.0.1) with the initial data $\mathcal{R}^n(\mu)f \in dom\, A^{n+1}$. Moreover, $\mathcal{R}^n(\mu)v(t) \in dom\, A^{n+1}$ for every $t \geq 0$. Hence

$$\frac{d}{ds}S_n^{(n)}(t-s)\mathcal{R}^n(\mu)v(s)$$

$$= -AS_n^{(n)}(t-s)\mathcal{R}^n(\mu)v(s) + S_n^{(n)}(t-s)A\mathcal{R}^n(\mu)v(s) = 0,$$

for $0 \leq s \leq t$. Therefore, $S_n^{(n)}(t-s)\mathcal{R}^n(\mu)v(s)$ as a function of s is constant on the segment $[0, t]$, in particular, it takes the same values as $s = t$ and $s = 0$:

$$S_n^{(n)}(0)\mathcal{R}^n(\mu)v(t) = S_n^{(n)}(t)\mathcal{R}^n(\mu)v(0).$$

20   *1. Semi-group methods for construction of solutions*

Commutativity of the semi-group operators with the resolvent of the generator implies

$$\mathcal{R}^n(\mu)S_n^{(n)}(0)v(t) = \mathcal{R}^n(\mu)S_n^{(n)}(t)f, \qquad f \in dom\, A^{n+1}, \quad t \geq 0,$$

hence $v = S_n^{(n)}f$, i.e., any solution of (1.0.1) is equal to $S_n^{(n)}f$. $\qquad\qquad\square$

Remark 1.2.1 As follows from the proof of Theorem 1.2.4, if $\overline{dom\, A} \neq \mathcal{X}$, then the operators $S_n(t)$, $S_n(t)f := v_n(t), t \geq 0$, are defined only on $\overline{dom\, A}$ and the operators $S_{n+1}(t) := \int_0^t S_n(s)\, ds$ can be extended to the whole space $\mathcal{X}$ and form an $(n+1)$-times integrated semi-group. That is why in the general case of a non-densely defined A the (n, ω)-well-posedness of the problem implies existence of only an $(n+1)$-times integrated semi-group defined by A. $\qquad\square$

Summarizing the results obtained in Theorem 1.2.4, we formulate them as the following theorem, an analog of the MFPHY theorem.

Theorem 1.2.5 *Let A be a linear densely defined on $\mathcal{X}$ operator with a non-empty resolvent set. Then the following assertions are equivalent:*

(i) the Cauchy problem (1.0.1) is uniformly (n, ω)-well-posed;

(ii) A is the generator of an n-times integrated exponentially bounded semi-group of operators $\{S_n(t),\, t \geq 0\}$;

(iii) the resolvent of A satisfies the condition (R1).

Under these equivalent conditions the solution of (1.0.1) has the form

$$u(t) = S_n^{(n)}(t)f, \qquad f \in dom\, A^{n+1}, \quad t \geq 0.$$

It is shown in [15, 16, 79] that the properties 1 and 2 of Theorem 1.2.2, where 2 is true on $\mathcal{X}$, are equivalent to the relation $(S_n 1)$. Thus (1.2.2) on $dom\, A$ and (1.2.3) on $\mathcal{X}$ can serve as an equivalent definition of an n-times integrated exponentially bounded semi-group generated by A, as well as of a local integrated semi-group, which we study further. This definition generalizes Definition 1.1.3 of a C_0-semi-group to the case of $n > 0$.

Definition 1.2.4 *Let A be a linear closed operator on $\mathcal{X}$ and let $n \in \mathbb{N}$. A strongly continuous exponentially bounded with respect to $t \geq 0$ family of bounded linear operators $\{S_n(t),\, t \geq 0\}$ acting in $\mathcal{X}$ and satisfying the equations*

$$S_n(t)Af = AS_n(t)f, \qquad f \in dom\, A, \quad t \geq 0, \tag{1.2.7}$$

$$A\int_0^t S_n(s)f\, ds = S_n(t)f - \frac{t^n}{n!}f, \qquad f \in \mathcal{X}, \quad t \geq 0, \tag{1.2.8}$$

is called an n-times integrated exponentially bounded semi-group of operators generated by A and A is called the generator of the family. The semi-group is called non-degenerate if $S_n(t)f = 0$ for every $t > 0$ implies $f = 0$.

1.2. The Cauchy problem with generators of regularized semi-groups — 21

1.2.2 The Cauchy problem with generators of local integrated semi-groups

The aims of this subsection are to determine the well-posedness of the Cauchy problem (1.0.1) with A satisfying the condition (R2), to find a "semi-group" family that allows us to obtain solution operators of such a problem, and to indicate the way of its regularization.

We show that further restriction of the set of regular points of A in the condition (R2) (compared with C_0-semi-groups and exponentially bounded integrated semi-groups) implies the property of the family to be local wrt t. In addition, the power growth of the resolvent, similar to the case of exponentially bounded integrated semi-groups, implies the family is integrated.

By analogy with Definition 1.2.4, we define a local integrated semi-group simultaneously with its generator.

Definition 1.2.5 *Let A be a linear closed operator in $\mathcal{X}$, $n \in \mathbb{N}$. A strongly continuous with respect to $t \in [0, \tau)$ family of linear bounded operators $\{S_n(t), t \in [0, \tau)\}$ that satisfies Equations (1.2.7) and (1.2.8) for $t \in [0, \tau)$ is called an n-times integrated semi-group generated by A and A is called the generator of the family; the semi-group is called local if $\tau < \infty$.*

To construct a semi-group family generated by A whose resolvent satisfies (R2), we neutralize the resolvent growth by multiplying it by $\lambda^{-(n+1)}$ and take the inverse Laplace transform of the product $\lambda^{-(n+1)} \mathcal{R}(\lambda)$, $\lambda \in \Lambda_{n,\nu,\omega}^{\ln}$. We show that a local $(n+1)$-times integrated semi-group is thus obtained.

Theorem 1.2.6 *Let the resolvent of A satisfy the condition (R2). Then A generates a local $(n+1)$-times integrated semi-group on $[0, \tau)$ for $\tau = \frac{1}{n\nu}$.*

Proof. Consider the inverse Laplace transform of the operator-function $\lambda^{-(n+1)} \mathcal{R}(\lambda)$, $\lambda \in \Lambda_{n,\nu,\omega}^{\ln}$:

$$S_n(t) := \int_{\Gamma} e^{\lambda t} \frac{1}{\lambda^{n+1}} \mathcal{R}(\lambda) \, d\lambda, \qquad \Gamma = \partial \Lambda_{n,\nu,\omega_1}^{\ln}, \quad \omega_1 > \omega.$$

Then, due to the estimate for $\mathcal{R}$, we have

$$\|S_n(t)\|_{\mathcal{L}(\mathcal{X})} \leq \int_{\Gamma} C \frac{e^{t \operatorname{Re}\lambda}}{|\lambda|} \, |d\lambda|.$$

Further, on the curve $\Gamma = \{\lambda \in \mathbb{C} : \operatorname{Re}\lambda = n\nu \ln |\lambda| + \omega_1\}$, we have

$$\frac{e^{t \operatorname{Re}\lambda}}{|\lambda|} = e^{\left(t - \frac{1}{n\nu}\right) \operatorname{Re}\lambda + \frac{\omega_1}{n\nu}}, \qquad \lambda \in \Gamma.$$

Hence, for each $t < \frac{1}{n\nu} =: \tau$, the operator $S_n(t)$ is bounded. We show that it

22   *1. Semi-group methods for construction of solutions*

satisfies the equality (1.2.8) from Definition 1.2.5. For each $f \in X$ we obtain:

$$
\begin{aligned}
A \int_0^t S_n(s) f \, ds &= \int_0^t \int_\Gamma \frac{e^{\lambda s}}{\lambda^{n+1}} (A \pm \lambda I) \mathcal{R}(\lambda) f \, d\lambda \, ds \\
&= \int_0^t \int_\Gamma \frac{e^{\lambda s}}{\lambda^{n+1}} \lambda \mathcal{R}(\lambda) f \, d\lambda \, ds - \int_0^t \int_\Gamma \frac{e^{\lambda s}}{\lambda^{n+1}} f \, d\lambda \, ds \\
&= S_n(t) f - \int_\Gamma \frac{1}{\lambda^{n+1}} \mathcal{R}(\lambda) f \, d\lambda - \frac{t^{n+1}}{(n+1)!} f.
\end{aligned}
$$

Since on the contour Γ we have

$$
\left\| \frac{1}{\lambda^{n+1}} \mathcal{R}(\lambda) \right\| \le \frac{C}{|\lambda|} = C e^{\frac{\omega_1 - Re\lambda}{n\nu}},
$$

the integral over Γ in the right-hand side is equal to zero for each $f \in \mathcal{X}$ and we obtain (1.2.8). Equation (1.2.7) can be derived in a similar manner using the equality $\mathcal{R}(\lambda)(\lambda I - A) f = f$, $f \in dom\, A$. □

For local integrated semi-groups, in contrast to exponentially bounded ones, the inverse result holds up to the value of the parameter n.

Theorem 1.2.7 [77] *Let a linear closed operator A generate a local n-times integrated semi-group $\{S_n(t),\ t \in [0, \tau)\}$, $n \in \mathbb{N}$. Then, for every $T \in (0, \tau)$, there exists $\omega \in \mathbb{R}$ such that*

$$
\Lambda^{\ln}_{n,\,1/T,\,\omega} = \left\{ \lambda \in \mathbb{C} : \ Re\lambda > \frac{n}{T} \ln |\lambda| + \omega \right\} \subseteq \rho(A),
$$

and the resolvent of A satisfies (R2) *in this region.*

Now let us see what type of regularization for the original Cauchy problem (1.0.1) is connected with the integrated semi-groups. Differentiating (1.2.8) on $dom\, A$, we obtain that $S_n(t) f$ provides a solution to the Cauchy problem for the inhomogeneous equation

$$
S'_n(t) f = A S_n(t) f + \frac{t^{n-1}}{(n-1)!} f, \quad f \in dom\, A, \quad t \in [0, \tau), \quad S_n(0) f = 0.
$$

This implies that the regularization of the Cauchy problem (1.0.1) with the generator of an n-times integrated semi-group is again due to n-times integration of (1.0.1).

To show the connection between integrated semi-groups and well-posedness of the Cauchy problem (1.0.1), we introduce the following definition of n-well-posedness.

Definition 1.2.6 *The Cauchy problem (1.0.1) is called n-well-posed, if, for each $f \in dom\, A^{n+1}$,*

(a) *there exists the unique solution of the problem*

$$
u \in C\left([0, \tau), dom\, A \right) \cap C^1\left([0, \tau), \mathcal{X} \right);
$$

1.2. The Cauchy problem with generators of regularized semi-groups   23

(b) *for each $T \in [0, \tau)$ there exists $C_T > 0$ such that*

$$\sup_{t \in [0,T]} \|u(t)\| \leq C_T \|f\|_n, \qquad (1.2.9)$$

where $\|f\|_n = \|f\| + \|Af\| + \ldots + \|A^n f\|$.

The following theorem gives an important characteristic of a solution of the n-well-posed Cauchy problem.

Theorem 1.2.8 *If for each $f \in dom\, A^{n+1}$ there exists the unique solution of the Cauchy problem (1.0.1) and $\rho(A) \neq \emptyset$, then the solution satisfies the stability condition (1.2.9).*

Proof. The proof is based on the Banach closed graph theorem applied to the spaces

$$[dom\, A^{n+1}] = \{f \in dom\, A^{n+1}, \|f\|_{n+1}\} \quad \text{and} \quad [dom\, A] = \{f \in dom\, A, \|f\|_1\}.$$

It is easy to check that both spaces introduced are complete. Consider the operator

$$U : [dom\, A^{n+1}] \to C\left([0,T], [dom\, A]\right),$$

which is defined everywhere on $[dom\, A^{n+1}]$:

$$U(t)f = u(t), \qquad f \in [dom\, A^{n+1}], \quad t \in [0, T].$$

Let us show that it is closed. Let $f_j \to f$ in $[dom\, A^{n+1}]$ and $u_j(t) \to v$ in $C\left([0,T], [dom\, A]\right)$. Then we get for each $t \in [0, T]$ the convergence in $\mathcal{X}$:

$$u_j'(t) = Au_j(t) \to Av(t) \quad \text{and} \quad u_j(t) - f_j \to v(t) - f = \int_0^t Av(s)\, ds,$$

hence, $v'(t) = Av(t)$, $v(0) = f$. By the Banach theorem, the operator U is continuous, i.e.,

$$\sup_{t \in [0,T]} \|u(t)\|_1 \leq C_T \|f\|_{n+1}.$$

Hence, due to the existence of a regular point for A, the n-stability estimate

$$\sup_{t \in [0,T]} \|u(t)\| \leq C_T \|f\|_n$$

holds. $\qquad\square$

On the base of this result and the properties of integrated semi-groups we have

Theorem 1.2.9 *If A is the generator of a local n-times integrated semi-group $\{S_n(t),\, t \in [0, \tau)\}$, then the Cauchy problem (1.0.1) is n-well-posed.*

24   *1. Semi-group methods for construction of solutions*

Under the condition $\rho(A) \neq \emptyset$ the reverse result holds true.

Theorem 1.2.10 *Let $\rho(A) \neq \emptyset$. If the Cauchy problem (1.0.1) is n-well-posed, then in the case $\operatorname{dom} A = \mathcal{X}$ the operator A generates a local n-times integrated semi-group on $\mathcal{X}$; in the general case A generates an $(n + 1)$-times integrated semi-group.*

Proof. By the condition, for each $f \in \operatorname{dom} A^{n+1}$ there exists a unique solution of the Cauchy problem (1.0.1), i.e., the solution operators

$$U(t) : \ \operatorname{dom} A^{n+1} \to \mathcal{X}, \qquad t \in [0, \tau),$$

are well defined, satisfy the equation

$$U'(t)f = AU(t)f, \qquad f \in \operatorname{dom} A^{n+1}, \quad t \in [0, \tau), \tag{1.2.10}$$

and satisfy the stability condition

$$\sup_{t \in [0,T] \subset [0,\tau)} \|U(t)f\| \leq C_T \|f\|_n, \qquad f \in \operatorname{dom} A^{n+1}.$$

The solution operators commute with the resolvent of A on the set $\operatorname{dom} A^{n+1}$. Indeed, for each $f \in \operatorname{dom} A^{n+1}$ the function $u = U\mathcal{R}(\lambda)f$ is a solution of the Cauchy problem (1.0.1) with the initial data $\mathcal{R}(\lambda)f$. On the other side the function $v = \mathcal{R}(\lambda)Uf$ is also a solution of (1.0.1) with the initial data $\mathcal{R}(\lambda)f$. It follows from the uniqueness of a solution of the problem that the solutions are equal and the commutativity

$$U(t)\mathcal{R}(\lambda)f = \mathcal{R}(\lambda)U(t)f, \qquad f \in \operatorname{dom} A^{n+1}, \quad t \in [0, \tau),$$

holds. Let us apply the resolvent to both parts of Equation (1.2.10):

$$\mathcal{R}(\lambda)U'(t)f = (\lambda I - A)^{-1}(A - \lambda I)U(t)f + \lambda \mathcal{R}(\lambda)U(t)f,$$

where $f \in \operatorname{dom} A^{n+1}$ and $t \in [0, \tau)$. Then

$$\mathcal{R}(\lambda)U'(t)f = \lambda \mathcal{R}(\lambda)U(t)f - U(t)f, \qquad f \in \operatorname{dom} A^{n+1}, \quad t \in [0, \tau).$$

By commutativity of the resolvent with the solution operators we obtain

$$U(t)f = \lambda U(t)\mathcal{R}(\lambda)f - U'(t)\mathcal{R}(\lambda)f, \qquad t \in [0, \tau).$$

Here the right-hand side is well defined for each $\mathcal{R}(\lambda)f \in \operatorname{dom} A^{n+1}$, hence for each $f \in \operatorname{dom} A^n$. Thus we can extend operators $U(t)$ to $\operatorname{dom} A^n$ and denote the extension by the same symbol. From (1.2.10) we obtain the solution operators in another form:

$$U(t)f = A \int_0^t U(s)f \, ds - f, \qquad t \in [0, \tau).$$

1.2. The Cauchy problem with generators of regularized semi-groups 25

Due to the closedness of A, the equality also can be extended to $dom A^n$.

Let us introduce the operators

$$
\begin{aligned}
S_1(t)f \quad &:= \quad \int_0^t U(s)f \, ds \\
&= \quad \lambda \int_0^t U(s)\mathcal{R}(\lambda)f \, ds - U(t)\mathcal{R}(\lambda)f + \mathcal{R}(\lambda)f \quad\quad (1.2.11) \\
&= \quad A \int_0^t S_1(s)f \, ds - tf, \quad\quad f \in dom\, A^n, \quad t \in [0,\tau). \quad (1.2.12)
\end{aligned}
$$

It is evident that the operators introduced satisfy the equation

$$
S_1{}'(t)f = AS_1(t)f - f, \quad\quad f \in dom\, A^n, \quad t \in [0,\tau).
$$

Applying the resolvent of A to it and acting by analogy with the previous case we obtain the equality

$$
S_1(t)f = \lambda S_1(t)\mathcal{R}(\lambda)f - S_1'(t)\mathcal{R}(\lambda)f - \mathcal{R}(\lambda)f, \quad\quad t \in [0,\tau),
$$

which allows us to continue the operators $S_1(t)$ and the equality (1.2.12) to $dom\, A^{n-1}$.

From (1.2.11) we obtain the estimate

$$
\|S_1(t)f\| \leq C_1 \|\mathcal{R}(\lambda)f\|_n \leq C_1' \|f\|_{n-1}, \quad\quad f \in dom\, A^n, \quad t \in [0,\tau),
$$

and integrating sequentially we obtain the operators

$$
S_n(t)f := \int_0^t S_{n-1}(s)f \, ds = A \int_0^t S_n(s)f \, ds - \frac{t^n}{n!}f, \quad f \in dom\, A, \; t \in [0,\tau),
$$

and their representation in the form

$$
S_n(t)f = \lambda S_n(t)\mathcal{R}(\lambda)f - S_n'(t)\mathcal{R}(\lambda)f - \frac{t^{n-1}}{(n-1)!}\mathcal{R}(\lambda)f, \quad\quad t \in [0,\tau).
$$

It allows us to continue these operators to the whole $\mathcal{X}$. Thus, we have built a family of operators satisfying the equation

$$
S_n(t)f = A \int_0^t S_n(s)f \, ds - \frac{t^n}{n!}f, \quad\quad f \in \mathcal{X}, \quad t \in [0,\tau),
$$

and the estimates

$$
\|S_n(t)f\| \leq C_n \|\mathcal{R}(\lambda)f\|_1 \leq \hat{C}_n \|f\|, \quad\quad f \in dom\, A, \quad t \in [0,\tau). \quad (1.2.13)
$$

If now $\overline{dom\, A} = \mathcal{X}$, then (1.2.13) can be continued to the whole $\mathcal{X}$. Hence, the family obtained is a family of bounded operators.

26   *1. Semi-group methods for construction of solutions*

If $\overline{dom\, A} \neq \mathcal{X}$, then integrating once more we obtain a family of operators satisfying the equation

$$S_{n+1}(t)f = A \int_0^t S_{n+1}(s)f\, ds - \frac{t^{n+1}}{(n+1)!}f, \qquad f \in \mathcal{X}, \quad t \in [0, \tau),$$

with the corresponding estimates (1.2.13) for $S_{n+1}(t)f$. Boundedness and the property of strong continuity with respect to t follow from here. $\qquad\square$

1.2.3   The Cauchy problem with generators of convoluted semi-groups

Here we investigate the Cauchy problem (1.0.1) with A satisfying the condition (R3):

$$\|\mathcal{R}(\lambda)\| \leq Ce^{\omega M(\gamma|\lambda|)}, \quad \lambda \in \Lambda^M_{\alpha,\,\gamma,\,\omega} = \{\lambda \in \mathbb{C}:\ Re\lambda > \alpha M(\gamma|\lambda|) + \omega\},$$

where $M(t)$, $t \geq 0$, is a positive function increasing as $t \to \infty$ and its growth rate does not exceed t^p, $p < 1$.

The aim of the subsection is to indicate the "semi-group" family generated by A and to find a way to regularize the problem (1.0.1) with such an operator.

In order to obtain the regularized family of operators, we neutralize the growth of the resolvent of A by multiplying it by an appropriate function $\widetilde{K}$, and then apply the inverse Laplace transform to the product $\mathcal{R}(\lambda)\widetilde{K}(\lambda)$, $\lambda \in \Lambda^M_{\alpha,\,\gamma,\,\omega}$. We will show that in this case a K-convoluted semi-group appears, namely, with $K = \mathcal{L}^{-1}[\widetilde{K}]$.

By analogy with the definition of integrated semi-groups we define a K-convoluted semi-group simultaneously with its generator via corresponding equations.

Definition 1.2.7 [16, 15] *Let A be a linear closed operator on a Banach space $\mathcal{X}$. Let K be a function continuous on $[0,\tau)$, $\tau \leq \infty$. A strongly continuous with respect to $t \in [0,\tau)$ family of bounded linear operators $\{S_K(t),\ t \in [0,\tau)\}$ on $\mathcal{X}$ satisfying the conditions*

$$S_K(t)Af = AS_K(t)f, \qquad f \in dom\, A, \quad t \in [0, \tau), \qquad (1.2.14)$$

$$A \int_0^t S_K(s)f\, ds = S_K(t)f - \int_0^t K(s)f\, ds, \qquad f \in \mathcal{X}, \quad t \in [0, \tau), \quad (1.2.15)$$

is called a K-convoluted semi-group of operators generated by A, and A is called the generator of the family.

Equations (1.2.14) and (1.2.15) imply the existence of a derivative for $\{S_K(t),\ t \in [0,\tau)\}$ on the domain of the generator and the next representation for it:

$$S_K'(t)f = AS_K(t)f + K(t)f, \qquad f \in dom\, A, \quad t \in [0, \tau).$$

1.2. The Cauchy problem with generators of regularized semi-groups   27

This means that $S_K(t)f$ is the solution of the inhomogeneous Cauchy problem:

$$v'(t) = Av(t) + K(t)f, \quad t \in [0, \tau), \qquad v(0) = 0. \tag{1.2.16}$$

If u is a solution of (1.0.1) with initial data $f \in dom\, A$, then its convolution $v = K * u$ is a solution of (1.2.16):

$$\frac{d}{dt}(K * u)(t) \;=\; \frac{d}{dt}\int_0^t K(s)u(t-s)\,ds = K(t)u(0) + \int_0^t K(s)u'(t-s)\,ds$$

$$=\; K(t)f + A(K*u)(t), \quad t \in [0, \tau), \qquad (K*u)(0) = 0.$$

This fact explains the name of the family: if $\{U(t),\, t \geq 0\}$ is a C_0-semi-group with the generator A, then its convolution with the function $K(t),\, t \geq 0$, as was just shown, forms a K-convoluted semi-group. In particular, if

$$K(t) = \frac{t^{n-1}}{(n-1)!}, \qquad t \geq 0, \tag{1.2.17}$$

the K-convoluted semi-group is an n-times integrated semi-group.

The semi-group property of a K-convoluted semi-group has the following form [16]:

$$S_K(t)S_K(s) \;=\; \int_s^{t+s} K(t+s-r)S_K(r)\,dr$$

$$=\; \int_0^{t+s} K(t+s-r)S_K(r)\,dr - \int_0^s K(t+s-r)S_K(r)\,dr$$

for $t, s, t+s \in [0, \tau)$. It is not difficult to show that for K having the form (1.2.17) the above equality coincides with the property (S_n1) of an n-times integrated semi-group.

In the following two theorems, which are extensions of Theorems 1.2.6 and 1.2.9, we establish a connection between the behavior of the resolvent of A and the existence of a K-convoluted semi-group with the generator A.

Theorem 1.2.11 *Let $M(t),\, t \geq 0$, be a positive function increasing as $t \to \infty$ with the growth rate not exceeding t^p, $p < 1$, and let* (R3) *hold. Then A generates a local K-convoluted semi-group $\{S_K(t),\, t \in [0, \tau)\}$ with $\tau = \frac{1}{\alpha}\left(\frac{\kappa}{\gamma} - \beta\right)$. The Laplace transform of the corresponding K satisfies the condition*

$$|\tilde{K}(\lambda)| = \mathcal{O}\left(e^{-M(\kappa|\lambda|)}\right) \qquad as \quad |\lambda| \to \infty \qquad for \quad \kappa > \beta\gamma.$$

Proof. Let us consider "nearly the inverse" Laplace transform

$$S_K(t) := \int_\Gamma e^{\lambda t}\,\tilde{K}(\lambda)\mathcal{R}(\lambda)\,d\lambda, \qquad t \geq 0, \tag{1.2.18}$$

28   *1. Semi-group methods for construction of solutions*

of the operator-function $\widetilde{K}\mathcal{R}$, where

$$\Gamma = \{\lambda \in \mathbb{C} : \ Re\lambda = \alpha M(\gamma|\lambda|) + \omega_1\} \subset \Lambda_{\alpha,\gamma,\omega}, \qquad \omega_1 > \omega.$$

According to assumptions of the theorem we have

$$\|S_K(t)\| \leq C_1 \int_\Gamma e^{tRe\lambda + \beta M(\gamma|\lambda|) - M(\kappa|\lambda|)} |d\lambda|.$$

On the contour Γ we have

$$M(\kappa|\lambda|) = M\left(\gamma \cdot \left|\tfrac{\kappa\lambda}{\gamma}\right|\right) = (Re\tfrac{\kappa\lambda}{\gamma} - \omega_1)/\alpha = (\tfrac{\kappa}{\gamma}Re\lambda - \omega_1)/\alpha.$$

Hence, for the integrand we obtain

$$e^{tRe\lambda + \beta M(\gamma|\lambda|) - M(\kappa|\lambda|)} = C_2 e^{\left(t + \frac{\beta}{\alpha} - \frac{\kappa}{\alpha\gamma}\right)Re\lambda}, \qquad \lambda \in \Gamma.$$

Therefore, $S_K(t)$ is defined for every $t < \left(\tfrac{\kappa}{\gamma} - \beta\right)\tfrac{1}{\alpha} =: \tau$.

Now we show that the operators $S_K(t)$, $t \in [0, \tau)$, satisfy (1.2.15). For any $f \in \mathcal{X}$ we obtain

$$\begin{aligned}
A \int_0^t S_K(s)f\,ds &= (A \pm \lambda I) \int_0^t \int_\Gamma e^{\lambda s}\mathcal{R}(\lambda)\widetilde{K}(\lambda)f\,d\lambda ds \\
&= \lambda \int_0^t \int_\Gamma e^{\lambda s}\mathcal{R}(\lambda)\widetilde{K}(\lambda)f d\lambda ds - \int_0^t \int_\Gamma e^{\lambda s}\widetilde{K}(\lambda)f\,d\lambda ds \\
&= \int_\Gamma \mathcal{R}(\lambda)\widetilde{K}(\lambda)\lambda \int_0^t e^{\lambda s}f\,ds d\lambda - \int_0^t K(s)f\,ds \\
&= S_K(t)f - \int_\Gamma \mathcal{R}(\lambda)\widetilde{K}(\lambda)f\,d\lambda - \int_0^t K(s)f\,ds.
\end{aligned}$$

For $\lambda \in \Gamma$ we have

$$\|\widetilde{K}(\lambda)\mathcal{R}(\lambda)\| \leq C_3 e^{\beta M(\gamma|\lambda|) - M(\kappa|\lambda|)} = C_4 e^{\frac{1}{\alpha}\left(\beta - \frac{\kappa}{\gamma}\right)Re\lambda},$$

where $\kappa > \beta\gamma$. Hence, the integral along Γ in the right-hand side is equal to zero for any $f \in \mathcal{X}$ and we obtain

$$A \int_0^t S_K(s)f\,ds = S_K(t)f - \int_0^t K(s)f\,ds, \qquad f \in \mathcal{X}.$$

A reader can easily check the condition (1.2.14) using the equality $\mathcal{R}(\lambda)(\lambda I - A)f = f$, $f \in dom\,A$, $\lambda \in \rho(A)$. $\qquad\square$

The following converse result is true.

Theorem 1.2.12 *Let linear closed operator A generate a K-convoluted semi-group $\{S_K(t), t \in [0, \tau)\}$ on $\mathcal{X}$, where K is an exponentially bounded*

1.2. The Cauchy problem with generators of regularized semi-groups   29

function: $|K(t)| \leq Ce^{\theta t}$, $t \geq 0$, $\theta \in \mathbb{R}$. *Let its Laplace transform satisfy the condition*

$$\left|\widetilde{K}(\lambda)\right| = O\left(e^{-M(\kappa|\lambda|)}\right) \quad as \ |\lambda| \to \infty,$$

where the function $M(t)$, $t \geq 0$, *is positive and increasing as* $t \to \infty$ *not faster than* t^p, $p < 1$. *Then, for any* $T \in (0, \tau)$ *there exists* $\omega \in \mathbb{R}$ *such that* (R3) *holds with* $\alpha = 1/T$ *and* $\gamma = \kappa$:

$$\|\mathcal{R}(\lambda)\| \leq Ce^{M(\kappa|\lambda|)}, \qquad \lambda \in \Lambda^M_{1/T, \kappa, \omega} = \left\{\lambda \in \mathbb{C} : \mathrm{Re}\lambda > \tfrac{1}{T}M(\kappa|\lambda|) + \omega\right\}.$$

Proof. Take $\lambda \in \mathbb{C}$ and consider the integral

$$\int_0^T e^{-\lambda s} S_K(s) \, ds, \qquad T \in (0, \tau).$$

Let us introduce an operator $\mathcal{R}_T(\lambda)$ via the local Laplace transform of the family $\{S_K(t), \ t \in [0, \tau)\}$, namely, by

$$\mathcal{R}_T(\lambda) := \widetilde{K}^{-1}(\lambda) \int_0^T e^{-\lambda s} S_K(s) \, ds, \qquad T \in (0, \tau).$$

We show that the operator is a "nearly resolvent" of A, i.e.,

$$\mathcal{R}(\lambda) = \mathcal{R}_T(\lambda)(I - B_T(\lambda)),$$

for all λ from a certain region $\Lambda^M_{1/T, \kappa, \omega}$ defined by $\|B_T(\lambda)\|_{\mathcal{L}(\mathcal{X})} \leq \delta < 1$.

It follows from (1.2.15) that for any $f \in \mathcal{X}$

$$\int_0^T S_K(s) f \, ds \in dom \, A.$$

Hence we can apply $(\lambda I - A)$ to $\mathcal{R}_T(\lambda)$:

$$(\lambda I - A)\mathcal{R}_T(\lambda) = \widetilde{K}^{-1}(\lambda) \left[\lambda \int_0^T e^{-\lambda s} S_K(s) \, ds - A \int_0^T e^{-\lambda s} S_K(s) \, ds\right].$$

Using (1.2.15) for the first term in the right-hand side and integrating the second one by parts we obtain

$$(\lambda I - A)\mathcal{R}_T(\lambda) = \widetilde{K}^{-1}(\lambda) \left[\lambda \int_0^T e^{-\lambda s} \int_0^s K(\xi) \, d\xi \, ds - e^{-\lambda T} A \int_0^T S_K(s) \, ds\right].$$

Now applying (1.2.15) to the second term and integrating the first one by parts we obtain

$$(\lambda I - A)\mathcal{R}_T(\lambda) = \widetilde{K}^{-1}(\lambda) \left[\int_0^T e^{-\lambda s} K(s) \, ds - e^{-\lambda T} S_K(T)\right].$$

1. Semi-group methods for construction of solutions

Let $\theta > 0$. Then, for $Re\lambda \geq \theta_1 > \theta$ we have

$$(\lambda I - A)\mathcal{R}_T(\lambda) = I - \widetilde{K}^{-1}(\lambda)\left[e^{-\lambda T}S_K(T) + \int_T^\infty e^{-\lambda s}K(s)\,ds\right] =: I - B_T(\lambda).$$

Let us estimate the operator $B_T(\lambda)$ for such λ. We have

$$\|B_T(\lambda)\| \leq |\widetilde{K}^{-1}(\lambda)|\left(e^{-TRe\lambda}\|S_K(T)\| + \int_T^\infty e^{-sRe\lambda}|K(s)|\,ds\right)$$

$$\leq C_1|\widetilde{K}^{-1}(\lambda)|\left(e^{-TRe\lambda} + \int_T^\infty e^{\theta s - \theta_1 s}\,ds\right)$$

$$\leq C_1|\widetilde{K}^{-1}(\lambda)|e^{T(\theta - Re\lambda)}\left(1 + \frac{1}{\theta_1 - \theta}\right) \leq C_2 e^{M(\kappa|\lambda|) - (Re\lambda - \theta)T}.$$

Now we find λ providing the inequality $C_2 e^{M(\kappa|\lambda|) - T(Re\lambda - \theta)} \leq \delta < 1$. It is valid if

$$M(\kappa|\lambda|) - T(Re\lambda - \theta) \leq \ln(\delta/C_2),$$

or, equivalently,

$$Re\lambda \geq \frac{M(\kappa|\lambda|) - \ln(\delta/C_2)}{T} + \theta =: \frac{M(\kappa|\lambda|)}{T} + \omega. \tag{1.2.19}$$

It follows from the condition imposed on M that this set is not empty. Thus for each λ satisfying (1.2.19) the estimate $\|B_T(\lambda)\|_{\mathcal{L}(\mathcal{X})} \leq \delta < 1$ holds. Hence there exists $(I - B_T(\lambda))^{-1}$. It follows from here that the resolvent $\mathcal{R}(\lambda)$ exists for $\lambda \in \Lambda_{1/T, \kappa, \omega} = \left\{\lambda \in \mathbb{C}: \ Re\lambda \geq \frac{M(\kappa|\lambda|)}{T} + \omega\right\}$ and satisfies the estimate

$$\|\mathcal{R}(\lambda)\| \leq |\widetilde{K}^{-1}(\lambda)|\left\|\int_0^T e^{-\lambda s}S_K(s)\,ds\right\|\frac{1}{1 - \delta} \leq C e^{M(\kappa|\lambda|)}. \qquad \square$$

We conclude this section with a note on uniqueness of solutions to the Cauchy problem (1.0.1). A few types of well-posedness (uniform well-posedness, (n, ω)-, and n-well-posedness) of the problem were established via semi-group techniques dependent on the resolvent behavior (R1)–(R3). The proof of uniqueness of the solution in each of these cases is based on the properties of the solution operators and is applicable for A with a non-empty resolvent set. There exists a general proof of uniqueness of the solution of the abstract Cauchy problem bypassing semi-group methods and connected with the Laplace transform instead. Namely, the following theorem holds true.

Theorem 1.2.13 [63] *If there exists such $\omega \in \mathbb{R}$ that*

$$\{\lambda \in \mathbb{C}: \ Re\lambda > \omega\} \subseteq \rho(A) \qquad and \qquad \overline{\lim}_{\lambda \to +\infty}\frac{\ln\|\mathcal{R}(\lambda)\|}{\lambda} < \infty,$$

then the Cauchy problem (1.0.1) on $[0, \infty)$ cannot have more than one solution in the Banach space $\mathcal{X}$.

If, in addition, the limit is equal to zero, then the local Cauchy problem cannot have more than one solution.

1.2. The Cauchy problem with generators of regularized semi-groups 31

If A does not have the resolvent, then the uniqueness of the solution in a Banach space can be proved by R-semi-group methods, which follow in the next subsection.

1.2.4 The Cauchy problem with generators of R-semi-groups. Regularized semi-groups

In the previous subsections we considered the Cauchy problem with the operator A being the generator of a semi-group of one of the classes C_0, C_1, or $\mathcal{A}$ or of a convoluted semi-group, in particular, an integrated one. Properties of these semi-groups were determined by the geometry of the set of regular points of the generator and by the estimates of its resolvent.

Now we consider the families of operators with the generators which can have no resolvent anywhere in the right half-plane. These are R-semi-groups introduced in [20, 25]. R-semi-groups, similar to convoluted and unlike strongly continuous ones, can be defined both on the whole semi-axis $[0, \infty)$ and locally. Further, following [27, 78, 79, 110], we formulate results on the properties of local R-semi-groups and on their connection with the Cauchy problem. We do not give detailed proofs here since many of them are conceptually closed and generalize the well-known "semi-group" proofs used in the previous subsections.

Let $R \in \mathcal{L}(\mathcal{X})$ be an injective operator with a dense $\operatorname{ran} R$ in $\mathcal{X}$.

Definition 1.2.8 *A family of bounded linear operators* $\{S(t), t \in [0, \tau)\}, \tau \leq \infty$, *on a Banach space* $\mathcal{X}$ *satisfying the conditions*

$(\mathbf{S}_R 1)$ $S(t + s)R = S(t)S(s), \quad s, t, s + t \in [0, \tau), \quad S(0) = R$

$(\mathbf{S}_R 2)$ *the operator-function* $S(\cdot)$ *is strongly continuous wrt t on* $[0, \tau)$

is called an R-*semi-group.*

An R-*semi-group is called exponentially bounded if* $\|S(t)\| \leq Ce^{\omega t}, t \geq 0$, *for some* $C > 0$, $\omega \in \mathbb{R}$, *and is called local if* $\tau < \infty$.

In the case $R = I$ the characteristic property $(S_R 1)$ of R-semi-groups coincides with the semi-group property (U1), which can be continued to all $t, s \geq 0$. Thus an I-semi-group is a C_0-semi-group.

As the basis for the definition of the generator of the family we take the definition of the infinitesimal generator. In the case of an R-semi-group there are a few such definitions.

Definition 1.2.9 *Let* $\{S(t), t \in [0, \tau)\}$ *be an* R-*semi-group. The operators* G *and* Z *defined by equalities*

$$Gf := \lim_{t \to 0} \frac{R^{-1}S(t) - I}{t} f, \qquad Zf := R^{-1} \lim_{t \to 0} \frac{S(t) - R}{t} f,$$

32   1. Semi-group methods for construction of solutions

with the domains $\operatorname{dom} G$, $\operatorname{dom} Z$ consisting of those f, for which the corresponding limits exist, are called infinitesimal generators of the family $\{S(t),\, t \in [0,\tau)\}$.

In [79] it is shown that these infinitesimal generators do not coincide in the general case. The inclusion $Z \subset G$ holds for them.

The operator $\overline{G}$ is called the (complete infinitesimal) generator of the family $\{S(t),\, t \in [0,\tau)\}$. It is easy to check that the operator G has a closure; therefore the definition of the complete infinitesimal generator is correct.

If $\{S(t),\, t \geq 0\}$ is an I-semi-group with the generator A, then, as we saw in Section 1.1, its Laplace transform coincides with the resolvent of A. In the case we now consider, first, if a semi-group is local, only the local Laplace transform can be defined, and, second, even for an exponentially bounded R-semi-group it is not necessarily equal to the resolvent of the generator. Nevertheless, in a certain sense that we will explain below, it plays the role of the resolvent.

Let us introduce the local Laplace transform of a family $\{S(t),\, t \in [0,\tau)\}$ by

$$\mathcal{R}_t(\lambda)f := \int_0^t e^{-\lambda\xi} S(\xi)f\, d\xi, \qquad t \in [0,\tau), \quad \lambda \in \mathbb{R}, \quad f \in \mathcal{X}. \qquad (1.2.20)$$

The main properties of R-semi-groups and of the operator-function $\mathcal{R}_t(\cdot)$ can be formulated as the following theorem.

Theorem 1.2.14 Let A be the complete infinitesimal generator of an R-semi-group $\{S(t),\, t \in [0,\tau)\}$ and the operator $\mathcal{R}_t(\lambda)$ be defined by the equality (1.2.20). Then

1) the operator G is densely defined in $\mathcal{X}$;

2) for any $f \in \operatorname{dom} A$, $t \in [0,\tau)$

$$S(t)x \in \operatorname{dom} A \quad and \quad AS(t)f = S(t)Af;$$

3) for any $f \in \operatorname{dom} A$, $t \in [0,\tau)$

$$A\mathcal{R}_t(\lambda)f = \mathcal{R}_t(\lambda)Af;$$

4) for any $f \in \mathcal{X}$, $t \in [0,\tau)$

$$\mathcal{R}_t(\lambda)f \in \operatorname{dom} A \quad and \quad (\lambda I - A)\mathcal{R}_t(\lambda)f = Rf - e^{-t\lambda}S(t)f;$$

5) for any $f \in \mathcal{X}$, $t \in [0,\tau)$

$$\mathcal{R}_t(\lambda)\mathcal{R}_t(\mu)f = \mathcal{R}_t(\mu)\mathcal{R}_t(\lambda)f;$$

1.2. The Cauchy problem with generators of regularized semi-groups   33

6) *for any* $t \in [0, \tau)$ *the operator-function* $\mathcal{R}_t(\cdot)$ *is differentiable with respect to* λ *on* $\mathbb{R}$ *and there exists a constant* $C_t > 0$ *such that*

$$\left\| \frac{\lambda^n}{(n-1)!} \frac{d^{n-1}}{d\lambda^{n-1}} \mathcal{R}_t(\lambda) \right\| \leq C_t, \quad \lambda > 0, \quad n \in \mathbb{N}. \tag{1.2.21}$$

The proof of the theorem generalizes the proofs of the corresponding properties of C_0-semi-groups and integrated ones.

Proceeding from properties 3–6 we introduce the following definition.

Definition 1.2.10 *Let A be a closed linear operator acting in $\mathcal{X}$ and let $\omega \in \mathbb{R}$. A family of bounded linear operators $\{\mathcal{R}_t(\lambda), \lambda > \omega\}$, $t \in [0, \tau)$, is called the asymptotical R-resolvent of A if*

(a) *for each $f \in \mathcal{X}$, $t \in [0, \tau)$ the function $\mathcal{R}_t(\lambda)f$, $\lambda > \omega$, is infinitely differentiable with respect to λ and*

$$\mathcal{R}_t(\lambda)\mathcal{R}_t(\mu)f = \mathcal{R}_t(\mu)\mathcal{R}_t(\lambda)f;$$

(b) *for any $f \in \mathcal{X}$, $t \in [0, \tau)$*

$$\mathcal{R}_t(\lambda)f \in \operatorname{dom} A \quad and \quad (\lambda I - A)\mathcal{R}_t(\lambda)f = Rf + B_t(\lambda)f,$$

where the function $B_t(\lambda)f$, $\lambda > \omega$, is infinitely differentiable with respect to λ and there exists $C_t > 0$ such that

$$\left\| \frac{d^{n-1}}{d\lambda^{n-1}} B_t(\lambda)f \right\| \leq C_t t^{n-1} e^{-t\lambda} \|f\|, \quad f \in \mathcal{X}, \quad \lambda > \omega, \quad n \in \mathbb{N};$$

(c) *for any $f \in \operatorname{dom} A$, $t \in [0, \tau)$,* $A\mathcal{R}_t(\lambda)f = \mathcal{R}_t(\lambda)Af$.

Similar to the case of C_0-semi-groups, integrated, and convoluted semi-groups, conditions for the existence of an R-semi-group with the generator A can be formulated in terms of the behavior of the asymptotical R-resolvent.

Theorem 1.2.15 *A linear closed operator A acting in a Banach space $\mathcal{X}$ is the complete infinitesimal generator of an R-semi-group $\{S(t), t \in [0, \tau)\}$ if and only if it satisfies the conditions*

1) $\operatorname{dom} A$ *is dense in $\mathcal{X}$ and*

$$\overline{A\big|_{R(\operatorname{dom} A)}} = A; \tag{1.2.22}$$

2) *for each $t \in [0, \tau)$ there exists the asymptotical R-resolvent $\mathcal{R}_t(\lambda)$ of the operator A and for $n/\lambda \in [0, t]$, $\lambda > \omega$, $n \in \mathbb{N}$, the estimates (1.2.21) hold true.*

34   *1. Semi-group methods for construction of solutions*

Now we show how R-semi-groups are connected with the Cauchy problem (1.0.1).

Definition 1.2.11 *The problem (1.0.1) is called R-well-posed on $[0, \tau)$ if for any $f \in R\,(dom\,A)$ there exists the unique solution u with $\|u(t)\| \le C_t \|R^{-1}f\|$, where C_t is bounded on each compact interval in $[0, \tau)$.*

We can formulate the main result on well-posedness of (1.0.1) in the following form.

Theorem 1.2.16 *Let A be a densely defined linear closed operator acting in $\mathcal{X}$. Let it commute with R on its domain and satisfy (1.2.22). Then A generates an R-semi-group if and only if the problem (1.0.1) is R-well-posed on $[0, \tau)$.*

Moreover, the solution of (1.0.1) has the form

$$u(t) = R^{-1}S(t)f, \qquad f \in R\,(dom\,A)\,, \quad t \in [0, \tau).$$

The following theorem shows the connections between integrated semi-groups and R-semi-groups.

Theorem 1.2.17 *Let A be a closed densely defined linear operator acting in $\mathcal{X}$. Let $\rho(A) \ne \emptyset$. Then the following conditions are equivalent:*

(i) A generates an n-times integrated semi-group $\{S_n(t),\, t \in [0, \tau)\}$;

(ii) A generates an R-semi-group $\{S(t),\, t \in [0, \tau)\}$ with $R = \mathcal{R}^n(\lambda_0)$, $\lambda_0 \in \rho(A)$.

Proof. The proof is constructive. The R-semi-group is built from the n-times integrated one and vice versa.

The R-semi-group is obtained from the integrated one by

$$S(t)f := \frac{d^n S_n(t)}{dt^n} \mathcal{R}^n(\lambda_0)f, \qquad f \in \mathcal{X}, \quad t \in [0, \tau). \qquad (1.2.23)$$

The definition is correct since the element $f_1 = \mathcal{R}^n(\lambda_0)f$ belongs to $dom\,A^n$ and the function $S_n(t)f_1,\, t \in [0, \tau)$, is n-times continuously differentiable with respect to t (Theorem 1.2.9).

The converse construction is given by

$$S_n(t)f = (\lambda_0 I - A)^n \int_0^t \int_0^{t_1} \ldots \int_0^{t_{n-1}} S(t_n)f\, dt_n \ldots dt_2\, dt_1, \qquad f \in \mathcal{X}.$$

$\square$

As follows from (1.2.23), the R-semi-group is the regularization by $R = \mathcal{R}^n(\lambda_0)$ of the solution operators

$$U(t) = d^n S_n(t)/dt^n, \quad t \in [0, \tau),$$

1.2. The Cauchy problem with generators of regularized semi-groups   35

of the Cauchy problem (1.0.1). Moreover, in the case of A generating an n-times integrated semi-group the regularization can be obtained by n-times integration of the solution operators since the primitives of the nth order of the family of solution operators can be extended from the set $dom\, A^n$ to the whole space $\mathcal{X}$ (Theorem 1.2.10).

Generally, this is not the case. If A is the generator of an R-semi-group, the resolvent does not generally exist, only the R-resolvent does. That is why regularization of solution operators is generally performed by R not necessarily equal to $\mathcal{R}^n(\lambda_0)$. Consider as an example the semi-groups of growth order α. These semi-groups have non-integrable singularities if $\alpha \geq 1$. The solution operators of the Cauchy problem with A being the generator of such a semi-group generally can be regularized neither via integration nor by a certain power of the resolvent $\mathcal{R}(\lambda_0)$. Nevertheless, as proved in [88], the generator of a semi-group of growth order α is the generator of a certain R-semi-group, namely, the R-semi-group with $R = (cI - A)^{-(n+1)}$, which is not generally the resolvent of A.

Theorem 1.2.18 *Let A be the generator of a semi-group $\{U(t), t > 0\}$ of growth order α and let $c > \omega_0$. Then A is the generator of an R-semi-group with R defined by*

$$Rf = \frac{1}{n!} \int_0^\infty t^n e^{-ct} U(t) f\, dt = (cI - A)^{-(n+1)}, \qquad f \in \mathcal{X},$$

where n is the integer part of α and ω_0 is the exponential type of the semi-group.

Summing up the assertions of Theorem 1.2.17 and Theorem 1.2.18 concerning the connections of the integrated semi-groups and the semi-groups of growth order α with R-semi-groups, we can say, that, in contrast to the relations (1.1.13) and (1.1.18) between the different classes of strongly continuous semi-groups in terms of embeddings, here we have relations which connect the semi-groups in terms of their generators, namely,

$$\{\text{class of generators of } n\text{-times integrated semi-groups}\} \subset$$
$$\subset \{\text{class of generators of } R\text{-semi-groups}\}.$$

Now we introduce regularized semi-groups. They serve as a generalization of integrated semi-groups, convoluted semi-groups, and R-semi-groups.

Definition 1.2.12 *Let A be a closed linear operator, $R(t), t \in [0, \tau), \tau \leq \infty$, be a continuous operator function with values in $\mathcal{L}(\mathcal{X})$ satisfying the conditions*

$$R(t)R(s)f = R(s)R(t)f, \qquad f \in \mathcal{X}, \quad t, s \in [0, \tau),$$
$$AR(t)f = R(t)Af, \qquad f \in dom\, A, \quad t \in [0, \tau).$$

36   *1. Semi-group methods for construction of solutions*

A strongly continuous with respect to $t \in [0, \tau)$ family of bounded linear operators $\{S(t), t \in [0, \tau)\}$ acting in $\mathcal{X}$ and satisfying the conditions

$$S(s)Af = AS(t)f, \qquad t \in [0, \tau), \quad f \in dom\, A,$$

$$A \int_0^t S(s)f\, ds = S(t)f - \int_0^t R(s)f\, ds, \qquad t \in [0, \tau), \quad f \in \mathcal{X},$$

is called a regularized (R-regularized) semi-group of operators generated by A. The operator A is called the generator of the family.

If the operator function $R(t)$, $t \in [0, \tau)$, is equal to the unit operator multiplied by a certain (smooth enough) function K: $R(\cdot) = K(\cdot)I$, then the semi-group $\{S(t), t \in [0, \tau)\}$ is K-convoluted. In particular, if $R(t) = \frac{t^{n-1}}{(n-1)!}I$, the semi-group is n-times integrated.

If we put the operator function $R(\cdot)$ equal to the delta-function multiplied by $R(t) \equiv R$ (in this case the equalities with $R(\cdot)$ must be understood in the sense of distributions), then $S(\cdot)$ is an R-semi-group.

The relations between the given classes of semi-groups can be written in the form of embeddings:

$$\{\text{integrated semi-groups}\} \subset \{\text{convoluted semi-groups}\} \subset$$

$$\subset \{\text{regularized semi-groups}\},$$

$$\{R\text{-semi-groups}\} \subset \{\text{regularized semi-groups}\}.$$

1.3 R-semi-groups and regularizing operators in the construction of approximated solutions to ill-posed problems

In this subsection we apply the regularizing methods developed in the theory of ill-posed problems to solving the Cauchy problem

$$u'(t) = Au(t), \quad t \in [0, T], \qquad u(0) = f. \qquad (1.3.1)$$

Using these methods we construct regularization which yields approximate solutions.

Let A be a closed linear operator in a Banach space $\mathcal{X}$. Suppose, as is conventional in the theory of ill-posed problems, that from a priori information it is known that a solution of (1.3.1) exists for certain initial data f, but the initial data is given with an error $\delta > 0$, i.e., instead of f we have f_δ and $\|f - f_\delta\| \leq \delta$. The error δ is called *the initial data error*. (This is typical for the most real-life situations where one knows not the exact initial data, but

1.3. R-semi-groups for construction of regularizing operators 37

only approximations. The error is usually caused by inaccuracy of measuring devices.)

In this setting of the problem one has to build a regularizing operator for the problem, i.e., an operator which assigns an approximate solution to a given f_δ in such a way that the approximate solutions converge to the exact one when the error in the initial data tends to zero.

Definition 1.3.1 *An operator* $\mathbf{R}_\varepsilon(t) : \mathcal{X} \to \mathcal{X}$ *is called a regularizing operator for (1.3.1) if the following conditions hold:*

(a) *for any* $\varepsilon > 0$ *and* $0 \le t \le T$ *the operator* $\mathbf{R}_\varepsilon(t)$ *is bounded and* $\mathbf{R}_\varepsilon f \in C([0,T], \mathcal{X})$ *for each* $f \in \mathcal{X}$;

(b) *there exists a function* $\varepsilon = \varepsilon(\delta)$ ($\varepsilon(\delta) \underset{\delta \to 0}{\longrightarrow} 0$) *such that*

$$\|\mathbf{R}_{\varepsilon(\delta)}(t)f_\delta - u(t)\| \underset{\delta \to 0}{\longrightarrow} 0, \qquad t \in [0, T].$$

The parameter ε *is called a regularizing parameter for (1.3.1).*

Like most of the ill-posed problems, the ill-posed Cauchy problem (1.3.1) can be considered as a problem of calculating the values of the unbounded operator

$$u(t) = U(t)f, \qquad t \in [0, T],$$

which, under the condition of uniqueness of the solution of the problem, can be written in the form of the operator equation of the first kind

$$Fu = v, \qquad F : W \to V, \tag{1.3.2}$$

where the operator F^{-1} is unbounded. To do this one has to take $W = C([0,T], \mathcal{X})$, $V = \mathcal{X}$, $F = U^{-1}$.

The main methods of regularization for equations of the first kind are variational methods (see, e.g., [47, 48, 112]. They are the Ivanov quasi-solutions method, the residual method, and the Tichonov method. As an approximate solution of (1.3.2) with a given v_δ ($\|v_\delta - v\|_V \le \delta$) one takes the solution of one of the following variational problems:

- for the Ivanov quasi-solutions method:

$$u_\varepsilon = \arg \inf\{ \|Fu - v_\delta\|_V : \|Lu\|_Z \le \varepsilon\}$$

- for the residual method:

$$u_\delta = \arg \inf \{ \|Lu\|_Z : \|Fu - v_\delta\|_V \le \delta\}$$

- for the Tichonov regularization method:

$$u_r = \arg \inf\{ \|Fu - v_\delta\|_V + r\|Lu\|_Z, \ r > 0\}$$

38   *1. Semi-group methods for construction of solutions*

with a stabilizing operator $L : W \to Z$.

All the above mentioned methods allow us to define regularizing operators of (1.3.2)

$$\mathbf{R}_\varepsilon v_\delta := u_{\varepsilon(\delta)} \qquad (\mathbf{R}_\delta v_\delta := u_\delta, \quad \mathbf{R}_r v_\delta := u_{r(\delta)})$$

so that $\varepsilon = \varepsilon(\delta) \underset{\delta \to 0}{\longrightarrow} 0 \quad (r = r(\delta) \underset{\delta \to 0}{\longrightarrow} 0)$ and

$$\|\mathbf{R}_{\varepsilon(\delta)} v_\delta - u\|_V \qquad (\|\mathbf{R}_\delta v_\delta - u\|_V, \quad \|\mathbf{R}_{r(\delta)} v_\delta - u\|_V) \to 0 \quad \text{as} \quad \delta \to 0.$$

Further, we show a connection between the existence of a family of regularizing operators and an R-semi-group depending on a regularizing parameter. This allows us to construct approximate solutions via a family of R-semi-groups depending on a regularizing parameter [66, 79, 85].

Theorem 1.3.1 *Suppose that the operator* $-A$ *generates a* C_0*-semi-group on a Banach space* $\mathcal{X}$*. Then the following statements are equivalent:*

(i) A generates a local R_ε*-semi-group* $\{S_\varepsilon(t), t \in [0, \tau)\}$ *with* R_ε *convergent to the identity operator on* $\operatorname{dom} A$ *as* $\varepsilon \to 0$;

(ii) for each $T < \tau$, *there exists a family of linear regularizing operators* $\{\mathbf{R}_\varepsilon(t), 0 < t \le T\}$ *for the problem (1.3.1). They are invertible and commute with* A.

Proof. (ii) $\Longrightarrow$ (i). The assumption that $-A$ generates a C_0-semi-group $\{U_{-A}(t), t \ge 0\}$ implies that the Cauchy problem

$$u'(t) = -Au(t), \quad t \in [0, T], \qquad u(0) = f, \tag{1.3.3}$$

and the equivalent one, $u'(t) = Au(t), t \in [0, T], u(T) = f$, are uniformly well-posed, while the problem (1.3.1) is generally not.

Let us denote by $U(t), t \in [0, T]$, the solution operators of (1.3.1) defined by $U(t)f := u(t), t \in [0, T]$. They are unbounded as solution operators of an ill-posed problem. Since (1.3.3) is well-posed, the semi-group $\{U_{-A}(t), t \ge 0\}$ is non-degenerate: $\ker U_{-A}(t) = \{0\}$. Therefore, there exist the inverse to the operators $U_{-A}(t)$, and the equality

$$[U_{-A}(t)]^{-1} = U(t)$$

holds. Commutativity of regularizing operators $\mathbf{R}_\varepsilon(t)$ with A implies commutativity of $\mathbf{R}_\varepsilon(t)$ with $U_{-A}(t)$ for $t \in [0, T]$. We define

$$S_\varepsilon(t) := \mathbf{R}_\varepsilon(T)U_{-A}(T - t) \;\; = \;\; \mathbf{R}_\varepsilon(T)U(t)U_{-A}(t)U_{-A}(T - t)$$
$$= \;\; \mathbf{R}_\varepsilon(T)U(t)U_{-A}(T), \qquad t \le T < \tau,$$

and

$$R_\varepsilon = S_\varepsilon(0) = \mathbf{R}_\varepsilon(T)U_{-A}(T).$$

We claim that $S_\varepsilon(\cdot)$ satisfies relations (R1)–(R2) of Definition 1.2.8. The equality

$$S_\varepsilon(t + s)R_\varepsilon = S_\varepsilon(t)S_\varepsilon(s), \qquad t, s, t + s \in [0, T],$$

1.3. R-semi-groups for construction of regularizing operators   39

is a consequence of commutativity of $\mathbf{R}_\varepsilon(T)$ with $U_{-A}(t)$ and the semi-group property of the family $U_{-A}(t)$:

$$
\begin{aligned}
S_\varepsilon(t+s)R_\varepsilon &= \mathbf{R}_\varepsilon(T)U_{-A}(T-t-s)\mathbf{R}_\varepsilon(T)U_{-A}(T) \\
&= \mathbf{R}_\varepsilon(T)U_{-A}(T-t)\mathbf{R}_\varepsilon(T)U_{-A}(T-s) = S_\varepsilon(t)S_\varepsilon(s).
\end{aligned}
$$

For any $T < \tau$ the operator $\mathbf{R}_\varepsilon(T)$ is defined on the whole $\mathcal{X}$ and is bounded. Moreover, the operators $U_{-A}(T-t)$ are strongly continuous with respect to t $(T-t \geq 0)$. Therefore, the functions $S_\varepsilon(t)f = \mathbf{R}_\varepsilon(T)U_{-A}(T-t)f$ are continuous with respect to $t \in [0,\tau)$ for all $f \in \mathcal{X}$. Thus, the operator family $\{S_\varepsilon(t), t \in [0,\tau)\}$ is a local R_ε-semi-group.

Let us show that R_ε is convergent to the identity operator as $\varepsilon \to 0$. For $f \in dom\,A$ we have

$$
\|R_\varepsilon f - f\| = \|\mathbf{R}_\varepsilon(T)U_{-A}(T)f - U(T)U_{-A}(T)f\|
$$
$$
= \|\mathbf{R}_\varepsilon(T)y - U(T)y\|, \qquad (1.3.4)
$$

where $y = U_{-A}(T)f$ is the solution of (1.3.3) with $t = T$ corresponding to the initial value f. The equality (1.3.4) and condition (b) of Definition 1.3.1 imply

$$
\|R_\varepsilon f - f\| \underset{\varepsilon \to 0}{\longrightarrow} 0, \qquad f \in dom\,A.
$$

Now we verify that for the complete infinitesimal generator $\overline{G}$ and for the infinitesimal generator Z of the obtained R_ε-semi-group (see Definition 1.2.9) the inclusions $\overline{G} \subset A \subset Z$ hold. These inclusions and the inclusion $\overline{Z} \subset \overline{G}$, which is always true for the generators introduced, imply $A = \overline{G}$.

By the definition of a complete infinitesimal generator we have

$$
Gf = \lim_{t \to 0} t^{-1}\left[R_\varepsilon^{-1}S_\varepsilon(t)f - f\right] = \lim_{t \to 0} t^{-1}[U(t)f - f] = U'(0)f = AU(0) = Af,
$$

for $f \in dom\,G$; therefore, $dom\,G \subseteq dom\,A$ and $A|_{dom\,G} = G$. This and closedness of A imply $\overline{G} \subset A$.

Let us show that $dom\,A \subseteq dom\,Z$ and $Z|_{dom\,A} = A$. For $f \in dom\,A$ we have

$$
\begin{aligned}
R_\varepsilon Af &= AR_\varepsilon f = A\mathbf{R}_\varepsilon(T)U_{-A}(T)f = \mathbf{R}_\varepsilon(T)AU_{-A}(T)f \\
&= \mathbf{R}_\varepsilon(T)\lim_{t \to 0} t^{-1}\left[U_{-A}(T-t) - U_{-A}(T)\right]f \\
&= \lim_{t \to 0} t^{-1}\left[S_\varepsilon(t)f - R_\varepsilon f\right] = R_\varepsilon Zf.
\end{aligned}
$$

Thus, $f \in dom\,Z$ and $Af = Zf$ for $f \in dom\,A$. Therefore, $A \subset Z$ and $A = \overline{Z} = \overline{G}$.

(i) $\Longrightarrow$ (ii). Suppose that A is the generator of a local R_ε-semi-group $\{S_\varepsilon(t), t \in [0,\tau)\}$ with R_ε convergent to the identity operator on $dom\,A$ as $\varepsilon \to 0$. We show that $\mathbf{R}_\varepsilon(t) := S_\varepsilon(t)$, $0 < t < \tau$, is a regularizing operator of (1.3.1), i.e., the conditions (a) and (b) of Definition 1.3.1 hold.

40   *1. Semi-group methods for construction of solutions*

According to the assumption, the linear operator $\mathbf{R}_\varepsilon(t)$ is defined on the whole $\mathcal{X}$ and is bounded; therefore, it is continuous. Suppose that for $f \in dom\, A$ there exists a solution u of (1.3.1). For a fixed $t \in [0,T]$, $T < \tau$, consider the error

$$\|\mathbf{R}_\varepsilon(t)f - u(t)\| = \|S_\varepsilon(t)R_\varepsilon^{-1}R_\varepsilon f - u(t)\|.$$

Since R_ε^{-1} commutes with $S_\varepsilon(t)$ on the range of R_ε we have

$$S_\varepsilon(t)R_\varepsilon^{-1}R_\varepsilon f = R_\varepsilon^{-1}S_\varepsilon(t)R_\varepsilon f.$$

By Theorem 1.2.16,

$$S_n(t) = \mathbf{R}_\varepsilon(t)f = R_\varepsilon^{-1}S_\varepsilon(t)R_\varepsilon f = R_\varepsilon^{-1}S_\varepsilon(t)y, \quad t \in [0,\tau),$$

where $y = R_\varepsilon f \in R_\varepsilon(dom\, A)$, is the unique solution of the Cauchy problem (1.3.1) with the generator of an R_ε-semi-group and initial value y:

$$v'(t) = Av(t), \quad 0 < t \le T, \quad v(0) = y.$$

On the other hand, $R_\varepsilon u$ is also a solution of (1.3.1) with the initial condition $R_\varepsilon u(0) = R_\varepsilon f = y$, hence $\mathbf{R}_\varepsilon(t)f = S_\varepsilon(t)f = R_\varepsilon u(t)$. Thus, if there exists a solution u for the initial value f, then

$$\|\mathbf{R}_\varepsilon(t)f - u(t)\| = \|R_\varepsilon u(t) - u(t)\| \underset{\varepsilon \to 0}{\longrightarrow} 0, \quad 0 \le t \le T,$$

that is, $\mathbf{R}_\varepsilon(t)$ is a regularizing operator of (1.3.1). $\qquad\square$

The condition that the operator $-A$ generates a C_0-semi-group is fulfilled for strongly ill-posed problems such as the Cauchy problem for the time-reversed heat equation. The most well-known examples of regularizing operators for such problems are the operators constructed by the *quasi-reversion* method and by the *auxiliary bounded conditions* (ABC) method [47, 62, 66, 79]. The results on regularization of ill-posed Cauchy problems by these methods with A belonging to quite a wide class of (R2) are presented in the following theorems.

Theorem 1.3.2 [47, 66] *Let A be a densely defined linear operator whose spectrum belongs to the region*

$$\Lambda_1 = \left\{ \lambda \in \mathbb{C} : |arg\, \lambda| < \beta < \frac{\pi}{4} \right\}.$$

Let the estimate of the resolvent of A,

$$\|\mathcal{R}(\lambda)\| \le C(1 + |\lambda|)^{-1},$$

hold for arbitrary $\lambda \notin \Lambda_1$ and some $C > 0$. Then the operator

$$\mathbf{R}_\varepsilon(t)f_\delta = u_{\varepsilon,\delta}(t) = U_{A_\varepsilon}(t)f_\delta = -\frac{1}{2\pi i}\int_{\partial\Lambda_1} e^{(\lambda - \varepsilon\lambda^2)t}\mathcal{R}(\lambda)f_\delta\, d\lambda,$$

1.3. R-semi-groups for construction of regularizing operators 41

constructed via the quasi-reversion method, i.e., as the solution operator of the Cauchy problem

$$u'_{\varepsilon,\delta}(t) = (A - \varepsilon A^2)u_{\varepsilon,\delta}(t) =: A_\varepsilon u_{\varepsilon,\delta}(t), \quad 0 < t \le T, \qquad u_{\varepsilon,\delta}(0) = f_\delta,$$

is a regularizing operator of the ill-posed problem (1.3.1). Here $\{U_{A_\varepsilon}(t), t \ge 0\}$ is a C_0-semi-group with the generator A_ε.

Theorem 1.3.3 [47, 66] *Let A be a densely defined linear operator whose spectrum belongs to the half-strip*

$$\Lambda_2 = \left\{ \lambda \in \mathbb{C} : \ |Im\lambda| < \alpha < \frac{\pi}{T}, \ Re\lambda > \omega, \ \omega \in \mathbb{R} \right\},$$

and the resolvent of A is bounded for $\lambda \notin \Lambda_2$. Then the operator

$$\widehat{\mathbf{R}}_\varepsilon(t)f_\delta = \widehat{u}_{\varepsilon,\delta}(t) = -\frac{1}{2\pi i} \int_{\partial\Lambda_2} \frac{e^{\lambda t}}{1 + \varepsilon e^{\lambda(T)}} \mathcal{R}(\lambda)f_\delta \, d\lambda, \qquad f_\delta \in X, \quad t < T,$$

constructed via the ABC method, i.e., as the solution operator of the boundary problem

$$\widehat{u}'_{\varepsilon,\delta}(t) = A\widehat{u}_{\varepsilon,\delta}(t), \qquad 0 < t < T,$$

$$\widehat{u}_{\varepsilon,\delta}(0) + \varepsilon\widehat{u}_{\varepsilon,\delta}(T) = f_\delta, \qquad \varepsilon > 0,$$

is a regularizing operator of the ill-posed problem (1.3.1).

Chapter 2

Distribution methods for construction of generalized solutions to ill-posed Cauchy problems

This chapter is devoted to generalized solutions for the abstract Cauchy problem

$$u'(t) = Au(t), \quad t \in [0, \infty), \quad u(0) = f, \tag{2.0.1}$$

where A is a closed linear operator with $dom\, A$ in a Banach space $\mathcal{X}$, in particular, for the problem with differential operators

$$\frac{\partial u(x;t)}{\partial t} = A\left(i\frac{\partial}{\partial x}\right)u(x;t), \quad t \in [0, T], \quad u(x;0) = f(x). \tag{2.0.2}$$

We explore solutions to (2.0.1) in dependence on the set of regular points and the behavior of the resolvent of A from classes (R1)–(R4). The classes were introduced to characterize different types of regularized semigroups generated by A.

Types of regularized semigroups generated by $A\,(i\partial/\partial x)$ are defined by solution operators $e^{tA(\sigma)}$ to the Fourier transformed Cauchy problem

$$\frac{\partial \widetilde{u}(\sigma;t)}{\partial t} = A(\sigma)\widetilde{u}(\sigma;t), \quad t \in [0, T], \quad \widetilde{u}(\sigma;0) = \widetilde{f}(\sigma), \quad \sigma \in \mathbb{R}^n,$$

and we explore solutions to (2.0.2) in dependence on the growth rate of $e^{tA(\cdot)}$.

In Section 2.1 we give the statement of (2.0.1) and construct generalized solutions in spaces of abstract distributions. Here the behavior of the resolvent under conditions (R1)–(R2) determines the choice of the spaces. We show that the Cauchy problem with A generating integrated semigroups is well-posed in spaces of distributions.

In Section 2.2 we construct generalized solutions to (2.0.1) in spaces of abstract ultra-distributions. Here the behavior of the resolvent under conditions (R3) determines the choice of the spaces. The Cauchy problem with A generating convoluted semigroups is well-posed in these spaces of ultra-distributions.

As for the class (R4) related to R-semigroups, we pay special attention to (2.0.2). In Section 2.3 we obtain a generalized solution to (2.0.2) using the generalized Fourier transform techniques in Gelfand–Shilov spaces:

$$u(x;t) = G_t(x) * f(x), \quad x \in \mathbb{R}^n.$$

43

Here the generalized (wrt x) Green function $G_t(x) = \mathcal{F}^{-1}\left[e^{tA(\sigma)}\right](x)$ is defined in a Gelfand–Shilov space, which depends on the growth rate of $e^{t\mathbf{A}(\cdot)}$.[1]

2.1 Solutions in spaces of abstract distributions

The section is devoted to generalized solutions for the Cauchy problem (2.0.1) with A generating integrated semigroups, exponentially bounded and local ones. As shown in Section 1.2, the resolvent of such an operator A satisfies conditions (R1) and (R2), respectively.

2.1.1 Statement of the generalized Cauchy problem. Abstract distribution spaces

We begin with definitions of abstract distribution spaces, where the generalized solutions to (2.0.1) can be found, and well-posedness of the problem in these spaces.

A space of abstract distributions $\Phi'(\mathcal{X})$ is the space of distributions on a test function space Φ taking their values in a Banach space $\mathcal{X}$. This means that $\Phi'(\mathcal{X})$ is the space of operators $\mathcal{L}(\Phi, \mathcal{X})$. Convergence in $\Phi'(\mathcal{X})$ is the uniform convergence on bounded subsets of Φ. These spaces are generalizations of $\mathbb{C}$-valued distribution spaces $\Phi' = \mathcal{L}(\Phi, \mathbb{C})$ introduced in Section 3.3 for various spaces of test functions Φ. For example, $\mathcal{L}(\mathcal{D}, \mathbb{C}) = \mathcal{D}'$ is the Schwartz distribution space, $\mathcal{L}(\mathcal{S}, \mathbb{C}) = \mathcal{S}'$ is the Schwartz slowly growing distribution space, $\mathcal{L}(\mathcal{S}_\alpha, \mathbb{C}) = \mathcal{S}'_\alpha$ is the Gelfand–Shilov exponentially growing distribution space, and so on.

Similar to the scalar case, *the support of an abstract distribution* $u \in \Phi'(\mathcal{X})$ is a minimal compact set $G \subset \mathbb{C}$ such that $u|_{\mathbb{C}\backslash G} = 0$. The last equality is understood in the sense of distributions:

$$\langle \varphi, u \rangle = 0 \quad \text{for all} \ \ \varphi \in \Phi \ \ \text{with} \ \ \operatorname{supp}\varphi \subset \mathbb{R}\backslash G.$$

According to notation accepted in the theory of distributions, we denote the subspace of distributions with supports in $[0, \infty)$ by $\Phi'_0(\mathcal{X})$ or $\Phi'_+(\mathcal{X})$.

In order to state the problem we have to define the concept of convolution for abstract distributions. Similar to the scalar case, we first define the convolution of a distribution $v_0 \in \Phi'(\mathcal{X})$ with a test function $\varphi \in \Phi$ by

$$(v_0 \overline{*} \varphi)(t) := \langle \varphi(t + \tau), v_0(\tau) \rangle. \tag{2.1.1}$$

If the function obtained possesses the properties of a test function with val-

---

[1] If we write a generalized function as a function of a variable, we mean that it is a distribution applied to test functions depending on this variable.

2.1. Solutions in spaces of abstract distributions

ues in the Banach space $\mathcal{X}$, then, according to scalar case terminology, the distribution v_0 is called *a convolutor* transforming Φ into $\Phi(\mathcal{X})$.

Now let us define *the convolution of two abstract distributions* in the following particular case. Let $\mathcal{X}, \mathcal{Y}, \mathcal{Z}$ be Banach spaces and suppose that there exists a multiplication operation

$$g \cdot f : \quad \mathcal{Y} \times \mathcal{X} \to \mathcal{Z}, \qquad f \in \mathcal{X}, \ g \in \mathcal{Y}.$$

For example, if $\mathcal{X} = \mathcal{L}(E_1, E_2)$ and $\mathcal{Y} = \mathcal{L}(E_2, E_3)$, then, in order to multiply the elements of these spaces, one should apply operators successively; the resultant operator is an element of $\mathcal{Z} = \mathcal{L}(E_1, E_3)$. Another example is $\mathcal{X} = E_1$ and $\mathcal{Y} = \mathcal{L}(E_1, E_2)$. Then multiplication of the elements of these spaces occurs through applying an operator to an element. It results in an element of $\mathcal{Z} = E_2$.

If $v_0 \in \Phi'(\mathcal{Y})$ is a *convolutor* in Φ, then we can define a convolution of $u \in \Phi'(\mathcal{X})$ with v_0 as follows:

$$\langle \varphi, v_0 * u \rangle := \langle v_0 \bar{*} \varphi, u \rangle, \qquad \varphi \in \Phi, \tag{2.1.2}$$

$$v_0 * u : \quad \Phi'(\mathcal{Y}) \times \Phi'(\mathcal{X}) \to \Phi'(\mathcal{Z}),$$

where $v_0 \bar{*} \varphi$ is obtained by (2.1.1). The mapping given by (2.1.2) is bounded and bilinear. In particular, the convolution of $v_0 \otimes g$ with $u \otimes f$,[2] where $f \in \mathcal{X}$, $g \in \mathcal{Y}$, and $v_0, u \in \Phi'$, is defined by

$$(v_0 \otimes g) * (u \otimes f) = (v_0 * u) \otimes (g \cdot f). \tag{2.1.3}$$

Note that one can find different definitions of convolutions of abstract distributions in [30, 49, 79]. They are based on structure theorems in spaces of distributions $\mathcal{D}'$, $\mathcal{S}'$, and $\mathcal{S}'_\omega$. (See details in Sections 3.3, 3.4.) If, for example, $v_0 \in \mathcal{D}'_0(\mathcal{Y})$, $u \in \mathcal{D}'_0(\mathcal{X})$, the structure theorem implies the existence of natural numbers m, p and continuous functions $g_0 : \mathbb{R} \to \mathcal{Y}$ and $g : \mathbb{R} \to \mathcal{X}$ equal to zero for $t < 0$ and satisfying

$$\langle \varphi, v_0 \rangle = \langle \varphi, g_0^{(p)} \rangle, \qquad \langle \varphi, u \rangle = \langle \varphi, g^{(m)} \rangle.$$

Then the convolution of u with v_0 is defined by

$$\langle \varphi, v_0 * u \rangle := \langle \varphi, (g_0 * g)^{(m+p)} \rangle, \qquad \varphi \in \mathcal{D},$$

and belongs to $\mathcal{D}'_0(\mathcal{Z})$. This definition does not depend on the choice of m, p, g_0, and g.

Now we are ready to state the Cauchy problem in a space of abstract distributions. As usually, the idea of the definition in the case of distributions comes from a formula obtained for classical functions.

---

[2] For any scalar-valued generalized function $u \in \Phi'$ and $f \in \mathcal{X}$ we define the abstract distribution $u \otimes f \in \Phi'(\mathcal{X})$ by $\langle \varphi, v \otimes f \rangle := \langle \varphi, v \rangle f$, $\varphi \in \Phi$.

46   2. *Distribution methods*

Suppose the Cauchy problem (2.0.1) is uniformly well-posed. In this case A is closed and densely defined in $\mathcal{X}$. Multiply the equation by a test function $\varphi \in \Phi$ and integrate both sides from 0 to ∞:

$$\int_0^\infty Au(t)\varphi(t)\,dt = \int_0^\infty u'(t)\varphi(t)\,dt = -f\varphi(0) - \int_0^\infty u(t)\varphi'(t)\,dt.$$

Since a solution u of a uniformly well-posed problem (2.0.1) is determined by the C_0-semi-group of bounded operators generated by A,

$$u(t) = U(t)f, \qquad t \geq 0, \quad f \in dom\,A,$$

by virtue of the closedness of A we obtain

$$A\int_0^\infty U(t)f\varphi(t)\,dt = -f\varphi(0) - \int_0^\infty U(t)f\varphi'(t)\,dt, \qquad f \in dom\,A. \quad (2.1.4)$$

The right-hand side of (2.1.4) is well defined for all $f \in \overline{dom\,A} = \mathcal{X}$. Furthermore, it continuously depends on $f \in \mathcal{X}$. Since A is closed, for any $f \in \mathcal{X}$ the integral in the left-hand side belongs to $dom\,A$. Thus the equality (2.1.4) holds on the whole space $\mathcal{X}$. In addition, if we keep the notation $u(t) = U(t)f$ on the whole $\mathcal{X}$, then (2.1.4) takes the form

$$A\int_0^\infty u(t)\varphi(t)\,dt = -f\varphi(0) - \int_0^\infty u(t)\varphi'(t)\,dt, \qquad f \in \mathcal{X}. \quad (2.1.5)$$

Here we consider $u(\cdot)$ as a distribution in $\Phi'(\mathcal{X})$ and it is natural to agree that any solution of (2.0.1) equals zero for $t < 0$. Therefore $supp\,u \subset [0, \infty)$; thus $u \in \Phi_0'(\mathcal{X})$. Hence we can rewrite (2.1.5) in the following form:

$$\langle \varphi, u' \rangle = -\langle \varphi', u \rangle = A\langle \varphi, u \rangle + \langle \varphi, \delta \rangle f, \qquad f \in \mathcal{X}, \quad \varphi \in \Phi, \quad (2.1.6)$$

where δ is the Dirac delta-function considered as an element of Φ_0' here. Let us introduce an operator-valued distribution

$$\mathbf{P} := \delta' \otimes I - \delta \otimes A, \quad \text{where} \quad \langle \varphi, \delta' \otimes I - \delta \otimes A \rangle := \langle \varphi, \delta' \rangle I - \langle \varphi, \delta \rangle A.$$

Denote by $[dom\,A]$ the space $dom\,A$ endowed with the graph-norm of A:

$$\|f\|_A = \|f\| + \|Af\|.$$

It is a Banach space and the operator $A : [dom\,A] \to \mathcal{X}$ is bounded: $A \in \mathcal{L}([dom\,A], \mathcal{X})$ and $\mathbf{P} \in \Phi_0'(\mathcal{L}([dom\,A], \mathcal{X}))$.

The Cauchy problem (2.0.1) and Equation (2.1.6), which is equivalent to (2.0.1) in the sense of distributions, take the following form in $\Phi_0'(\mathcal{X})$:

$$\mathbf{P} * u = \delta \otimes f, \qquad f \in \mathcal{X}. \quad (2.1.7)$$

Recall that $\delta * u = u$ and $\delta' * u = u'$ for $u \in \Phi_0'(\mathcal{X})$.

2.1. Solutions in spaces of abstract distributions

Definition 2.1.1 *An element $u \in \Phi_0'(\mathcal{X})$ satisfying (2.1.7) is called a generalized solution of the (generalized) Cauchy problem (2.0.1).*

Definition 2.1.2 *The problem (2.1.7) is called well-posed or, in other words, the generalized Cauchy problem (2.0.1) is called well-posed in a space $\Phi_0'(\mathcal{X})$ if for each $f \in \mathcal{X}$ there exists a unique solution of (2.1.7) stable in $\Phi_0'(\mathcal{X})$, i.e., for any sequence f_n converging to zero in $\mathcal{X}$, the corresponding sequence u_n converges to zero in $\Phi_0'(\mathcal{X})$.*

The aim of the present section is to reveal the conditions that provide the generalized well-posedness of (2.0.1) in a certain $\Phi_0'(\mathcal{X})$ and to obtain its generalized solution. The main tool applied is the generalized Laplace transform (see Section 3.4).

Applying the generalized Laplace transform to (2.1.7) in a space $\Phi_0'(\mathcal{X})$ and using the property of the Laplace transform of a derivative,

$$\mathcal{L}\left[du(t)/dt\right](\lambda) = \lambda \mathcal{L}u(\lambda) - f,$$

we obtain

$$(\lambda I - A)\mathcal{L}u(\lambda) = f, \quad f \in \mathcal{X}.$$

If the resolvent of A exists, we apply it to both sides of this equation and obtain

$$\mathcal{L}u(\lambda) = \mathcal{R}(\lambda)f, \quad f \in \mathcal{X}, \quad \lambda \in \rho(A).$$

It follows that the generalized solution u of (2.0.1) must be equal to the (generalized) inverse Laplace transform of the resolvent:

$$u = \mathcal{L}^{-1}\left[\mathcal{R}(\lambda)f\right], \quad f \in \mathcal{X}.$$

In order to obtain a generalized solution by this method, one should select such a space Φ of test functions that the generalized inverse Laplace transform of the resolvent exists in the abstract conjugate space $\Phi'(\mathcal{X})$.

Thus the resolvent behavior determines the choice of the test function space. For the case when $\mathcal{R}$ satisfies (R1), we consider the Cauchy problem (2.0.1) in spaces of abstract exponentially bounded distributions $\mathcal{S}_\omega'(\mathcal{X})$ and for the case (R2) we use the space of abstract distributions $\mathcal{D}'(\mathcal{X})$. The Cauchy problem with operators satisfying (R3) will be considered in the next section in spaces of abstract ultra-distributions.

In both cases (R1) and (R2) we first show that the existence of a unique solution to (2.1.7) and the corresponding behavior of the resolvent of A are equivalent to the existence of a generalized operator family U solving the equations

$$U * \mathbf{P} = \delta \otimes I_{[\text{dom }A]}, \qquad \mathbf{P} * U = \delta \otimes I_{\mathcal{X}}, \tag{2.1.8}$$

in the corresponding space. Second, we show that the solution of (2.1.8) exists if and only if (2.0.1) is well-posed in the generalized sense.

48   *2. Distribution methods*

2.1.2   Generalized solutions to the Cauchy problem with generators of integrated semigroups

We begin with the case of the Cauchy problem (2.0.1) with the generator of an exponentially bounded integrated semi-group and the resolvent of A satisfying (R1). We consider the problem in the space of abstract exponentially bounded distributions $\mathcal{S}'_\omega(\mathcal{X})$ and show that (R1) is the necessary and sufficient condition for generalized well-posedness of (2.0.1) in this space.

The space $\mathcal{S}'_\omega(\mathcal{X})$ is the space of $\mathcal{X}$-valued distributions f satisfying the condition $\{e^{-\omega t}\}f(t) \in \mathcal{S}'(\mathcal{X})$,[3] where $\{e^{-\omega t}\}$ is a smoothed exponential function, namely, an infinitely differentiable function

$$\{e^{-\omega t}\} = \left\{ \begin{array}{ll} e^{-\omega t}, & t \geq 0, \\ 0, & t \leq c < 0. \end{array} \right. \tag{2.1.9}$$

According to Theorem 1.2.5, the problem (2.0.1) is (n, ω)-well-posed if and only if the resolvent of A satisfies conditions (R1), which is in turn equivalent to the existence of an exponentially bounded n-times integrated semi-group of operators generated by A. The connection between the generalized well-posedness of such a problem and the estimates of the resolvent of A based on the connection with integrated semigroups is given in the next theorem.

Theorem 2.1.1 *Let A be a linear operator in a Banach space $\mathcal{X}$ densely defined and with a nonempty set of regular points. The following assertions are equivalent:*

(i) *the Cauchy problem (2.0.1) is well-posed in the space of exponentially bounded distributions $\mathcal{S}'_\omega(\mathcal{X})$;*

(ii) *there exists an operator-valued distribution $U \in \mathcal{S}'_\omega(\mathcal{L}(\mathcal{X}))$ with support in $[0, \infty)$ solving (2.1.8);*

(iii) *(R1) holds true for the resolvent of A.*

Proof. (i)$\Longrightarrow$(ii). Define for every $\varphi \in \mathcal{S}$ a linear operator $\langle \varphi, \{e^{-\omega t}\}U \rangle$ acting in $\mathcal{X}$ by

$$\langle \varphi, \{e^{-\omega t}\}U \rangle f := \langle \varphi, \{e^{-\omega t}\}u \rangle, \qquad f \in \mathcal{X},$$

where $u \in \mathcal{S}'_\omega(\mathcal{X})$ is the solution of the well-posed problem (2.1.7) and $\{e^{-\omega t}\}$ is defined by (2.1.9).

We show that $U \in \mathcal{S}'_\omega(\mathcal{L}(\mathcal{X}))$, i.e.,

1) $\{e^{-\omega t}\}U$ is a linear operator from $\mathcal{S}$ to $\mathcal{L}(\mathcal{X})$,

2) for any sequence $\varphi_n \in \mathcal{S}$ convergent to zero in $\mathcal{S}$ it holds that

$$\|\langle \varphi_n, \{e^{-\omega t}\}U \rangle\|_{\mathcal{L}(\mathcal{X})} \to 0,$$

i.e., $\langle \varphi_n, \{e^{-\omega t}\}U \rangle f \to 0$ in $\mathcal{X}$ uniformly with respect to f from any bounded set in $\mathcal{X}$.

---

[3]Recall that if we write a distribution $u \in \Phi'$ as $u(t)$, we mean u is applied to test functions $\varphi \in \Phi$ of the variable t.

2.1. Solutions in spaces of abstract distributions

The linearity of U follows from that of the solution. The stability of the solution implies

$$\|\langle\varphi, \{e^{-\omega t}\}U\rangle f_n\| = \|\langle\varphi, \{e^{-\omega t}\}u_n\rangle\| \to 0 \quad \text{as} \quad \|f_n\| \to 0$$

uniformly with respect to φ from any bounded set in $\mathcal{S}$. Thus, for each $\varphi \in \mathcal{S}$, we have $\langle\varphi, \{e^{-\omega t}\}U\rangle \in \mathcal{L}(\mathcal{X})$.

Consider a bounded set $\{f : \|f\| \le C\}$ in $\mathcal{X}$. It generates the set $\mathcal{B} = \{Uf : \|f\| \le C\} \subset \mathcal{S}'_\omega(\mathcal{X})$, which is bounded in $\mathcal{S}'_\omega(\mathcal{X})$ due to the stability of the solution: for each $\varphi \in \mathcal{S}$ it holds that

$$\sup_{Uf \in \mathcal{B}} \|\langle\varphi, \{e^{-\omega t}\}U\rangle f\| = \sup_{\|f\| \le C} \|\langle\varphi, \{e^{-\omega t}\}u\rangle\| < \infty.$$

Then, by the structure theorem for elements from $\mathcal{S}'(\mathcal{X})$, there exists $m \in \mathbb{N}$, independent of elements of $\mathcal{B}$, and for every $Uf \in \mathcal{B}$ there exists a continuous primitive $g : \mathbb{R} \to \mathcal{X}$ such that

$$\langle\varphi, \{e^{-\omega t}\}Uf\rangle = \langle\varphi, g^{(m)}\rangle, \quad \varphi \in \mathcal{S}.$$

Moreover, all the primitives have the same power of growth: $|g(t)| \le C|t|^r$ as $|t| \to \infty$. Hence,

$$
\begin{aligned}
\|\langle\varphi, \{e^{-\omega t}\}U\rangle f\| &= \|\langle\varphi, g^{(m)}\rangle\| \\
&= \|\langle(-1)^m \varphi^{(m)}, g\rangle\| \le C|t|^r \sup_{t \in \mathbb{R}} |\varphi^{(m)}(t)|, \quad \varphi \in \mathcal{S}.
\end{aligned}
$$

In accordance with the definition of convergence in $\mathcal{S}$, this estimate implies

$$\|\langle\varphi_n, \{e^{-\omega t}\}U\rangle f\| \to 0 \quad \text{as} \quad \varphi_n \to 0 \quad \text{in} \quad \mathcal{S}.$$

Since $\operatorname{supp} u \subset [0, \infty)$, the support of the operator-valued distribution belongs to $[0, \infty)$ as well. It only remains to prove that U satisfies Equations (2.1.8) in $\mathcal{S}'_\omega(\mathcal{L}(\mathcal{X}))$. Using the notation $\varphi_\omega := \{e^{-\omega t}\}\varphi$, it is sufficient to prove

$$\langle\varphi_\omega, \mathbf{P} * U\rangle = \langle\varphi_\omega, \delta \otimes I_\mathcal{X}\rangle, \quad \langle\varphi_\omega, U * \mathbf{P}\rangle = \langle\varphi_\omega, \delta \otimes I_{[\operatorname{dom} A]}\rangle.$$

By (2.1.2) and (2.1.3), for any $f \in \mathcal{X}$ we obtain

$$
\begin{aligned}
\langle\varphi_\omega, \mathbf{P} * U\rangle f &= \langle\mathbf{P}\bar{*}\varphi_\omega, U\rangle f = \langle(\delta' \otimes I_{[\operatorname{dom} A]} - \delta \otimes A)\bar{*}\varphi_\omega, U\rangle f \\
&= \langle\varphi_\omega, U'\rangle f - \langle\varphi_\omega, AU\rangle f = \langle\varphi_\omega, (U' - AU)f\rangle.
\end{aligned}
$$

For any $U \in \mathcal{S}'_\omega(\mathcal{L}(\mathcal{X}))$ and $f \in \mathcal{X}$ we have $U'f = (Uf)'$. Therefore,

$$
\begin{aligned}
\langle\varphi_\omega, \mathbf{P} * U\rangle f &= \langle\varphi_\omega, (Uf)' - AUf\rangle = \langle\varphi_\omega, u' - Au\rangle \\
&= \langle\varphi_\omega, \mathbf{P} * u\rangle = \langle\varphi_\omega, \delta \otimes f\rangle = \langle\varphi_\omega, \delta \otimes I_\mathcal{X}\rangle f.
\end{aligned}
$$

This completes the proof of the first equality in (2.1.8).

50   *2. Distribution methods*

Let us prove the second equality. We have

$$\langle \varphi_\omega, \mathbf{P} * (U'f) \rangle = \langle \varphi_\omega, \mathbf{P} * (Uf)' \rangle = \langle \varphi_\omega, \mathbf{P} * u' \rangle = I_\mathcal{X} \langle \varphi_\omega, u'' \rangle - A \langle \varphi_\omega, u' \rangle$$

$$= -I_\mathcal{X} \langle \varphi_\omega', u' \rangle + A \langle \varphi_\omega', u \rangle = -\langle \varphi_\omega', (\delta' \otimes I_\mathcal{X} - A) * u \rangle.$$

It follows that

$$\langle \varphi_\omega, \mathbf{P} * (U'f) \rangle = -\langle \varphi_\omega', \mathbf{P} * u \rangle = -\langle \varphi_\omega', \delta \otimes f \rangle = \langle \varphi_\omega, \delta' \otimes f \rangle, \quad f \in \mathcal{X}.$$

Thus we have the equality $\mathbf{P} * (U'f) = \delta' \otimes f$, $f \in \mathcal{X}$. In a similar manner we have

$$(\mathbf{P} * U)Af = \delta \otimes Af, \qquad \mathbf{P} * (\delta \otimes f) = \delta' \otimes f - \delta \otimes Af, \quad f \in [dom\, A].$$

Then, for any $f \in [dom\, A]$,

$$\mathbf{P} * (U'f - UAf - \delta \otimes f) = 0$$

and by the uniqueness of the solution we obtain

$$U' - UA = \delta \otimes I_{[dom\, A]}.$$

Further,

$$\langle \varphi_\omega, U * \mathbf{P} \rangle = \langle U * \varphi_\omega, \delta' \otimes I_{[dom\, A]} \rangle - \langle U * \varphi_\omega, \delta \otimes A \rangle$$

$$= \langle \varphi_\omega, U' \rangle - \langle \varphi_\omega, UA \rangle = \langle \varphi_\omega, U' - UA \rangle$$

and the two last equations provide $U * \mathbf{P} = \delta \otimes I_{[dom\, A]}$, which proves the assertion.

(ii)$\Longrightarrow$(i). For $f \in \mathcal{X}$ consider the distribution $u := Uf$. By the first equation in (2.1.8) we have

$$\langle \varphi, \{e^{-\omega t}\} u \rangle \in [dom\, A], \quad \varphi \in \mathcal{S},$$

and by the second one we conclude that the distribution $\{e^{-\omega t}\} u$ is a solution of (2.1.7):

$$\mathbf{P} * \left(\{e^{-\omega t}\} u \right) = \mathbf{P} * \{e^{-\omega t}\} Uf = \delta \otimes f.$$

The associativity of convolution and the first equality in (2.1.8) imply the uniqueness of the solution. To prove the stability of the solution, we show that $u_n \to 0$ in $\mathcal{S}'_\omega(\mathcal{X})$ (i.e., uniformly with respect to φ on a bounded set in $\mathcal{S}$) as $f_n \to 0$ in $\mathcal{X}$.

Let $\mathcal{B}$ be a bounded set in $\mathcal{S}$. Then, for any $q, k \in \mathbb{N}_0$, there exists a constant $C_{k,q}$ independent of the elements of $\mathcal{B}$, such that the following inequality holds:

$$|t^k \varphi^{(q)}(t)| \leq C_{k,q}, \quad t \in \mathbb{R}, \quad \varphi \in \mathcal{B}.$$

2.1. Solutions in spaces of abstract distributions

For any $\varphi \in \mathcal{B}$ the operator $\langle \varphi, U_\omega := \{e^{-\omega t}\}U \rangle$ belongs to $\mathcal{L}(\mathcal{X})$; therefore,

$$\|\langle \varphi, \{e^{-\omega t}\}u_n \rangle\| = \|\langle \varphi, U_\omega \rangle f_n\| \leq \|\langle \varphi, U_\omega \rangle\|_{\mathcal{L}(\mathcal{X})} \cdot \|f_n\|.$$

Using the structure theorem for $\mathcal{S}(\mathcal{L}(\mathcal{X}))$ we obtain

$$\langle \varphi, U_\omega \rangle = \langle \varphi, g^{(m)} \rangle, \qquad \varphi \in \mathcal{S}.$$

Therefore, for every $\varphi \in \mathcal{B}$,

$$\|\langle \varphi, U_\omega \rangle\|_{\mathcal{L}(\mathcal{X})} = \|\langle (-1)^{(m)} \varphi^{(m)}, g \rangle\|_{\mathcal{L}(\mathcal{X})} \leq C_{r,m}.$$

Hence,

$$\|\langle \varphi, \{e^{-\omega t}\}u_n \rangle\| \leq C_{r,m}\|f_n\| \to 0 \quad \text{as} \quad f_n \to 0.$$

(ii) $\implies$ (iii). Let $U \in \mathcal{S}'_\omega(\mathcal{L}(\mathcal{X}))$ be a solution of Equations (2.1.8). Applying to them the generalized Laplace transform, we obtain

$$(\lambda I - A)\mathcal{L}U(\lambda) = I_{[\mathrm{dom}\,A]}, \qquad \mathcal{L}U(\lambda)(\lambda I - A) = I_{\mathcal{X}}, \quad \mathrm{Re}\,\lambda > \omega. \quad (2.1.10)$$

These equalities imply that the half-plane $\Re\lambda > \omega$ lies in the resolvent set of A and $(\lambda I - A)^{-1} = \mathcal{L}U(\lambda)$. In order to prove (R1) let us construct a continuous primitive of the family U and show its connection with the resolvent of A. The primitive obtained turns out to be the exponentially bounded integrated semi-group generated by A.

Since $U_\omega \in \mathcal{S}'(\mathcal{L}(\mathcal{X}))$ and the space $\mathcal{S}$ is the intersection of spaces $\mathcal{S}^p$ with the norms

$$\|\varphi\|_p = \sup_{k,q \leq p} \sup_{t \in \mathbb{R}} |x^k \varphi^{(q)}(t)|, \quad p \in \mathbb{N},$$

the distribution U_ω belongs to the space which is dual to some $\mathcal{S}^p$. Hence,

$$\|\langle \varphi, U_\omega \rangle\|_{\mathcal{L}(\mathcal{X})} \leq C\|\varphi\|_p, \qquad \varphi \in \mathcal{S}, \qquad (2.1.11)$$

and, by the density of $\mathcal{S}$ in $\mathcal{S}^p$, one can extend U_ω to $\mathcal{S}^p$.

Let us consider the functions

$$\chi(s) = \begin{cases} 0, & s \leq -1, \\ 1, & s \geq 0, \end{cases} \qquad \chi(\cdot) \in C^\infty(\mathbb{R}), \qquad \eta_p(t) = \begin{cases} t^p/p!, & t \geq 0, \\ 0, & t < 0. \end{cases}$$

The product $\chi(s)\eta_p(t-s)$ considered as a function of s belongs to $\mathcal{S}^p$. Now introduce the function

$$\langle \chi(s)\eta_p(t-s), U \rangle := \langle e^{\omega s}\chi(s)\eta_p(t-s), U_\omega \rangle =: S_{p+2}(t)$$

and prove that S_{p+2} is a $(p+2)$-times integrated semigroup. It is easy to verify that the mapping

$$t \to e^{\omega s}\chi(s)\eta_p(t-s) : \quad \mathbb{R} \to \mathcal{S}^p$$

is continuous; therefore, $S_{p+2}(\cdot)$ is continuous in $\mathcal{L}(\mathcal{X})$. Since $\operatorname{supp} U \subseteq [0, \infty)$ and $\operatorname{supp} e^{\omega s} \chi(s) \eta_p(t-s) \subseteq [-1, t]$, we have $S_{p+2}(t) = 0$ for $t \leq 0$. In addition, (2.1.11) implies

$$\|S_{p+2}(t)\|_{\mathcal{L}(\mathcal{X})} = \|\langle e^{\omega s} \chi(s) \eta_p(t-s), U_\omega \rangle\|_{\mathcal{L}(\mathcal{X})} \leq C' e^{\omega_1 t},$$

for $\omega_1 > \omega$ and $t \geq 0$. Let us show that U is the generalized derivative of $S_{p+2}(\cdot)$ of order $p+2$. For an arbitrary $\varphi \in \mathcal{S}$ we have

$$\begin{aligned}
\langle \varphi, S_{p+2}^{(p+2)} \rangle &= \langle (-1)^{p+2} \varphi^{(p+2)}, S_n \rangle = (-1)^{p+2} \int_0^\infty \varphi^{(p+2)}(t) S_n(t) \, dt \\
&= \left\langle \chi(s)(-1)^{p+2} \int_0^\infty \varphi^{(p+2)}(t) \eta_p(t-s) \, dt, U(s) \right\rangle \\
&= \langle \chi(s) \varphi(s), U \rangle = \langle \varphi, U \rangle.
\end{aligned}$$

Thus we have $S_{p+2}^{(p+2)} = U$. Together with (2.1.10) it implies

$$(\lambda I - A)^{-1} = \lambda^{p+2} \int_0^\infty e^{-\lambda t} S_{p+2}(t) \, dt, \qquad Re\lambda > \omega_1.$$

Therefore, (R1) takes place:

$$\begin{aligned}
\left\| \frac{d^k}{d\lambda^k} \left(\frac{(\lambda I - A)^{-1}}{\lambda^{p+2}} \right) \right\| = \left\| \frac{d^k}{d\lambda^k} \mathcal{L} S_{p+2}(\lambda) \right\| &\leq \int_0^\infty e^{-t Re\lambda} t^k \|S_{p+2}(t)\| \, dt \\
&\leq C' \int_0^\infty t^k e^{t(\omega_1 - Re\lambda)} \, dt = \frac{C' k!}{(Re\lambda - \omega_1)^{k+1}}
\end{aligned}$$

for $Re\lambda > \omega_1$, $k = 0, 1, \dots$.

(iii) $\implies$ (ii). The estimate (R1) with $k = 0$ provides that there exists the classical inverse Laplace transform of the analytical in the half-plane $\Re\lambda \geq \omega_1$ function $\lambda^{-p-4} \mathcal{R}(\lambda)$:

$$g(t) = \frac{1}{2\pi i} \int_{\omega_1 - i\infty}^{\omega_1 + i\infty} e^{\lambda t} \lambda^{-p-4} \mathcal{R}(\lambda) \, d\lambda.$$

According to the abstract Cauchy theorem, $g(t) = 0$ for $t \leq 0$ and $g(\cdot)$ is exponentially bounded for $t \geq 0$:

$$\|g(t)\|_{\mathcal{L}(\mathcal{X})} \leq \frac{1}{2\pi} \int_{\omega_1 - i\infty}^{\omega_1 + i\infty} e^{Re\lambda t} |\lambda|^{-p-4} \|\mathcal{R}(\lambda)\| \cdot |d\lambda| \leq C e^{\omega_1 t}.$$

Therefore, its Laplace transform exists and

$$\mathcal{L} g^{(p+4)}(\lambda) = \lambda^{p+4} \mathcal{L} g(\lambda) = \lambda^{p+4} \lambda^{-p-4} \mathcal{R}(\lambda) = \mathcal{R}(\lambda).$$

Thus, we have proved that the distribution $U = g^{(p+4)} \in \mathcal{S}'_\omega(\mathcal{L}(\mathcal{X}))$ satisfies the equality

$$\mathcal{L} U(\lambda) = \mathcal{R}(\lambda).$$

2.1. Solutions in spaces of abstract distributions

By the properties of the Laplace transform, we conclude that the U obtained is the unique solution to

$$\mathcal{L}(\mathbf{P} * U)(\lambda) = (\lambda I - A)\mathcal{R}(\lambda) = I_{\mathcal{X}}, \quad \mathcal{L}(U * \mathbf{P})(\lambda) = \mathcal{R}(\lambda)(\lambda I - A) = I_{[dom\, A]}.$$

$\square$

Now consider the Cauchy problem in the Schwartz space of abstract distributions $\mathcal{D}'(\mathcal{X})$. Let the resolvent of A satisfy (R2), i.e.,

$$\|\mathcal{R}(\lambda)\| \leq C|\lambda|^n, \qquad \lambda \in \Lambda_{n,\,\nu,\,\omega}^{\ln} = \{\lambda \in \mathbb{C} : \ Re\lambda > n\nu \ln|\lambda| + \omega\},$$

for some $n \in \mathbb{N}$, $\nu > 0$, $C > 0$, $\omega \in \mathbb{R}$.

It was mentioned above that the generalized solution of (2.0.1) can be obtained as the generalized inverse Laplace transform of the resolvent:

$$\langle \varphi, \mathcal{L}^{-1}\mathcal{R} \rangle = \frac{1}{2\pi i} \int_{b-i\infty}^{b+i\infty} \mathcal{R}(\lambda)\mathcal{L}\varphi(-\lambda)\, d\lambda, \qquad \varphi \in \Phi,$$

where the choice of Φ is determined by the properties of the resolvent. If (R2) holds, we can obtain the generalized inverse transform of the resolvent by integrating the product $\mathcal{R}(\lambda)\mathcal{L}\varphi(-\lambda)$ over a contour

$$\Gamma = \partial \Lambda_{n,\,\nu,\,\omega_1}^{\ln} = \{\lambda \in \mathbb{C} : \ Re\lambda = n\nu \ln|\lambda| + \omega_1\}, \qquad \omega_1 > \omega.$$

To define such a transform, it is sufficient to choose test functions φ in such a way that the Laplace transform $\mathcal{L}\varphi$ neutralizes the resolvent growth on Γ. Therefore, it is sufficient for $\mathcal{L}\varphi$ to decrease on Γ faster than any power of $1/|\lambda|$. It turns out that φ from $\mathcal{D}$ are suitable here.

The next three theorems are in a sense analogs of Theorem 2.1.1 to the case of test functions $\varphi \in \mathcal{D} \subset \mathcal{S}_\omega$.

Theorem 2.1.2 *If the resolvent of A satisfies (R2), then there exists an operator-valued distribution $U \in \mathcal{D}_0'(\mathcal{L}(\mathcal{X}))$ solving (2.1.8).*

Theorem 2.1.3 *Let A be a closed linear operator in a Banach space $\mathcal{X}$. Then the following statements are equivalent:*

(i) the Cauchy problem (2.0.1) is well-posed in the space of distributions $\mathcal{D}_0'(\mathcal{X})$;

(ii) there exists a generalized family of operators $U \in \mathcal{D}_0'(\mathcal{L}(\mathcal{X}))$ solving Equations (2.1.8).

Theorem 2.1.4 *If there exists an operator-valued distribution $U \in \mathcal{D}_0'(\mathcal{L}(\mathcal{X}))$ solving Equations (2.1.8), then the resolvent of A satisfies the condition (R2).*

The proofs of these theorems are carried out in the same manner as the proof of Theorem 2.1.1. They are even less complicated compared with the case of

54 2. Distribution methods

solutions from $\mathcal{S}'_\omega(\mathcal{X})$ because here the solution operator distribution U should be applied to test functions from $\mathcal{D}$ without multiplication by an exponent (see, e.g., [77]).

In addition, we give below the proof for the case of test functions from $\mathcal{D}^{\{M_q\}}$, which are the functions from $\mathcal{D}$ with additional conditions on the growth of $\varphi^{(k)}$.

2.2 Solutions in spaces of abstract ultra-distributions

In this section for the case (R3) we show that the existence of a unique solution to (2.1.7) and the corresponding behavior of the resolvent of A are equivalent to the existence of a generalized operator family U solving Equations (2.1.8):

$$U * \mathbf{P} = \delta \otimes I_{[dom\,A]}, \qquad \mathbf{P} * U = \delta \otimes I_{\mathcal{X}},$$

in the space of ultra-distributions. The solution of (2.1.8) exists if and only if (2.0.1) is well-posed in the generalized sense.

Now let the resolvent of A satisfy the condition (R3), i.e.,

$$\|\mathcal{R}(\lambda)\| \le C e^{\beta M(\gamma|\lambda|)}, \qquad \lambda \in \Lambda^M_{\alpha,\gamma,\omega} := \{\lambda \in \mathbb{C} : \ Re\lambda > \alpha M(\gamma|\lambda|) + \omega\},$$

where $\gamma > 0$, $\beta > 0$, $C > 0$, $\omega \in \mathbb{R}$, and M is an *associated function* with a certain sequence $\{M_q\}$:

$$M(x) := \sup_{q \in \mathbb{N}_0} \ln \frac{x^q M_0}{M_q}, \qquad x > 0, \tag{2.2.1}$$

or equivalently

$$e^{-M(x)} = \inf_{q \in \mathbb{N}_0} \frac{M_q}{M_0 x^q}, \qquad x > 0.$$

We consider the generalized well-posedness of (2.0.1) with the operator A satisfying (R3) on the space of ultra-differentiable test functions $\Phi = \mathcal{D}^{\{M_q\}}$, which is defined by the choice of the sequence $\{M_q\}$. We show that (2.0.1) is well-posed in the abstract dual space $\left(\mathcal{D}_a^{\{M_q\},\,B}\right)'(\mathcal{X})$, where the parameter B is defined by parameters a, α, β. (See definitions and properties of the spaces in Section 3.3.)

As we shall see below in the course of studying generalized well-posedness, the definition of M in (2.2.1) is consistent with the definition of M in the theory of K-convoluted semigroups and the generalized solutions obtained are closely related to K-regularized solutions.

It is known that the Fourier transform of a test function $\varphi \in \mathcal{D}_a^{\{M_q\},\,B}$ decreases as $e^{-M(B\,a\,|\lambda|)}$, $\lambda \in \mathbb{C}$ (Section 3.4). We show that the Laplace

2.2. Solutions in spaces of ultra-distributions

transform of a test function $\varphi \in \mathcal{D}_a^{\{M_q\}, B}$ behaves the same. As a consequence, we obtain the solution operators to (2.1.7) as the generalized inverse Laplace transform of the resolvent of A in the abstract dual space $\left(\mathcal{D}_a^{\{M_q\}, B}\right)'(\mathcal{L}(\mathcal{X}))$.

Thus we consider the generalized inverse Laplace transform of the resolvent:

$$\langle \varphi, \mathcal{L}^{-1}\mathcal{R} \rangle = \frac{1}{2\pi i} \int_\Gamma \mathcal{R}(\lambda)\mathcal{L}\varphi(-\lambda)\,d\lambda, \qquad \varphi \in \mathcal{D}_a^{\{M_q\}, B},$$

$$\Gamma = \partial \Lambda_{\alpha, \gamma, \omega_1}^M. \qquad (2.2.2)$$

We show that under the condition (R3) it defines an operator

$$\langle \varphi, U \rangle := \langle \varphi, \mathcal{L}^{-1}\mathcal{R} \rangle,$$

which is bounded in $\mathcal{X}$ for each $\varphi \in \mathcal{D}_a^{\{M_q\}, B}$. The U obtained is a family of generalized solution operators to (2.0.1).

Comparing this generalized solution with the solution regularized by means of convolution, i.e., with the K-convoluted semigroup, one can see that the role of $\widetilde{K}$ in (1.2.18) is played here by the test functions $\mathcal{L}\varphi$ for $\varphi \in \mathcal{D}_a^{\{M_q\}, B}$.

Theorem 2.2.1 *If the resolvent of A satisfies* (R3), *then, for any $a > 0$ and some $B = B(a, \alpha, \beta)$, there exists $U \in \left(\mathcal{D}_a^{\{M_q\}, B}\right)'(\mathcal{L}(\mathcal{X}))$ solving (2.1.8).*

Proof. We prove that the equality (2.2.2) defines a bounded operator on $\mathcal{X}$ by the formula

$$\langle \varphi, U \rangle := \langle \varphi, \mathcal{L}^{-1}\mathcal{R} \rangle, \qquad \varphi \in \mathcal{D}_a^{\{M_q\}, B}.$$

Let us take $a > 0$. To begin with, we estimate the behavior of the Laplace transform of a test function $\varphi \in \mathcal{D}_a^{\{M_q\}, B}$:

$$\mathcal{L}\varphi(-\lambda) = \int_0^{+\infty} e^{\lambda t}\varphi(t)\,dt. \qquad (2.2.3)$$

The supports of functions of $\mathcal{D}_a^{\{M_q\}, B}$ lie in the segment $[-a, a]$. Integrating q times by parts, we obtain the estimate

$$|\mathcal{L}\varphi(-\lambda)| = \left| \frac{(-1)^q}{\lambda^q} \int_0^a e^{\lambda t}\varphi^{(q)}(t)\,dt \right| \le \frac{\|\varphi^{(q)}\|_{C[-a,a]}}{|\lambda|^q} \int_0^a e^{\mathrm{Re}\lambda\, t}\,dt.$$

By the definition of the space $\mathcal{D}_a^{\{M_q\}, B}$, the inequality (3.3.14)

$$\|\varphi^{(q)}\|_{C[-a,a]} \le \|\varphi\|_m M_q(B + 1/m)^q,$$

holds for every $m \in \mathbb{N}$. If m is fixed, multiplying the numerator and the denominator by γ^q, we obtain

$$|\mathcal{L}\varphi(-\lambda)| \le C\|\varphi\|_m \frac{M_q\gamma^q(B + 1/m)^q}{M_0|\gamma\lambda|^q} e^{a\,\mathrm{Re}\lambda} = C\|\varphi\|_m \frac{M_q B_1^q}{M_0|\gamma\lambda|^q} e^{a\,\mathrm{Re}\lambda}$$

56   2. *Distribution methods*

for arbitrary $q \in \mathbb{N}_0$ and $B_1 := \gamma(B + 1/m)$. Now we can pass to the infimum with respect to all $q \in \mathbb{N}_0$, which, in accordance with the definition of the associated function (2.2.1), is equal to

$$\inf_{q \in \mathbb{N}_0} \frac{M_q B_1^q}{M_0 |\gamma \lambda|^q} = e^{-M(|\gamma \lambda|/B_1)}.$$

As a result we obtain the estimate

$$|\mathcal{L}\varphi(-\lambda)| \leq C \|\varphi\|_m e^{a \, \mathrm{Re}\lambda - M(|\gamma\lambda|/B_1)}, \qquad \lambda \in \mathbb{C}, \quad \varphi \in \mathcal{D}_a^{\{M_q\}, B}.$$

On the contour Γ it takes the form

$$|\mathcal{L}\varphi(-\lambda)| \leq C \|\varphi\|_m e^{\left(a - \frac{1}{\alpha B_1}\right) \mathrm{Re}\lambda + \omega_1/\alpha}, \qquad \lambda \in \Gamma.$$

Now we estimate the integral (2.2.2):

$$\left\| \frac{1}{2\pi i} \int_\Gamma \mathcal{R}(\lambda) \mathcal{L}\varphi(-\lambda) \, d\lambda \right\| \leq \frac{1}{2\pi} \int_\Gamma \|\mathcal{R}(\lambda)\| |\mathcal{L}\varphi(-\lambda)| \cdot |d\lambda|.$$

The condition (R3) implies the following estimate of the resolvent on Γ:

$$\|\mathcal{R}(\lambda)\|_{\mathcal{L}(\mathcal{X})} \leq C e^{\beta(\mathrm{Re}\lambda - \omega_1)/\alpha}, \qquad \lambda \in \Gamma.$$

Hence,

$$\left\| \frac{1}{2\pi i} \int_\Gamma \mathcal{R}(\lambda) \mathcal{L}\varphi(-\lambda) \, d\lambda \right\| \leq e^{\omega_1(1-\beta)/\alpha} C_1 \|\varphi\|_m \int_\Gamma e^{\left(\beta\alpha + a - \frac{1}{\alpha B_1}\right) \mathrm{Re}\lambda} |d\lambda|.$$

For given β, α, and arbitrary $a > 0$ one can choose B_1 in such a way that $\beta/\alpha + a - 1/\alpha B_1 < 0$, which provides convergence of the integral in the right-hand side. As a result, we have the estimate

$$\left\| \frac{1}{2\pi i} \int_\Gamma \mathcal{R}(\lambda) \mathcal{L}\varphi(-\lambda) \, d\lambda \right\| \leq C_2 \|\varphi\|_m, \tag{2.2.4}$$

which proves that the generalized inverse Laplace transform of the resolvent satisfying the condition (R3) defines a bounded on $\mathcal{X}$ operator $\langle \varphi, U \rangle$ for every $\varphi \in \mathcal{D}_a^{\{M_q\}, B}$. Here B in chosen in the following way. From the inequality $\beta/\alpha + a - \alpha/B_1 < 0$ we find B_1 and for the chosen and fixed $m \in \mathbb{N}$ we define $B = B_1/\gamma - 1/m$.

Besides, (2.2.4) demonstrates that $U = \mathcal{L}^{-1} \mathcal{R} \in \left(\mathcal{D}_a^{\{M_q\}, B} \right)' (\mathcal{L}(\mathcal{X}))$, and by (2.2.3), it follows that $\langle \varphi, U \rangle = 0$ for φ with supports in $(-\infty, 0]$. Hence, $\mathrm{supp}\, U$ lies in the non-negative semi-axis.[4]

---

[4] This fact, in contrast to the case of $\mathcal{D}_0'$, we do not reflect in notations of ultra-distribution spaces $\left(\mathcal{D}_a^{\{M_q\}, B} \right)'$ because of many indexes present here.

2.2. Solutions in spaces of ultra-distributions 57

The proof of equations (2.1.8) is done similarly to their proof in the spaces of distributions. $\qquad\square$

Now we connect the equalities (2.1.8) in $\left(\mathcal{D}_a^{\{M_q\},\,B}\right)'(\mathcal{X})$ with the generalized well-posedness of (2.0.1) in this space.

Theorem 2.2.2 *Let A be a closed linear operator in a Banach space $\mathcal{X}$. Then the following statements are equivalent:*

(i) the Cauchy problem (1.0.1) is well-posed in the space of ultra-distributions $\left(\mathcal{D}_a^{\{M_q\},\,B}\right)'(\mathcal{X})$;

(ii) there exists a generalized family of operators $U \in \left(\mathcal{D}_a^{\{M_q\},\,B}\right)'(\mathcal{L}(\mathcal{X}))$ solving (2.1.8).

Proof. (i)$\Longrightarrow$(ii). For every $\varphi \in \mathcal{D}_a^{\{M_q\},\,B}$ we define

$$\langle \varphi, U \rangle f := \langle \varphi, u \rangle, \qquad f \in \mathcal{X},$$

where $u \in \left(\mathcal{D}_a^{\{M_q\},\,B}\right)'(\mathcal{X})$ is the solution of a well-posed problem (2.1.7). Stability of the solution implies

$$\|\langle \varphi, U \rangle f_n\| = \|\langle \varphi, u_n \rangle\| \to 0 \quad \text{as} \quad \|f_n\| \to 0$$

uniformly with respect to φ in every bounded set in $\mathcal{D}_a^{\{M_q\},\,B}$. Therefore, $\langle \varphi, U \rangle \in \mathcal{L}(\mathcal{X})$ for any $\varphi \in \mathcal{D}_a^{\{M_q\},\,B}$.

It follows from the definition of U that $Uf \in \left(\mathcal{D}_a^{\{M_q\},\,B}\right)'(\mathcal{X})$ for any $f \in \mathcal{X}$. We show that $U \in \left(\mathcal{D}_a^{\{M_q\},\,B}\right)'(\mathcal{L}(\mathcal{X}))$ with $\operatorname{supp} U \subset [0,\infty)$, and that for any sequence $\varphi_n \in \mathcal{D}_a^{\{M_q\},\,B}$ convergent to zero in $\mathcal{D}_a^{\{M_q\},\,B}$, it holds that $\|\langle \varphi_n, U \rangle\| \to 0$.

Consider the set

$$\mathcal{B} = \{Uf : \|f\| \le C\} \subset \left(\mathcal{D}_a^{\{M_q\},\,B}\right)'(\mathcal{X}).$$

According to the structure theorems, the set $\mathcal{B}$ is bounded in $\left(\mathcal{D}_a^{\{M_q\},\,B}\right)'(\mathcal{X})$ if and only if for any compact set Υ there exist measures $u_k = u_k \in C'(\Upsilon, \mathcal{X})$ satisfying the estimate

$$\|u_k\|_{C'(\Upsilon, \mathcal{X})} \le CB^k/M_k \tag{2.2.5}$$

for any element Uf of $\mathcal{B}$ and such that

$$Uf = \sum_{k=0}^{\infty} d^k u_k/dt^k.$$

58   *2. Distribution methods*

Therefore, for each sequence $\varphi_n \to 0$ and each $Uf \in \mathcal{B}$, we have

$$\|\langle \varphi_n, U \rangle f\| = \left\| \sum_{k=0}^{\infty} \langle \varphi_n, D^k u_k \rangle \right\| = \left\| \sum_{k=0}^{\infty} \langle (-1)^k \varphi_n^{(k)}, u_k \rangle \right\|$$

$$\leq \sum_{k=0}^{\infty} \|u_k\|_{C'(\Upsilon, \mathcal{X})} \|\varphi_n^{(k)}\|_{C([-a,a])}.$$

Multiplying the numerator and the denominator by an arbitrary number h^k and taking into account (2.2.5), we obtain

$$\|\langle \varphi_n, U \rangle f\| \leq C_1 \sum_{k=0}^{\infty} \frac{B^k (h + \frac{1}{m})^k \|\varphi_n^{(k)}\|_{C([-a;a])}}{M_k h^k} \leq C_1 \sum_{k=0}^{\infty} h^k B^k \|\varphi_n\|_m$$

for each $m \in \mathbb{N}$, where $\|\varphi\|_m$ is defined by (3.3.14):

$$\|\varphi\|_m = \sup_{q \in \mathbb{N}_0} \sup_{|x| \leq a} \frac{|\varphi^{(q)}(x)|}{\left(B + \frac{1}{m}\right)^q M_q}.$$

Now, setting $h = 1/2B$, we obtain

$$\|\langle \varphi_n, U \rangle f\| \leq 2C_1 \|\varphi_n\|_m \underset{\varphi_n \to 0}{\longrightarrow} 0, \qquad Uf \in \mathcal{B}.$$

This means that $\langle \varphi_n, U \rangle$ is convergent to zero in $\mathcal{L}(\mathcal{X})$.

One can verify that U satisfies (2.1.8) in the same way as done in the case of distributions.

(ii)$\Longrightarrow$(i). For $f \in \mathcal{X}$ we consider the element $u := Uf \in \left(\mathcal{D}_a^{\{M_q\}, B} \right)' (\mathcal{X})$. The first equation in (2.1.8) implies that u is a solution of the problem (2.1.7).

To prove the stability of the solution introduced we show that $u_n \to 0$ in $\left(\mathcal{D}_a^{\{M_q\}, B} \right)' (\mathcal{X})$ if $f_n \to 0$ in $\mathcal{X}$. Again for a bounded set $\mathcal{B}$ in $\mathcal{D}_a^{\{M_q\}, B}$ the operator $\langle \varphi, U \rangle$ is linear and bounded; therefore,

$$\|\langle \varphi, u_n \rangle\| = \|\langle \varphi, U \rangle f_n\| \leq \|\langle \varphi, U \rangle\|_{\mathcal{L}(\mathcal{X})} \cdot \|f_n\|$$

and there exists $m_0 \in \mathbb{N}$ such that U is an element of the space which is dual to the normed space $\left(\mathcal{D}_a^{\{M_q\}, B}, \| \cdot \|_{m_0} \right)$. Hence,

$$\|\langle \varphi, U \rangle\|_{\mathcal{L}(\mathcal{X})} \leq C \|\varphi\|_{m_0}, \qquad \varphi \in \mathcal{D}_a^{\{M_q\}, B}.$$

Then we have

$$\|\langle \varphi, u_n \rangle\| \leq C \|\varphi\|_{m_0} \|f_n\| = C_1 \|f_n\|, \qquad \varphi \in \mathcal{B},$$

for all the functions $\varphi \in \mathcal{B}$, which proves convergence of $\|\langle \varphi, u_n \rangle\|$ to zero uniformly with respect to $\varphi \in \mathcal{B}$.

2.3. Differential systems in Gelfand–Shilov spaces 59

Prove that the solution is unique. Suppose v is another solution of (2.1.7); then we have

$$U * \mathbf{P} * v = U * (\delta \otimes f) = Uf = u \in \left(\mathcal{D}_a^{\{M_q\},\, B} \right)' (\mathcal{X})$$

and the second equation in (2.1.8) implies

$$u = U * \mathbf{P} * v = (\delta \otimes I_{[dom\, A]}) * v = v. \qquad \square$$

The inverse to the result of Theorem 2.2.1 is valid [77].

In Part II the results on regularized and generalized (wrt t and wrt x) solutions to (2.0.1) will be used for construction of corresponding solutions to stochastic Cauchy problems. In addition, a few more results on generalized solutions will be available immediately for stochastic problems with A generating R-semigroups, without preliminary results for the corresponding deterministic ones. Among the problems with the generators of R-semigroups the important place take the problems with differential operators $A = \mathbf{A}\,(i\partial/\partial x)$, whose generalized well-posedness we begin to study in the next section.

2.3 Solutions to the Cauchy problem for differential systems in Gelfand–Shilov spaces

The present section is devoted to study of the special case, important for applications, of the abstract Cauchy problem (2.0.1) with $A = \mathbf{A}\,(i\partial/\partial x)$,[5] i.e., the Cauchy problem for the system of differential equations (2.0.2):

$$\frac{\partial u(x;t)}{\partial t} = \mathbf{A}\left(i\frac{\partial}{\partial x}\right) u(x;t), \quad t \in [0,T], \quad x \in \mathbb{R}^n, \qquad (2.3.1)$$

$$u(x;0) = f(x). \qquad (2.3.2)$$

We obtain generalized solutions $u = (u_1, \ldots, u_m)$ to the problem in spaces of distributions Φ' for initial data f from some other spaces of distributions Ψ', which are chosen according to the properties of the operator $\mathbf{A}\,(i\partial/\partial x)$, in particular for f from $L_m^2(\mathbb{R}^n) := L^2(\mathbb{R}^n) \times \ldots \times L^2(\mathbb{R}^n)$.

These generalized solutions are obtained via the generalized Fourier transform techniques developed for the systems which are Petrovsky correct, conditionally correct, and incorrect according to the Gelfand–Shilov classification. These results will be used in Part II in construction of generalized solutions to stochastic Cauchy problems with $A = \mathbf{A}\,(i\partial/\partial x)$ proved to be the generator of an R-semi-group in $H = L_m^2(\mathbb{R}^n)$.

---

[5]In this section (and in some sections later) we use the designation $\mathbf{A}\,(i\partial/\partial x)$ for the differential operator in order to distinguish the operator symbol from designations of constants A, B customary in the theory of Gelfand–Shilov generalized functions used here.

60   2. Distribution methods

2.3.1   The Gelfand–Shilov classification for the Cauchy problem with differential operators

We consider the Cauchy problem (2.3.1)–(2.3.2), where $\mathbf{A}\left(i\partial/\partial x\right)$ is a matrix operator:

$$\mathbf{A}\left(i\frac{\partial}{\partial x}\right) = \left\{A_{jk}\left(i\frac{\partial}{\partial x}\right)\right\}_{j,k=1}^{m}, \qquad x = (x_1,\dots,x_n) \in \mathbb{R}^n,$$

$A_{jk}\left(i\frac{\partial}{\partial x}\right)$ are linear differential operators of orders not exceeding $p > 0$, and

$$A_{jk}\left(i\frac{\partial}{\partial x}\right) = \sum C_\alpha^{j,k}\left(i\frac{\partial}{\partial x}\right)^\alpha, \qquad C_\alpha^{j,k} \in \mathbb{R}, \quad |\alpha| = \alpha_1 + \cdots + \alpha_n \leq p.$$

For every $x \in \mathbb{R}^n$, a solution of (2.3.1)–(2.3.2) is an m-dimensional vector:

$$u(x;t) = (u_1(x;t),\dots,u_m(x;t)) \in \mathbb{R}^m, \qquad t \in [0,T].$$

We begin with some notations and preliminary information on operators connected with the systems of differential equations on linear topological spaces.[6]

Let Φ be the direct product of linear topological spaces Φ_k, $k = \overline{1,m}$, i.e., $\Phi = \Phi_1 \times \dots \times \Phi_m$ with ordinary operations of addition and multiplication by a scalar. This implies that the elements $\varphi \in \Phi$ are m-dimensional vector-functions: $\varphi(x) = (\varphi_1(x),\dots,\varphi_m(x)) \in \mathbb{C}^m$, $x \in \mathbb{R}^n$, with the elements φ_k belonging to linear scalar topological spaces Φ_k.

Linear continuous functionals on Φ are defined as vector-valued distributions $f = (f_1,\dots,f_m) \in \Phi'$ with $f_k \in \Phi_k'$, $k = \overline{1,m}$. A functional f acts on a test function $\varphi \in \Phi$ as follows:

$$\langle \varphi, f \rangle = \langle \varphi_1, f_1 \rangle + \dots + \langle \varphi_m, f_m \rangle,$$

where $\langle \varphi_k, f_k \rangle$ denotes the result of applying a scalar distribution $f_k \in \Phi_k'$ to a scalar test function $\varphi_k \in \Phi_k$. Multiplication of a distribution by a scalar is defined as follows:

$$\langle \varphi, \lambda f \rangle = \overline{\lambda} \langle \varphi, f \rangle, \qquad \lambda \in \mathbb{C}, \quad f \in \Phi', \quad \varphi \in \Phi.$$

Now we construct generalized (wrt x) solutions of the Cauchy problem (2.3.1)–(2.3.2) in spaces of distributions Φ' for $f \in \Psi'$ which are defined by the properties of $\mathbf{A}\left(i\partial/\partial x\right)$. In contrast to generalized (wrt t) solutions constructed in the previous section by use of the generalized Laplace transform defined for distributions with supports in $[0,\infty)$, here we use the Fourier transform techniques in the spaces of test functions $\varphi(x)$, $x \in \mathbb{R}^n$, in particular $\varphi(x)$, $x \in \mathbb{R}$ [36, 37, 77] (see also Sections 3.3–3.4).

---

[6]For readers, convenience, some useful information on specific topological spaces used as test function spaces and operators in the spaces is presented in Section 3.3.

2.3. Differential systems in Gelfand–Shilov spaces

Let $\widetilde{\Phi}$ be a space formed by classical Fourier transforms of functions from Φ and let $\widetilde{\Phi}'$ denote the space of generalized Fourier transforms of distributions from Φ'. The consideration is restricted to the common case of spaces Φ of test functions in which there exists a one-to-one correspondence between Φ and $\widetilde{\Phi}$ and $\left(\widetilde{\Phi}\right)' = \widetilde{\Phi}'$.

The definition of the generalized Fourier transform of $f \in \Psi'$ denoted by $\mathcal{F}[f]$, $\tilde{f}$, or $\mathcal{F}f \in \widetilde{\Psi}'$ is given by

$$\langle \tilde{\psi}, \mathcal{F}f \rangle := (2\pi)^n \langle \mathcal{F}^{-1}[\tilde{\psi}], f \rangle, \qquad \tilde{\psi} = \mathcal{F}\psi \in \widetilde{\Psi}, \qquad \psi \in \Psi.$$

It is based on the Parseval equality for the Fourier transform in $L^2(\mathbb{R}^n)$, where it has the form of equality for scalar products:

$$\langle \mathcal{F}\varphi, \mathcal{F}f \rangle = (2\pi)^n \langle \varphi, f \rangle, \qquad f, \varphi \in L^2(\mathbb{R}^n).$$

Let us apply the generalized Fourier transform to the problem (2.3.1)–(2.3.2) in the spaces Ψ' and Φ' and consider solutions to the dual one in $\widetilde{\Psi}'$ and $\widetilde{\Phi}'$:

$$\langle \tilde{\varphi}, \frac{\partial \tilde{u}(\cdot;t)}{\partial t} \rangle = \langle \tilde{\varphi}, \mathbf{A}(\cdot)\tilde{u}(\cdot;t) \rangle = \langle \mathbf{A}^*(\cdot)\tilde{\varphi}, \tilde{u}(\cdot;t) \rangle, \qquad t \in [0,T],$$

$$\langle \tilde{\psi}, \tilde{u}(\cdot;0) \rangle = \langle \tilde{\psi}, \tilde{f}(\cdot) \rangle, \qquad \tilde{\varphi} \in \widetilde{\Phi}, \quad \tilde{\psi} \in \widetilde{\Psi}. \tag{2.3.3}$$

Here the matrix function $\mathbf{A}(s)$, $s = \sigma + i\tau$, defines an operator of multiplication by the matrix $\{A_{jk}(s)\}_{j,k=1}^m$, which has polynomials of powers not exceeding p as its elements and

$$\mathbf{A}^*(s) = \{\overline{A_{kj}(s)}\}_{j,k=1}^m \quad : \quad \widetilde{\Phi} \to \widetilde{\Phi} \quad (\widetilde{\Phi}' \to \widetilde{\Phi}'). \tag{2.3.4}$$

Taking into consideration the isomorphism of Φ and $\widetilde{\Phi}$ and the isomorphism of Φ' and $\widetilde{\Phi}'$, which are carried out by the direct and inverse Fourier transforms, we go from exploring (2.3.1)–(2.3.2) on Ψ', Φ' to the study of the Fourier transformed (2.3.3) on $\widetilde{\Psi}'$, $\widetilde{\Phi}'$ with the solution having the form

$$\tilde{u}(s;t) = e^{t\mathbf{A}(s)}\tilde{f}(s), \qquad t \in [0,T], \quad s = \sigma + i\tau \in \mathbb{C}^n.$$

It follows that

$$u(x;t) = \mathcal{F}^{-1}\left[e^{t\mathbf{A}(\cdot)}\tilde{f}(\cdot)\right](x) = (G_t * f)(x), \qquad t \in [0,T], \quad x \in \mathbb{R}^n, \tag{2.3.5}$$

where the Green function is defined by

$$G_t(x) := \mathcal{F}^{-1}\left[e^{t\mathbf{A}(\cdot)}\right](x), \qquad t \in [0,T], \quad x \in \mathbb{R}^n.\,[7]$$

---

[7] Recall that if we write a distribution f as $f(s)$, it means that the distribution is applied to test functions of variable s. The same applies to the equalities for distributions of type (2.3.5) written without applying to test functions.

The existence and stability of solutions to (2.3.3) in $\widetilde{\Phi}'$ for $\tilde{f} \in \widetilde{\Psi}'$ hold for those $\widetilde{\Psi}'$, $\widetilde{\Phi}'$, where the solution operators $e^{t\mathbf{A}(s)}$ are bounded as multiplication operators acting from $\widetilde{\Psi}'$ to $\widetilde{\Phi}'$ and hence $e^{t\mathbf{A}^*(s)}$, where $\mathbf{A}^*(s)$ is defined by (2.3.4), are bounded as multiplication operators acting from $\widetilde{\Phi}$ to $\widetilde{\Psi}$. In this case $e^{t\mathbf{A}(s)}$ is called a *multiplier* from $\widetilde{\Psi}'$ to $\widetilde{\Phi}'$ and $e^{t\mathbf{A}^*(s)}$ is a *multiplier* from $\widetilde{\Phi}$ to $\widetilde{\Psi}$. The estimates of $e^{t\mathbf{A}(s)}$ for $s = \sigma + i\tau \in \mathbb{C}^n$ and especially for $s = \sigma$ determine the choice of these spaces. The corresponding solution $u(\cdot\,;t) = \mathcal{F}^{-1}[e^{t\mathbf{A}(s)}\tilde{f}(s)](\cdot) \in \Phi'$ to the problem (2.3.1)–(2.3.2) exists and is stable wrt $f \in \Psi'$.

For $e^{t\mathbf{A}(s)}$ the following obvious estimates take place:

$$\left\|e^{t\mathbf{A}(s)}\right\|_m = \left\|\sum_{k=0}^{\infty} \frac{t^k}{k!}\mathbf{A}^k(s)\right\|_m \leq \sum_{k=0}^{\infty} \frac{t^k}{k!}\|\mathbf{A}(s)\|_m^k$$

$$\leq \sum_{k=0}^{\infty} \frac{t^k}{k!}(mC)^k|s|^{kp} = e^{mCt|s|^p}.$$

The estimates imply that the components of the matrix $e^{t\mathbf{A}(s)}$ are analytical functions of $s \in \mathbb{C}^n$ whose growth orders do not exceed p.

In order to make the estimates more accurate we introduce $\lambda_1(s)$, ..., $\lambda_m(s)$, the roots of the characteristic equation

$$\det\left(\lambda I - \mathbf{A}(s)\right) = 0, \qquad s \in \mathbb{C}^n,$$

which are called *characteristic roots of the system* (2.3.1). The way the characteristic roots are constructed implies that they are polynomials of powers not exceeding p. Let

$$\Lambda(s) = \max_{1 \leq k \leq m} Re\lambda_k(s), \qquad s \in \mathbb{C}^n.$$

The following significant theorem holds true.

Theorem 2.3.1 [36] *For any $m \times m$ matrix $\mathbf{A}(s)$, $s \in \mathbb{C}^n$, with the components that are polynomials of powers not exceeding p, the following estimate holds:*

$$e^{t\Lambda(s)} \leq \left\|e^{t\mathbf{A}(s)}\right\|_m \leq C(1+|s|)^{p(m-1)}e^{t\Lambda(s)}, \qquad t \geq 0, \quad s \in \mathbb{C}^n. \qquad (2.3.6)$$

This theorem implies the following estimate of $e^{t\mathbf{A}(\cdot)}$ in terms of the reduced order p_0:

$$\left\|e^{t\mathbf{A}(s)}\right\|_m \leq C(1+|s|)^{p(m-1)}e^{b_0t|s|^{p_0}}, \qquad b_0 \in \mathbb{R}, \quad t \geq 0, \quad s \in \mathbb{C}^n. \qquad (2.3.7)$$

A number p_0 defined by $p_0 = \inf\{\rho : |\Lambda(s)| \leq C_\rho(1+|s|)^\rho, \ s \in \mathbb{C}^n\}$ is called

2.3. Differential systems in Gelfand–Shilov spaces

the exact power growth order of $\Lambda(\cdot)$ and the reduced order of the system (2.3.1). It follows from the definition that $p_0 \leq p$ and that the conjugate system has the same reduced order as the initial system.

The behavior of $\Lambda(\cdot)$ for real values of the variable underlies distinguishing of the following classes of systems (2.3.1) in the Gelfand–Shilov classification [36].

Definition 2.3.1 *The system (2.3.1) is called*
Petrovsky correct *if there exists such a constant $C > 0$ that*

$$\Lambda(\sigma) \leq C, \qquad \sigma \in \mathbb{R}^n, \tag{2.3.8}$$

in particular, the important subclasses of Petrovsky correct systems are:
parabolic systems *for which there exist such constants $C > 0$, $h > 0$, $C_1 > 0$ that*

$$\Lambda(\sigma) \leq -C|\sigma|^h + C_1, \qquad \sigma \in \mathbb{R}^n;$$

hyperbolic systems *for which $p_0 \leq 1$, i.e.,*

$$\Lambda(s) \leq C_1|s| + C_2, \qquad s \in \mathbb{C}^n,$$

and the property (2.3.8) holds;
conditionally correct *if there exist such constants $C > 0$, $0 < h < 1$, $C_1 > 0$ that*

$$\Lambda(\sigma) \leq C|\sigma|^h + C_1, \qquad \sigma \in \mathbb{R}^n; \tag{2.3.9}$$

incorrect *if the function $\Lambda(\cdot)$ grows for real $s = \sigma$ in the same way as for the complex ones:*

$$\Lambda(\sigma) \leq C|\sigma|^{p_0} + C_1, \qquad \sigma \in \mathbb{R}^n.$$

Further, following to [36, 77], we first prove the general theorem on the well-posedness of the Cauchy problem (2.3.3) in spaces $\widetilde{\Psi}'$, $\widetilde{\Phi}'$ (i.e., the theorem stating that for each $\widetilde{f}(\cdot) \in \widetilde{\Psi}'$ there exists the unique solution $\widetilde{u}(\cdot; t) \in \widetilde{\Phi}'$ which is stable wrt $\widetilde{f}$) and as a consequence we obtain the result on the well-posedness of the Cauchy problem (2.3.1)–(2.3.2) in the spaces Ψ', Φ'. Then we introduce more precise results on well-posedness for each type of system in the Gelfand–Shilov classification. In the study of our main object, the stochastic problems in Part II, we will use a simpler version of these results for the case of $\widetilde{f} \in L^2_m(\mathbb{R})$ and hence $f \in L^2_m(\mathbb{R})$.

As mentioned above, the choice of the spaces $\widetilde{\Psi}'$, $\widetilde{\Phi}'$ is based on the estimates of $e^{t\mathbf{A}(\cdot)}$. It follows from (2.3.4) that $\|e^{t\mathbf{A}^*(\cdot)}\|_m$ has the same estimates as $\|e^{t\mathbf{A}(\cdot)}\|_m$. To obtain solutions for different estimates corresponding to different classes of $\mathbf{A}$, we will use the conjugate spaces to $\mathcal{S}^\beta_\alpha$, W^Ω_M as well as their subspaces

$$W^{\Omega,b}_{M,a}, \quad S^{\beta,B}_{\alpha,A}, \quad \dots$$

The spaces are considered in detail in Section 3.3. Test functions from the

64   *2. Distribution methods*

spaces decrease exponentially at infinity and the corresponding spaces of distributions are called distributions of exponential growth. (Actually, these spaces are used as Ψ'_k and Φ'_k, i.e., coordinate spaces of Ψ' and Φ', but we do not stress it in notation to avoid overloading.)

Before the estimation of $\|e^{t\mathbf{A}(\cdot)}\|_m$, we recall some spaces of test functions used as Ψ, Φ: W_M and W^Ω. Comparing with the well-known spaces of functions with compact supports $\mathcal{D}$ and rapidly decreasing functions $\mathcal{S}$, the spaces W_M (as well as S_α and S_α^β used later) allow us to trace more subtle exponential decay at infinity.

Let μ and ω be increasing continuous functions on $[0, \infty)$ under conditions $\mu(0) = \omega(0) = 0$ and $\lim_{\xi \to \infty} \mu(\xi) = \lim_{\xi \to \infty} \omega(\xi) = \infty$. Let

$$M(x) := \int_0^x \mu(\xi)d\xi, \qquad \Omega(x) := \int_0^x \omega(\xi)d\xi, \qquad x \geq 0,$$
$$M(-x) := M(x), \qquad \Omega(-x) := \Omega(x).$$

The space W_M consists of all infinitely differentiable functions φ satisfying the condition

$$|\varphi^{(q)}(x)| \leq C_q e^{-M(ax)}, \qquad x \in \mathbb{R},$$

with some constants $C_q = C_q(\varphi)$, $a = a(\varphi)$. Since $M(x)$ increases as $|x| \to \infty$ faster than any linear function, the functions from W_M decrease at infinity faster than any exponent of type $e^{-a|x|}$.[8] The space W^Ω consists of all functions $\varphi(z)$ of the variable $z = x + iy \in \mathbb{C}$ satisfying the inequality

$$|z^k \varphi(z)| \leq C_k e^{\Omega(by)}, \qquad z = x + iy \in \mathbb{C},$$

with some constants $C_k = C_k(\varphi)$, $b = b(\varphi)$. There exists the relation between spaces W_M and $\widetilde{W^\Omega}$: they coincide if M and $\Omega(\cdot)$ are dual by Young functions. (In more detail these spaces, as well as W_M^Ω, $W_{M,a}^{\Omega,b}$, S_α^β, $S_{\alpha,A}^{\beta,B}$, and some other spaces of test functions and their dual are presented in Section 3.3.)

Below we formulate the promised results on the well-posedness of the Cauchy problem limited to the case $p_0 > 1$. (For the case of $0 < p_0 \leq 1$, see, e.g., [36, 37, 77].)

2.3.2   Generalized well-posedness of the Cauchy problem in Gelfand–Shilov spaces

Theorem 2.3.2 *Let the reduced order of the system (2.3.1) be $p_0 > 1$ and let the matrix exponent $e^{t\mathbf{A}(\cdot)}$ satisfy (2.3.7). Consider $a > 0$, $b > 0$, and $\theta < a$. Then for each $t \in [0, T]$ the matrix exponent $e^{t\mathbf{A}^*(\cdot)}$ defines a bounded multiplication operator from $W_{M,a}^{\Omega,b}$ to $W_{M,a-\theta}^{\Omega,b+\theta}$, where $M(x) = \Omega(x) = \frac{|x|^{p_0}}{p_0}$, $x \in \mathbb{R}$, and $T = T(p_0; q; \theta)$.*

---

[8]Note that M in the notations of W_M has different growth as $x \to \infty$ comparing with M in (2.2.1). For simplicity, further on we will mainly work with the case of $n = 1$, i.e., $x \in \mathbb{R}$.

2.3. Differential systems in Gelfand–Shilov spaces

Proof. We begin by noting that for $q > 0$, $p_0 > 1$ and for an arbitrary $\theta > 0$ there exists such $T > 0$ that

$$qT < \frac{1}{p_0}\theta^{p_0} 2^{-p_0}. \tag{2.3.10}$$

This is indeed the case since for each $a > 0$ and $x > 1$ there always exists $0 < C < \frac{a^x}{x}$ that implies the inequality $Cx < a^x$. In other words, $(qT)p_0 < (\theta/2)^{p_0}$ for $p_0 > 1$. Therefore, (2.3.10) implies the existence of a constant $C_1 > 0$ such that (2.3.7) results in

$$\left\| e^{t\mathbf{A}(s)} \right\|_m \leq C(1+|s|)^{p(m-1)} e^{qT|s|^{p_0}} \leq C_1 e^{\frac{1}{p_0}\theta^{p_0} 2^{-p_0}|s|^{p_0}}.$$

Applying the estimate of binomial coefficients

$$|s|^{p_0} = |\sigma + i\tau|^{p_0} = (\sigma^2 + \tau^2)^{p_0/2} \leq (|\sigma|^{p_0} + |\tau|^{p_0})2^{p_0},$$

we obtain

$$\left\| e^{t\mathbf{A}^*(s)} \right\|_m = \left\| e^{t\mathbf{A}(s)} \right\|_m \leq C_1 e^{\frac{(\theta|\sigma|)^{p_0}}{p_0} + \frac{(\theta|\tau|)^{p_0}}{p_0}}. \tag{2.3.11}$$

Let us take $M(x) = \Omega(x) = \frac{|x|^{p_0}}{p_0}$, $x \in \mathbb{R}$, and consider the space $W_{M,a}^{\Omega,b}$ consisting of entire functions ψ satisfying the condition

$$|\psi(\sigma + i\tau)|_m \leq Ce^{-M(\bar{a}\sigma) + \Omega(\bar{b}\tau)},$$

for all $\bar{a} < a$, $\bar{b} > b$. According to (3.3.19), each entire function f satisfying

$$|f(\sigma + i\tau)|_m \leq Ce^{M(a_1\sigma) + \Omega(b_1\tau)}$$

defines a bounded multiplication operator acting from $W_{M,a}^{\Omega,b}$ to $W_{M,a-a_1}^{\Omega,b+b_1}$ for $a_1 < a$ and $b_1 > 0$. Thus the matrix $e^{t\mathbf{A}^*(s)}$, $s \in \mathbb{C}^n$, satisfying

$$\left\| e^{t\mathbf{A}^*(s)} \right\|_m \leq C_1 e^{M(\theta\sigma) + \Omega(\theta\tau)},$$

defines a bounded multiplication operator acting from $W_{M,a}^{\Omega,b}$ to $W_{M,a-\theta}^{\Omega,b+\theta}$ for all $b > 0$ and $0 < \theta < a$.

Now it is easy to complete the proof. Let us take an arbitrary $a > 0$ and choose $0 < \theta < a$. For the chosen θ and $p_0 > 1$, $q > 0$ given, we can find $T = T(p_0; q; \theta)$ in such a way that the inequalities (2.3.10) and (2.3.11) hold. Hence $e^{t\mathbf{A}^*(s)}$ defines a bounded multiplication operator acting from $\widetilde{\Phi} = W_{M,a}^{\Omega,b}$ to $\widetilde{\Psi} = W_{M,a-\theta}^{\Omega,b+\theta}$ for all $t \in [0,T]$. $\qquad\square$

As a consequence, we obtain the existence of solution operators for (2.3.3) from $\left(\widetilde{\Psi}\right)' = \left(W_{M,a-\theta}^{\Omega,b+\theta}\right)'$ to $\left(\widetilde{\Phi}\right)' = \left(W_{M,a}^{\Omega,b}\right)'$.

Taking into account the fact that the uniqueness of the solution to the Cauchy problem with operator $\mathbf{A}^*(\cdot)$ holds in the spaces obtained [36], we see

2. Distribution methods

that Theorem 2.3.2 implies well-posedness of the Cauchy problem with $\mathbf{A}^*(\cdot)$ from $\widetilde{\Phi} = W_{M,a}^{\Omega,b}$ to $\widetilde{\Psi} = W_{M,a-\theta}^{\Omega,b+\theta}$. It follows the generalized well-posedness of (2.3.3) from $(W_{M,a-\theta}^{\Omega,b+\theta})'$ to $(W_{M,a}^{\Omega,b})'$.

Now we point out such spaces Φ', Ψ' that there exists a generalized solution $u(\cdot\,;t) \in \Phi'$ of the Cauchy problem (2.3.1)–(2.3.2) for $f(\cdot) \in \Psi'$.

In order to define these spaces we apply the generalized Fourier transform to the problem. Then the problem (2.3.1)–(2.3.2) turns into the problem (2.3.3). The operator of multiplication by a matrix exponent $e^{t\mathbf{A}(\cdot)}$ solves this problem. It means that the solution of (2.3.3) has the form $\widetilde{u}(\cdot\,;t) = e^{t\mathbf{A}(\cdot)}\widetilde{f}(\cdot)$ and exists in spaces $\widetilde{\Phi}'$, $\widetilde{\Psi}'$ such that $e^{t\mathbf{A}(\cdot)}$ defines a bounded multiplication operator from $\widetilde{\Psi}'$ to $\widetilde{\Phi}'$ or, equivalently, $e^{t\mathbf{A}^*(\cdot)}$ defines a bounded multiplication operator from $\widetilde{\Phi}$ to $\widetilde{\Psi}$; the generalized solution of the Cauchy problem (2.3.1)–(2.3.2) has the form (2.3.5):

$$u(\cdot\,;t) = (G_t * f)(\cdot), \quad t \in [0,T], \qquad G_t(\cdot) = \mathcal{F}^{-1}[e^{t\mathbf{A}(s)}](\cdot). \qquad (2.3.12)$$

Thus the problem of constructing a solution to the problem (2.3.1)–(2.3.2) turns into the problem of choosing a pair of spaces $\widetilde{\Psi}'$, $\widetilde{\Phi}'$, where $e^{t\mathbf{A}(\cdot)}$ defines a bounded multiplication operator. Namely, the following theorem holds.

Theorem 2.3.3 *Suppose the function $e^{t A(\cdot)}$ defines a bounded multiplication operator acting on $\widetilde{\Psi}'$ to $\widetilde{\Phi}'$. Then for any $f \in \Psi'$ the distribution (2.3.12) is the generalized solution of (2.3.1)–(2.3.2) belonging to Φ'. In addition, if $f_n \to 0$ in Ψ', then the corresponding sequence of solutions $u_n(\cdot\,;t) \to 0$ in Φ' for any $t \in [0,T]$.*

It follows from Theorem 2.3.2 that the spaces Ψ and Φ can be taken such that $\widetilde{\Phi} = W_{M,a}^{\Omega,b}$ and $\widetilde{\Psi} = W_{M,a-\theta}^{\Omega,b+\theta}$. (For spaces dual to $W_{M,a}^{\Omega,\beta}$ wrt the Fourier transform see below and Section 3.4).

Now we present more special results for Petrovsky correct, conditionally correct, and incorrect systems.

For **Petrovsky correct systems**, according to (2.3.6) and (2.3.8), the following estimate holds:

$$\left\|e^{t\mathbf{A}(\sigma)}\right\|_m \leq C(1 + |\sigma|)^h, \qquad t \geq 0, \quad \sigma \in \mathbb{R}^n, \qquad (2.3.13)$$

where h is the minimal natural number l, providing the inequality

$$\left\|e^{t\mathbf{A}(\sigma)}\right\|_m \leq C(1 + |\sigma|)^l, \quad t \geq 0, \quad \sigma \in \mathbb{R}^n, \quad \text{i.e.,} \quad h \leq p(m-1).$$

If an entire function satisfies (2.3.7) for each $s \in \mathbb{C}^n$, then, according to Theorem 3.4.6, the estimate (2.3.13) can be extended to a certain neighborhood of real values $s = \sigma$. Namely, there exists a region

$$H_\mu = \{s = \sigma + i\tau : \ |\tau| \leq c(1 + |\sigma|)^\mu\}, \quad 1 - p_0 \leq \mu \leq 1, \quad c = c(b_1, h),$$

2.3. Differential systems in Gelfand–Shilov spaces

where

$$\left\|e^{t\mathbf{A}(s)}\right\|_m \leq C(1+|\sigma|)^h, \qquad t \geq 0, \quad s \in H_\mu. \tag{2.3.14}$$

Suppose that $p_0 > 1$; then μ determining H_μ can be either positive or negative.

a. If $0 < \mu \leq 1$, then, by Theorem 3.4.8, the inequalities (2.3.7) and (2.3.14) imply that the matrix exponent satisfies the inequality

$$\left\|e^{t\mathbf{A}(s)}\right\|_m \leq C(1+|\sigma|)^h e^{bt|\tau|^{p_0/\mu}}, \qquad t \geq 0, \quad s = \sigma + i\tau, \quad b = b(b_1, h, c).$$

Hence, for each $t \in [0, T]$, we have

$$\left\|e^{t\mathbf{A}(s)}\right\|_m \leq C(1+|\sigma|)^h e^{bT|\tau|^{p_0/\mu}}, \qquad s = \sigma + i\tau.$$

Applying the substitute

$$bT = \frac{\rho^{p_0/\mu}}{p_0/\mu} \qquad \Longleftrightarrow \qquad \rho = \left(\frac{bTp_0}{\mu}\right)^{\mu/p_0},$$

we obtain

$$\left\|e^{t\mathbf{A}(s)}\right\|_m \leq C(1+|\sigma|)^h \exp\left(\frac{(\rho|\tau|)^{p_0/\mu}}{p_0/\mu}\right), \quad t \in [0, T], \quad T = \frac{\mu\rho^{p_0/\mu}}{bp_0}, \quad s = \sigma + i\tau.$$

Introducing $\Omega(\tau) = \frac{|\tau|^{p_0/\mu}}{p_0/\mu}, \tau \in \mathbb{R}^n$, we turn the latter inequality into

$$\left\|e^{t\mathbf{A}(s)}\right\|_m \leq C(1+|\sigma|)^h e^{\Omega(\rho\tau)}, \qquad t \in [0, T], \quad s = \sigma + i\tau. \tag{2.3.15}$$

Let us take an arbitrary $\beta = (\beta_1, \ldots, \beta_n)$ with $\beta_j > 0$ and consider the spaces $W^{\Omega, \beta}$, $W^{\Omega, \beta+\rho}$ where $\rho = (\rho_1, \ldots, \rho_n)$, $\rho_j > 0$. From (2.3.15) it follows that $e^{t\mathbf{A}^*(\cdot)}$ defines a bounded multiplication operator (multiplier) acting from $W^{\Omega, \beta}$ to $W^{\Omega, \beta+\rho}$ and, respectively, $e^{t\mathbf{A}(\cdot)}$ from $\left(W^{\Omega, \beta+\rho}\right)'$ to $\left(W^{\Omega, \beta}\right)'$.

The space $\widetilde{W}^{\Omega, \beta}$ is equal to $W_{M, 1/\beta}$, where M is the Young dual to Ω. For the introduced function $\Omega(\tau) = \frac{|\tau|^{p_0/\mu}}{p_0/\mu}, \tau \in \mathbb{R}^n$

$$M(x) = \frac{|x|^{p_1}}{p_1}, \quad x \in \mathbb{R}^n, \quad \text{where} \quad \frac{1}{p_1} + \frac{1}{p_0/\mu} = 1.$$

Therefore, $G_t(\cdot)$ in (2.3.12) defines a continuous convolution operator (convolutor) acting from $\left(W_{M, 1/(\beta+\rho)}\right)'$ to $\left(W_{M, 1/\beta}\right)'$. According to Theorem 2.3.3, this implies that for each $f \in \left(W_{M, 1/(\beta+\rho)}\right)'$ there exists a generalized solution $u(\cdot, t) \in \left(W_{M, 1/\beta}\right)'$ of (2.3.1)–(2.3.2) in the case of a Petrovsky correct system and $p_0 > 1$.

Since the elements of the space $\mathcal{S}'$ are known as distributions of slow growth and the elements of $\mathcal{D}'$ are known as distributions of arbitrary growth, the elements of $\left(W_M\right)'$ (and of its subspaces $\left(W_{M, a}\right)'$) are called *the distributions*

68   2. *Distribution methods*

of exponential growth. Given the function $M(x) = \frac{|x|^{p_1}}{p_1}$, $x \in \mathbb{R}^n$, we obtain $u(\cdot; t)$ in spaces of distributions of exponential growth of order $p_1 = \frac{p_0}{p_0 - \mu}$.

b. Now consider the case of $1 - p_0 \leq \mu \leq 0$. According to Theorem 3.4.11, the condition (2.3.14) on H_μ for $\mu \leq 0$ implies that the estimates

$$\left\| \left(\frac{\partial}{\partial \sigma} \right)^q e^{t\mathbf{A}(\sigma)} \right\|_m \leq C_q (1 + |\sigma|)^{h - \mu|q|}, \quad t \geq 0, \quad q = (q_1, \ldots, q_n), \quad q_j \in \mathbb{N},$$

hold for real values of $s = \sigma$. According to the condition (3.3.18), this inequality shows that $e^{t\mathbf{A}(\cdot)}$ generates a bounded multiplication operator on $\mathcal{S}$ and hence on $\mathcal{S}'$. This means that $G_t(\cdot)$ is a convolutor on $\mathcal{S}'$.

According to Theorem 2.3.3, for each $f \in \mathcal{S}'$ there exists a generalized solution $u(\cdot, t) \in \mathcal{S}'$ of (2.3.1)–(2.3.2) defined by (2.3.12). Hence the generalized solution of the Cauchy problem (2.3.1)–(2.3.2) is a distribution of slow growth in the case of a Petrovsky correct system and $1 - p_0 \leq \mu \leq 0$.

For **conditionally correct systems**, according to (2.3.6) and (2.3.9), the following estimate holds:

$$\left\| e^{t\mathbf{A}(\sigma)} \right\|_m \leq C e^{a_0 t |\sigma|^h}, \quad a_0 > 0, \quad t \geq 0, \quad \sigma \in \mathbb{R}^n, \quad 0 < h < 1, \quad (2.3.16)$$

for the real values of $s = \sigma$. According to Theorem 3.4.5, under the condition (2.3.7) the estimate (2.3.16) can be extended to a certain neighborhood of the real values of $s = \sigma$. Namely, for each $a_1 > a_0$ there exists a region

$$H_\mu = \{ s = \sigma + i\tau : \ |\tau| \leq c(1 + |\sigma|)^\mu \}, \quad 1 - (p_0 - h) \leq \mu \leq 1,$$

with $c = c(b_1; a_0; a_1)$, where

$$\left\| e^{t\mathbf{A}(s)} \right\|_m \leq C e^{a_1 t |\sigma|^h}, \quad t \geq 0, \quad a_1 > a_0, \quad s \in H_\mu. \quad (2.3.17)$$

Due to $h < 1 \leq p_0$, the number μ determining H_μ may turn out to be either positive or negative.

a. If $0 < \mu \leq 1$, then, according to Theorem 3.4.7, the inequalities (2.3.7) and (2.3.17) imply that the matrix exponent satisfies the condition

$$\left\| e^{t\mathbf{A}(s)} \right\|_m \leq C e^{a_1 t |\sigma|^h + bt|\tau|^{p_0/\mu}}, \quad t \geq 0, \quad s = \sigma + i\tau, \quad b = b(b_1; a_1; c).$$

Hence, supposing $t \in [0, T]$ for all $a_1 T = \eta$, $bT = \rho$, we obtain

$$\left\| e^{t\mathbf{A}(s)} \right\|_m \leq C e^{a_1 T |\sigma|^h + bT|\tau|^{p_0/\mu}} = e^{\eta|\sigma|^h + \rho|\tau|^{p_0/\mu}}, \quad s = \sigma + i\tau.$$

Since $h \leq p_0/\mu$, in this case Theorem 3.4.9 provides the following estimates for the derivatives on the real axis:

$$\left\| \left(\frac{\partial}{\partial \sigma} \right)^q e^{t\mathbf{A}(\sigma)} \right\|_m \leq C B_0^{|q|} |q|^{|q|(1 - \frac{\mu}{p_0})} e^{a_2|\sigma|^h}, \quad a_2 > a_1.$$

2.3. Differential systems in Gelfand–Shilov spaces

Suppose $\alpha = 1/h$, $\beta = 1 - \mu/p_0$. Let us take arbitrary $B > 0$ and $a > a_2$. Then the inequality (3.3.17) is true. The estimates obtained imply that $e^{t\mathbf{A}^*(\cdot)}$ defines a multiplier from $\mathcal{S}_{\alpha, A}^{\beta, B}$ to $\mathcal{S}_{\alpha, A_1}^{\beta, B+B_0}$ with

$$A = (h\,e\,a)^{-1/h}, \qquad A_1 = (h\,e\,(a - a_2))^{-1/h}$$

and $e^{t\mathbf{A}(\cdot)}$ defines a multiplier from $\left(\mathcal{S}_{\alpha, A_1}^{\beta, B+B_0}\right)'$ to $\left(\mathcal{S}_{\alpha, A}^{\beta, B}\right)'$. Therefore, $G_t(\cdot)$ defines a continuous convolution operator acting from $\left(\mathcal{S}_{\beta, B+B_0}^{\alpha, A_1}\right)'$ to $\left(\mathcal{S}_{\beta, B}^{\alpha, A}\right)'$.

It follows from Theorem 2.3.3 that for each $f \in \left(\mathcal{S}_{\beta, B+B_0}^{\alpha, A_1}\right)'$ there exists a generalized solution $u(\,\cdot\,;t) \in \left(\mathcal{S}_{\beta, B}^{\alpha, A}\right)'$ defined by (2.3.12). Therefore, in the case of conditionally correct systems and $0 < \mu \leq 1$ the generalized solution of the Cauchy problem (2.3.1)–(2.3.2) is a distribution increasing exponentially with order $p_1 = \frac{p_0}{p_0 - \mu}$ for $p_0 > \mu$ and with arbitrary order for $p_0 = \mu$.

b. Let us now consider the case of $1 - p_0 \leq \mu \leq 0$.

Due to Theorem 3.4.10, it follows from (2.3.17), which holds true on H_μ with $\mu \leq 0$, that there exists such $B_0 = B_0(a_1; \mu)$ that for the real values of $s = \sigma$ the following estimates hold:

$$\left\|\left(\frac{\partial}{\partial\sigma}\right)^q e^{t\mathbf{A}(\sigma)}\right\|_m \leq CB_0^{|q|}|q|^{|q|(1-\frac{\mu}{h})}e^{a_2 t|\sigma|^h}, \qquad t \geq 0, \quad a_2 > a_1.$$

Hence, supposing $a_2 T = \rho$, we obtain

$$\left\|\left(\frac{\partial}{\partial\sigma}\right)^q e^{t\mathbf{A}(\sigma)}\right\|_m \leq CB_0^{|q|}|q|^{|q|(1-\frac{\mu}{h})}e^{\rho|\sigma|^h}, \qquad t \in [0, T], \quad \sigma \in \mathbb{R}^n,$$

for all $t \in [0, T]$. Let us take $a > \rho$ and $B > 0$. Suppose $A = \frac{1}{(h\,e\,a)^{1/h}}$ and $\alpha = 1/h$, $\beta = (1 - \mu)/h$. Then, according to (3.3.17), this inequality shows that the exponent $e^{t\mathbf{A}^*(\cdot)}$ generates a bounded multiplication operator from $\mathcal{S}_{\alpha, A}^{\beta, B}$ to $\mathcal{S}_{\alpha, A_1}^{\beta, B+B_0}$ with $A_1 = (1/h\,e\,(a - \rho))^{1/h}$ and hence $e^{t\mathbf{A}^*(\cdot)}$ from $\left(\mathcal{S}_{\alpha, A_1}^{\beta, B+B_0}\right)'$ to $\left(\mathcal{S}_{\alpha, A}^{\beta, B}\right)'$. It follows that $G_t(\cdot)$ is a convolutor acting from $\left(\mathcal{S}_{\beta, B+B_0}^{\alpha, A_1}\right)'$ to $\left(\mathcal{S}_{\beta, B}^{\alpha, A}\right)'$.

It follows from Theorem 2.3.3 that for each $f \in \left(\mathcal{S}_{\beta, B+B_0}^{\alpha, A_1}\right)'$ there exists a generalized solution $u(\cdot\,;t) \in \left(\mathcal{S}_{\beta, B}^{\alpha, A}\right)'$ defined by (2.3.12). Therefore, in this case the generalized solution of the Cauchy problem (2.3.1)–(2.3.2) is a distribution increasing exponentially with order $1/\beta = h/(h - \mu)$.

An **incorrect system** satisfies (2.3.7) in the whole complex plane; therefore, denoting $a = (b_1 T)^{1/p_0}$, we obtain

$$\left\|e^{t\mathbf{A}(s)}\right\|_m \leq Ce^{b_1 T|s|^{p_0}} = Ce^{(a|s|)^{p_0}}, \qquad s \in \mathbb{C}^n, \tag{2.3.18}$$

70   2. *Distribution methods*

for $t \in [0, T]$. Such $e^{t\mathbf{A}(\cdot)}$ generates a functional in $\mathcal{D}'$, hence its generalized inverse Fourier transform $G_t(\cdot) \in \mathcal{Z}'$. In addition, for incorrect systems we can use the general results obtained in Theorems 2.3.2–2.3.3 and consider the generalized inverse Fourier transform $G_t(\cdot)$ as a convolutor acting from Ψ' to Φ', where $\widetilde{\Phi} = W_{M,a}^{\Omega,b}$ and $\widetilde{\Psi} = W_{M,a-\theta}^{\Omega,b+\theta}$ with $M(x) = \Omega(x) = |x|^{p_0}/p_0$, $x \in \mathbb{R}$.

The spaces constructed for Petrovsky correct, conditionally correct, and incorrect systems, as in the general case of the systems, provide uniqueness of a solution to the Cauchy problem (2.3.1)–(2.3.2) [36]. Therefore, the results obtained are the results on the generalized well-posedness of (2.3.1)–(2.3.2) in corresponding spaces Φ', Ψ'.

Now we can prove results on the well-posedness of the Cauchy problem (2.3.3) for the important special case $\widetilde{\Psi}' = L_m^2(\mathbb{R}) := L^2(\mathbb{R}) \times \ldots \times L^2(\mathbb{R})$ and hence on the well-posedness of the Cauchy problem (2.3.1)–(2.3.2) for the case $\Psi' = L_m^2(\mathbb{R})$.[9]

Theorem 2.3.4 *Let the matrix-function $e^{t\mathbf{A}(\cdot)}$ satisfy the estimate (2.3.7). Then, for a Petrovsky correct system $e^{t\mathbf{A}(\cdot)}$ defines a multiplier from $\widetilde{\Psi}' = L_m^2(\mathbb{R})$ to $\widetilde{\Phi}' = \mathcal{S}_m'$. For a conditionally correct system it defines a multiplier from $\widetilde{\Psi}' = L_m^2(\mathbb{R})$ to $\widetilde{\Phi}' = (\mathcal{S}_{\alpha,A})_m'$ with $\alpha = \frac{1}{h}$, and $\frac{1}{h\,e\,A^h} > a_0$, where the constants a_0, h are from the estimate (2.3.16). For an incorrect system it defines a multiplier from $\widetilde{\Psi}' = L_m^2(\mathbb{R})$ to $\widetilde{\Phi}' = (\mathcal{S}_{\alpha,A})_m'$ with $\alpha = \frac{1}{p_0}$, $\frac{1}{p_0\,e\,A^{p_0}} > b_1$, where b_1, p_0 are from (2.3.18).*

Proof. For a Petrovsky correct system $e^{t\mathbf{A}(\cdot)}$ and $e^{t\mathbf{A}^*(\cdot)}$ satisfy the estimate

$$\left\| e^{t\mathbf{A}(\sigma)} \right\|_m = \left\| e^{t\mathbf{A}^*(\sigma)} \right\|_m \leq C(1 + |\sigma|)^h, \qquad t \geq 0, \quad \sigma \in \mathbb{R}.$$

It follows that for $\widetilde{\varphi} \in \mathcal{S}_m := \mathcal{S} \times \ldots \times \mathcal{S}$ we have $e^{t\mathbf{A}^*(\cdot)}\widetilde{\varphi} \in \mathcal{S}_m$ and for $\widetilde{f} \in L_m^2(\mathbb{R})$ we have

$$\langle \widetilde{\varphi}(\cdot),\, \widetilde{u}(\cdot, t) \rangle = \langle e^{t\mathbf{A}^*(\sigma)}\widetilde{\varphi}(\sigma),\, \widetilde{f}(\sigma) \rangle.$$

Hence, the generalized solution to (2.3.3) $\widetilde{u}(\cdot, t) = e^{t\mathbf{A}(\cdot)}\widetilde{f}(\cdot) \in \widetilde{\Phi} = \mathcal{S}_m'$ for any $t \in [0, T]$.

For a conditionally correct system the operator-matrixes $e^{t\mathbf{A}(\cdot)}$ and $e^{t\mathbf{A}^*(\cdot)}$ satisfy the estimate (2.3.16):

$$\left\| e^{t\mathbf{A}(\sigma)} \right\|_m = \left\| e^{t\mathbf{A}^*(\sigma)} \right\|_m \leq C e^{a_0 t |\sigma|^h}, \qquad h < 1, \quad a_0 > 0, \quad t \geq 0, \quad \sigma \in \mathbb{R}.$$

We show that in this case one can take $\widetilde{\Phi} = (\mathcal{S}_{\alpha,A})_m$ with specially chosen α

---

[9]Note that in contrast to the notations above, where we did not stress that all the spaces of test and generalized functions are actually spaces of vector-functions, we will further denote by $L_m^2(\mathbb{R})$ ($\mathcal{S}_m, \ldots$) the space of all vector-functions $f : t \mapsto (f_1(t), \ldots, f_m(t))$, where $f_i \in L^2(\mathbb{R})$ ($f_i \in S(\mathbb{R}), \ldots$). We will use the results in these notations for stochastic equations in Part II.

2.3. Differential systems in Gelfand–Shilov spaces

and A. By the definition of $(\mathcal{S}_{\alpha,A})$ $(\alpha \geq 0,\ A > 0)$, the space consists of all infinitely differentiable functions $\widetilde{\varphi}$ satisfying the inequalities

$$|x^k \widetilde{\varphi}(x)| \leq C_\varepsilon(\widetilde{\varphi})(A + \varepsilon)^k k^{k\alpha}, \qquad k \in \mathbb{N}_0, \quad x \in \mathbb{R},$$

for any $\varepsilon > 0$. Equivalently, by the structure theorem, it consists of all functions satisfying the inequalities

$$|\widetilde{\varphi}(x)| \leq C_\rho(\widetilde{\varphi}) e^{-(a-\rho)|x|^{1/\alpha}}, \qquad x \in \mathbb{R}, \quad a = \frac{\alpha}{e\,A^{1/\alpha}},$$

for any $\rho > 0$.

As shown above, while studying conditionally correct systems, for $\widetilde{\varphi} \in \mathcal{S}_{\alpha,A}$ we have $e^{t\mathbf{A}^*(\cdot)}\widetilde{\varphi} \in (\mathcal{S}_{\alpha,A_1})$, where $A_1 = \left(\frac{\alpha}{e\,(a-a_0)}\right)^\alpha$. Hence, for $\widetilde{f} \in L^2_m(\mathbb{R}) \subset (\mathcal{S}_{\alpha,A_1})'_m$ the generalized solution $\widetilde{u}(\cdot,t) = e^{t\mathbf{A}(\cdot)}\widetilde{f}(\cdot) \in (\mathcal{S}_{\alpha,A})'_m$, where $\alpha = 1/h$ and A is taken such that $a = \frac{1}{h\,e\,A^h} > a_0$.

The similar proof is true for incorrect systems. Here we deal with $\widetilde{\Phi}' = (\mathcal{S}_{\alpha,A})'_m$, where $\alpha = \frac{1}{p_0}$ and $\frac{1}{p_0\,e\,A^{p_0}} > b_1$, the constants b_1, p_0 are from the estimate (2.3.18). $\qquad\square$

As a consequence of these results we obtain results on the well-posedness of (2.3.1)–(2.3.2) for initial data from $L^2_m(\mathbb{R})$.

Theorem 2.3.5 *Suppose the matrix-function $e^{t\mathbf{A}(\cdot)}$ satisfies the estimate (2.3.7). Then, for a Petrovsky correct system, a unique stable solution of (2.3.1)–(2.3.2) with $f \in L^2_m(\mathbb{R})$ exists and belongs to $\Phi' = \mathcal{S}'_m$; for a conditionally correct system, $u(\cdot,t) \in \Phi' = \left(\mathcal{S}^{\alpha,A}\right)'_m$ with $\alpha = 1/h$ and $\frac{1}{h\,e\,A^h} > a_0$, where the constants a_0, h are from (2.3.16); for an incorrect system, $u(\cdot,t) \in \Phi' = \left(\mathcal{S}^{\alpha,A}\right)'_m$ with $\alpha = 1/p_0$, $\frac{1}{p_0\,e\,A^{p_0}} > b_1$, where b_1, p_0 are from (2.3.18).*

It is important to note that the spaces obtained in Theorems 2.3.4–2.3.5, by taking into account the behavior of $e^{t\mathbf{A}(\cdot)}$ only on the real axis, turned out to be ordered wrt the behavior of $e^{t\mathbf{A}(\cdot)}$ on $\mathbb{R}$, while the more subtle results taking into account the behavior of $e^{t\mathbf{A}(\cdot)}$ in different regions of a complex plane do not.

2.3.3 Regularization of solutions in a broad sense

In conclusion of Chapters 1 and 2, let us sum up the methods and approaches used in the study of solutions to the problems (I.2) and (I.3). We show that, generally, all of the methods are based on constructing regularized solutions. Regularization is understood here in a broad sense. This means constructing corrected (smoothed) solutions, which generally are not approximated solutions. (Approximate solutions are those constructed by virtue of regularizing operators.)

72   *2. Distribution methods*

Let us discuss the ideas of regularization used in solving the Cauchy problem for homogeneous equations in more detail. The abstract Laplace and Fourier transforms play an important role in constructing solutions to the Cauchy problems (I.2) and (I.3). Application of the Laplace transform to (I.2) turns it into the equation

$$(\lambda I - A)\widetilde{u}(\lambda) = f, \quad \lambda \in \rho(A).$$

As a result the solution operators $U(t), t \geq 0$, of the problem become equal to the inverse Laplace transform of $\mathcal{R}(\lambda) = (\lambda I - A)^{-1}$, the resolvent of A:

$$u(t) = U(t)f = \mathcal{L}^{-1}[\mathcal{R}(\cdot)f](t), \quad t \geq 0. \tag{2.3.19}$$

The solution operators generally are not defined on the whole $\mathcal{X}$. We show that the regularization used within semi-group methods, namely, in construction of K-convoluted (in the particular case, integrated) semigroups $\{S_K(t), t \in [0, \tau)\}$, is performed by multiplying the resolvent in (2.3.19) by a function $\widetilde{K}$. This function allows us to obtain the regularized solution:

$$u_K(t) = S_K(t)f = \mathcal{L}^{-1}[\widetilde{K}\mathcal{R}](t)f = (U * K)(t)f, \quad t \in [0, \tau), \quad f \in \mathcal{X}. \tag{2.3.20}$$

In the case of R-semigroups $\{S_R(t), t \in [0, \tau)\}$ the regularization is performed by smoothing $(\lambda I - A)^{-1}f$ by the operator R:

$$u_R = S_R(t)f = \mathcal{L}^{-1}\left[(\lambda I - A)^{-1}Rf\right](t), \quad t \in [0, \tau), \quad f \in \mathcal{X}, \tag{2.3.21}$$

where $(\lambda I - A)^{-1}$ usually is not the resolvent in the case of R-semigroups.

The methods of abstract distributions provide the regularization by applying (unbounded) solution operators to test functions φ, which allows constructing a generalized solution to (I.2) as the generalized inverse Laplace transform:

$$\langle \varphi, u \rangle = \langle \varphi, Uf \rangle = \langle \varphi, \mathcal{L}^{-1}[\mathcal{R}f] \rangle := \langle \widetilde{\varphi}(-\overline{\lambda}), \mathcal{R}(\lambda)f \rangle, \quad f \in \mathcal{X},$$

where $\varphi \in \mathcal{D}$ in the case of integrated semigroups and $\varphi \in \mathcal{D}^{\{M_q\}}$ in the case of K-convoluted semigroups. Here, instead of the decreasing function $\widetilde{K}$ in (2.3.20), test functions φ play the role of "regularizing factors".

In solving the problem (I.3) by the Fourier transform method, the regularization is absolutely analogous to the regularization (2.3.20) and (2.3.21). Here there are also two possible variants of regularization. The first one is multiplication of $e^{tA(\sigma)}, \sigma \in \mathbb{R}^n$, the solution operators to (2.3.3), by a certain regularizing function $K(\sigma)$. This gives rise to an R-semi-group with R defined as

$$Rf(x) = (\mathcal{F}^{-1}[K] * f)(x), \quad x \in \mathbb{R}^n.$$

Such R-semi-groups are constructed and used for solving stochastic problems in Section 5.1. The second variant is applying $e^{tA(\cdot)}$ to test functions $\widetilde{\varphi}(\cdot)$ in the construction of generalized solutions. In this case again a space of test

2.3. Differential systems in Gelfand–Shilov spaces

functions is chosen according to the growth rate of $e^{tA(\sigma)}$ as $\sigma \to \infty$ in such a way that $e^{tA(\cdot)}$ defines a bounded multiplication operator in the space.

While studying stochastic problems in Part II we will see that the same ideas work in the regularization of Wiener processes and solutions by a trace class operator Q as well as in the construction of generalized solutions to stochastic problems in different spaces of distributions.

Chapter 3

Examples. Supplements

3.1 Examples of regularized semi-groups and their generators

Example 3.1.1 *of an operator which depends on a parameter γ and generates a C_0-semi-group, an integrated semi-group, a semi-group of growth order α, or an R-semi-group in dependence on the value of γ.*

Let $\mathcal{X} = L^p(\mathbb{R}) \times L^p(\mathbb{R})$, $1 \leq p < \infty$, be the space of vector-functions $f(\cdot) = (f_1(\cdot), f_2(\cdot))$ with the norm $\|f\| = \|f_1\|_p + \|f_2\|_p$. Consider the operator A of multiplication by the matrix

$$\begin{pmatrix} -h & 0 \\ -g & -h \end{pmatrix}, \tag{3.1.1}$$

where
$$h(x) = 1 + x^2, \qquad g(x) = x^{2\gamma}, \qquad x \in \mathbb{R}, \quad \gamma \geq 0,$$

with $dom\, A = \{f \in \mathcal{X} : hf_1,\ gf_1 + hf_2 \in L^p(\mathbb{R})\}$.

Note that this matrix can be obtained as the Fourier transform of a class of differential systems depending on a parameter, in particular, fractional differential systems (see Example 3.2.4).

We construct the operator-function

$$U(t) = e^{At} = I + tA + \frac{t^2 A^2}{2!} + \ldots + \frac{t^n A^n}{n!} + \ldots, \qquad t \geq 0,$$

formally. The corresponding matrix-function calculated by components has the form

$$e^{-t(1+x^2)} \begin{pmatrix} 1 & 0 \\ -tx^{2\gamma} & 1 \end{pmatrix}, \qquad t \geq 0, \quad x \in \mathbb{R}.$$

Let us investigate whether the operators obtained are bounded. To do this we calculate the norm of $U(t)$:

$$\|U(t)\|_{\mathcal{L}(\mathcal{X})} = \max \left\{ \max_{x \in \mathbb{R}} e^{-t(1+x^2)}; \ \max_{x \in \mathbb{R}} t|x|^{2\gamma} e^{-t(1+x^2)} \right\}.$$

76   *3. Examples. Supplements*

The maximum value of the first function is evidently achieved at $x = 0$. The extreme points of the second function are $\pm\sqrt{\frac{\gamma}{t}}$, $t > 0$. Hence, we obtain

$$\|U(t)\| = \max\left\{e^{-t};\ \gamma^\gamma t^{1-\gamma} e^{-t-\gamma}\right\}. \tag{3.1.2}$$

For the norms of the operators with $\gamma = 0$, we have

$$\|U(t)\| = \max\left\{e^{-t},\ te^{-t}\right\} = \begin{cases} e^{-t}, & 0 \le t \le 1, \\ te^{-t}, & t > 1. \end{cases}$$

If $\gamma = 1$, we obtain

$$\|U(t)\| = \max\left\{e^{-t}; e^{-t-1}\right\} = e^{-t}, \quad t \ge 0.$$

In order to find the maximum of two functions in (3.1.2) with $\gamma > 0$, $\gamma \ne 1$, we solve the inequality

$$\gamma^\gamma t^{1-\gamma} e^{-\gamma} < 1 \qquad \Longleftrightarrow \qquad t^{1-\gamma} < \left(\frac{e}{\gamma}\right)^\gamma. \tag{3.1.3}$$

If we denote $t_\gamma := \left(\frac{e}{\gamma}\right)^{\gamma/1-\gamma}$, then the solutions of (3.1.3) have the form

$$t < t_\gamma, \quad \text{for } 0 < \gamma < 1, \qquad t > t_\gamma, \quad \text{for } \gamma > 1.$$

Therefore, if $0 < \gamma < 1$ and $t \le t_\gamma$, then the first argument is maximal, and if $t > t_\gamma$, then the second one is maximal. For $\gamma > 1$, it is vice versa.

To sum up, we have obtained, in the case of $0 < \gamma < 1$,

$$\|U(t)\| = \max\left\{e^{-t};\ \gamma^\gamma t^{1-\gamma} e^{-t-\gamma}\right\} = \begin{cases} e^{-t}, & 0 \le t \le t_\gamma, \\ \gamma^\gamma t^{1-\gamma} e^{-t-\gamma}, & t > t_\gamma, \end{cases}$$

and in the case of $\gamma > 1$, we have

$$\|U(t)\| = \max\left\{e^{-t};\ \gamma^\gamma t^{1-\gamma} e^{-t-\gamma}\right\} = \begin{cases} e^{-t}, & t = 0, \\ \gamma^\gamma t^{1-\gamma} e^{-t-\gamma}, & 0 < t \le t_\gamma, \\ e^{-t}, & t > t_\gamma. \end{cases} \tag{3.1.4}$$

This means that, if $0 \le \gamma \le 1$, then the operators $U(t)$ are bounded on the whole semi-axis $t \ge 0$: the norm of $U(t)$ is equal to e^{-t} for every $0 \le \gamma \le 1$ and for sufficiently small t; for $t \to \infty$, the norm of $U(t)$ is equal to e^{-t} in the case of $\gamma = 1$, to te^{-t} in the case of $\gamma = 0$, and to $t^{1-\gamma} e^{-t-\gamma}$ in the case of $0 < \gamma < 1$; therefore, it is bounded.

Thus, if $0 \le \gamma \le 1$, we have obtained a family of linear bounded operators.

In the case of $\gamma > 1$ the norms of $U(t)$ increase proportionally to $\frac{1}{t^{\gamma-1}}$ as $t \to 0$. Hence the family of operators $\{U(t),\ t \ge 0\}$ is unbounded in a neighborhood of $t = 0$ for such γ.

1. Let us show that for $0 \le \gamma \le 1$ the constructed family $\{U(t),\ t \ge 0\}$ of bounded operators forms a C_0-semi-group.

3.1. Examples of regularized semi-groups

The semi-group property follows from the corresponding equality for matrices:

$$e^{-t(1+x^2)} \begin{pmatrix} 1 & 0 \\ -tx^{2\gamma} & 1 \end{pmatrix} e^{-\tau(1+x^2)} \begin{pmatrix} 1 & 0 \\ -\tau x^{2\gamma} & 1 \end{pmatrix} =$$

$$= e^{-(t+\tau)(1+x^2)} \begin{pmatrix} 1 & 0 \\ -(t+\tau)x^{2\gamma} & 1 \end{pmatrix}.$$

The strong continuity of the family can be easily checked directly. We investigate the existence and behavior of the resolvent of A as well as its derivatives. We have

$$\frac{d^k}{d\lambda^k}(\lambda I - A)^{-1} = \frac{(-1)^k k!}{(\lambda + h)^{k+2}} \begin{pmatrix} \lambda + h & 0 \\ -(k+1)g & \lambda + h \end{pmatrix}, \qquad k \in \mathbb{N}_0,$$

To check the conditions of the MFPHY theorem we estimate the norm of these derivatives for arbitrary $f \in \mathcal{X}$:

$$\left\| \frac{1}{k!} \cdot \frac{d^k}{d\lambda^k}(\lambda I - A)^{-1} f \right\| \leq \left\| (\lambda + h)^{-k-1} f_1 \right\|_p$$

$$+ \left\| (k+1)(\lambda + h)^{-k-2} g f_1 \right\|_p + \left\| (\lambda + h)^{-k-1} f_2 \right\|_p.$$

If $\lambda > 0$ and $i = 1, 2$, we obtain the following estimates for the first and the last terms:

$$\left\| (\lambda + h)^{-k-1} f_i \right\|_p = \left(\int_{-\infty}^{\infty} \frac{|f_i(x)|^p}{|\lambda + 1 + x^2|^{(k+1)p}} \, dx \right)^{1/p} \leq \frac{1}{\lambda^{k+1}} \| f_i \|_p. \quad (3.1.5)$$

Let us estimate the second term:

$$\left\| (k+1)(\lambda + h)^{-k-2} g f_1 \right\|_p = \left(\int_{-\infty}^{\infty} \frac{(k+1)^p |x|^{2p\gamma}}{|\lambda + 1 + x^2|^{(k+2)p}} \cdot |f_1(x)|^p \, dx \right)^{1/p}$$

$$= \left(\int_{-\infty}^{\infty} \frac{(k+1)^p}{|\lambda + 1 + x^2|^{(k+1)p}} \cdot \frac{|x|^{2p\gamma}}{|\lambda + 1 + x^2|^p} \cdot |f_1(x)|^p \, dx \right)^{1/p}$$

$$= \frac{1}{|\lambda|^{k+1}} \left(\int_{-\infty}^{\infty} \frac{(k+1)^p}{|1 + \frac{1+x^2}{\lambda}|^{(k+1)p}} \cdot \frac{|x|^{2p\gamma}}{|\lambda + 1 + x^2|^p} \cdot |f_1(x)|^p \, dx \right)^{1/p}.$$

Note that

$$\frac{k+1}{(1 + \frac{1+x^2}{\lambda})^{k+1}} \leq \frac{\lambda}{1+x^2}, \qquad x \in \mathbb{R}, \quad \lambda > 0. \quad (3.1.6)$$

This is indeed the case since the inequality

$$ny \leq 1 + ny \leq (1+y)^n, \qquad y > -1, \quad n \in \mathbb{N},$$

78   3. *Examples. Supplements*

is easy to prove by the Taylor expansion of the right-hand side. The change of n by $k+1$ and y by $\frac{1+x^2}{|\lambda|}$ yields (3.1.6). Then, for $\lambda > 0$, we have

$$\left\|(k+1)(\lambda+h)^{-k-2}gf_1\right\|_p$$

$$\leq \frac{1}{\lambda^{k+1}}\left(\int_{-\infty}^{\infty}\frac{\lambda^p}{(1+x^2)^p}\cdot\frac{|x|^{2p\gamma}}{(\lambda+1+x^2)^p}\cdot|f_1(x)|^p\,dx\right)^{1/p}$$

$$= \frac{1}{\lambda^{k+1}}\left(\int_{-\infty}^{\infty}\frac{\lambda^p}{(1+x^2)^p}\cdot\frac{|x|^{2p\gamma}}{\lambda^p\left(1+\frac{1+x^2}{\lambda}\right)^p}\cdot|f_1(x)|^p\,dx\right)^{1/p}. \qquad (3.1.7)$$

Further, for every $x \in \mathbb{R}$, the estimates

$$\frac{|x|^{2\gamma}}{1+x^2}\leq 1, \quad\text{as } \gamma\in[0,1], \qquad \frac{1}{1+\frac{1+x^2}{\lambda}}\leq 1, \quad\text{as }\lambda>0,$$

hold true. Therefore,

$$\left\|(k+1)(\lambda+h)^{-k-2}gf_1\right\|_p \leq \frac{1}{\lambda^{k+1}}\left(\int_{-\infty}^{\infty}|f_1(x)|^p\,dx\right)^{1/p}=\frac{1}{\lambda^{k+1}}\|f_1\|_p.$$

Finally, we obtain

$$\left\|\frac{1}{k!}\cdot\frac{d^k}{d\lambda^k}(\lambda I-A)^{-1}f\right\|\leq\frac{1}{\lambda^{k+1}}\|f_1\|_p+\frac{1}{\lambda^{k+1}}\|f_1\|_p+\frac{1}{\lambda^{k+1}}\|f_2\|_p$$

$$\leq\frac{2}{\lambda^{k+1}}\|f\|, \qquad k\in\mathbb{N}\cup\{0\},$$

for $\lambda > 0$ and $0 \leq \gamma \leq 1$. This estimate implies that for $0 \leq \gamma \leq 1$ the operator $(\lambda I - A)^{-1}$ is the resolvent of A and satisfies the MFPHY condition.

Thus, for $0 \leq \gamma \leq 1$, the family of operators $\{U(t),\ t \geq 0\}$ forms the C_0-semi-group.

2. Now let us consider the values $\gamma > 1$. In this case, as seen from the estimates obtained the MFPHY condition is not true for $(\lambda I - A)^{-1}$. However, this operator is bounded for $1 < \gamma \leq 2$. Indeed, the estimate (3.1.5) is independent on γ. The estimate (3.1.7) also does not depend on γ and in the case of $k = 0$ we have

$$\left\|(\lambda+h)^{-2}gf_1\right\|_p\leq\frac{1}{\lambda}\left(\int_{-\infty}^{\infty}\frac{\lambda^p}{(1+x^2)^p}\cdot\frac{|x|^{2p\gamma}}{(\lambda+1+x^2)^p}\cdot|f_1(x)|^p\,dx\right)^{1/p}.$$

$$(3.1.8)$$

The function under the integral sign with $\gamma \leq 2$ satisfies the estimates

$$\frac{|x|^{\gamma}}{1+x^2}\leq 1, \qquad \frac{|x|^{\gamma}}{(\lambda+1+x^2)^p}\leq 1, \quad\text{as }\lambda>0,$$

3.1. Examples of regularized semi-groups

for each $x \in \mathbb{R}$; therefore,

$$\left\|(\lambda + h)^{-2} g f_1\right\|_p \leq \left(\int_{-\infty}^{\infty} |f_1(x)|^p \, dx\right)^{1/p} = \|f_1\|_p.$$

Thus we obtain the estimate

$$\left\|(\lambda I - A)^{-1} f\right\| \leq \frac{1}{\lambda} \|f_1\|_p + \|f_1\|_p + \frac{1}{\lambda} \|f_2\|_p \leq 2\|f\|, \quad \lambda > 0, \quad 1 < \gamma \leq 2.$$

Let us emphasize that the condition (3.1.8) implies the operator $(\lambda I - A)^{-1}$ to be unbounded for $\gamma > 2$; therefore, the resolvent of A exists for $\gamma \leq 2$ only.

As noted in the case of $\gamma > 1$, the relation (3.1.4) implies that the family $\{U(t), t \geq 0\}$ has a peculiarity of the kind

$$\|U(t)\| = \mathcal{O}(t^{1-\gamma}) \qquad \text{as } t \to 0. \tag{3.1.9}$$

If, in addition, $\gamma < 2$, then the peculiarity is integrable. We show that in the case of $1 < \gamma < 2$ the operator A generates a one-time integrated semi-group.

Consider a family of primitives of $\{U(t), t \geq 0\}$, i.e.,

$$S_1(t) = \int_0^t U(s) \, ds = \frac{1}{h} \begin{pmatrix} 1 - e^{-ht} & 0 \\ tge^{-ht} - (1 - e^{-ht})g/h & 1 - e^{-ht} \end{pmatrix}, \qquad t \geq 0.$$

These operators are bounded:

$$
\begin{aligned}
\|S_1(t)\| &\leq \int_0^t \|U(s)\| \, ds \leq \gamma^\gamma \int_0^{t_\gamma} s^{1-\gamma} e^{-s-\gamma} \, ds + \int_{t_\gamma}^t e^{-s} \, ds \\
&= \gamma^\gamma e^{-\xi-\gamma} \int_0^{t_\gamma} s^{1-\gamma} \, ds + e^{-t_\gamma} - e^{-t} \\
&\leq \gamma^\gamma e^{-\gamma} \frac{t_\gamma^{2-\gamma}}{2-\gamma} + e^{-t_\gamma} - e^{-t} \leq C.
\end{aligned}
$$

The same reason implies that the family constructed is exponentially bounded (with $\omega = 0$). Further, the operators $S_1(t)$, $t \geq 0$, being primitives of the family $\{U(t), t \geq 0\}$ that possesses the semi-group property, satisfy the condition $(S_n 1)$ of Definition 1.2.1 with $n = 1$. Finally, it is easy to see that the operator-function $S_1(\cdot)$ is strongly continuous wrt $t \geq 0$. Therefore, the family $\{S_1(t), t \geq 0\}$ with $1 < \gamma < 2$ satisfies all conditions of Definition 1.2.1 of an integrated semi-group.

We show that A is the generator of this semi-group. Indeed, the matrix

$$\lambda I - \begin{pmatrix} \frac{1}{\lambda+h} & -\frac{f}{(\lambda+h)^2} \\ 0 & \frac{1}{\lambda+h} \end{pmatrix}^{-1} = \begin{pmatrix} -h & 0 \\ -g & -h \end{pmatrix}$$

corresponds to the operator

$$\lambda I - \mathcal{R}(\lambda)^{-1} = \lambda I - \left(\int_0^\infty \lambda e^{-\lambda t} S_1(t) \, dt\right)^{-1}.$$

80   3. Examples. Supplements

Hence, in accordance with Definition 1.2.2, A is the generator of the constructed family.

3. Let us show that for the case $\gamma > 1$ the relation (3.1.9) implies the values $\|t^\alpha U(t)\|$ are bounded as $t \to 0$ for any $\alpha \geq \gamma - 1$. The set $\mathcal{X}_0 := \bigcup_{t>0} U(t)(\mathcal{X})$ is dense in $\mathcal{X}$ due to the property of any operator $U(t)$ with $t > 0$ to map a set dense in $\mathcal{X}$ into a dense set. In addition, it follows from the construction of operators $U(t)$, $t \geq 0$, that this family is non-degenerate: if $U(t)f = 0$ for each $t \geq 0$, then $f = 0$. Therefore, the conditions of Definition 1.1.6 of a semi-group of growth order α hold and for each $\gamma > 1$ the operator family $\{U(t), t \geq 0\}$ forms a semi-group of growth order $\alpha \geq \gamma - 1$.

Then, according to Theorem 1.2.18, the operator A generates an R-semi-group with $R = (\lambda I - A)^{-n}$, $n = [\alpha] + 1$. Note that here the operator $(\lambda I - A)^{-n}$ is an nth power of the resolvent only in the case of $\gamma \leq 2$. For another γ, as we have shown above, the resolvent does not exist and only the operator $(\lambda I - A)^{-n}$ is bounded.

Example 3.1.2 *of a global one-time integrated semi-group which is not exponentially bounded.*

Let $\mathcal{X} = l_2$. Consider the operator A of multiplication by a sequence $a = (a_1, a_2, \ldots, a_m, \ldots)$ defined as follows:

$$Af := (a_1 f_1, a_2 f_2, \ldots, a_m f_m, \ldots),$$

where

$$a_m = m + i\sqrt{e^{2m^2} - m^2}, \qquad m \in \mathbb{N},$$

with $dom\, A = \{f \in \mathcal{X} : \ Af \in \mathcal{X}\}$.

Consider the operators $U(t)$, $t \geq 0$, formally defined as operators of multiplication by the sequence

$$e^{At} = (e^{a_1 t}, e^{a_2 t}, \ldots, e^{a_m t}, \ldots).$$

They are not bounded as $t > 0$ since the sequence $|e^{a_m t}| = e^{mt}$ is unbounded. However, they possess the semi-group property on their domain (this is a consequence of the corresponding property for the real-valued exponents), and operators $S_1(t) := \int_0^t U(s)\, ds$ of multiplication by the vector

$$\left(\frac{e^{a_1 t} - 1}{a_1}, \frac{e^{a_2 t} - 1}{a_2}, \ldots, \frac{e^{a_m t} - 1}{a_m}, \ldots \right)$$

are bounded for all $t \geq 0$. Indeed,

$$\|S_1(t)\| = \sup_m \left| \frac{e^{a_m t} - 1}{a_m} \right| \leq \sup_m \frac{e^{mt}}{e^{m^2}} + \sup_m \frac{1}{e^{m^2}}.$$

Here $\sup_m \frac{1}{e^{m^2}} = e^{-1}$. As for the second term, for an even number t the supremum of $\frac{e^{mt}}{e^{m^2}}$ is reached as $m = \frac{t}{2}$:

$$\sup_m \frac{e^{mt}}{e^{m^2}} = e^{t^2/2 - t^2/4} = e^{t^2/4}, \qquad t = 2k,$$

3.1. Examples of regularized semi-groups

81

in the general case $t \geq 0$ the estimate $\sup_m e^{mt}/e^{m^2} \leq e^{t^2/4}$ holds. Thus the operator $S_1(t)$ is bounded for every fixed $t \geq 0$ and the behavior of the operator function $S_1(t)$ as $t \to \infty$ is determined by the relationship

$$\|S_1(t)\| = \mathcal{O}\left(e^{\frac{t^2}{4}}\right).$$

It follows that the family of operators $\{S_1(t),\, t \geq 0\}$ does not possess the property of exponential boundedness. This means that there are no such $C > 0$ and $\omega \in \mathbb{R}$ that

$$\|S_1(t)\| \leq Ce^{\omega t}, \qquad t \geq 0.$$

Nevertheless, this family satisfies the characteristic property of integrated semi-groups $(S_n 1)$ with $n = 1$ as a primitive of the family $\{U(t),\, t \geq 0\}$ which possesses the semi-group property. In addition, the defined family $\{S_1(t),\, t \geq 0\}$ is strongly continuous wrt t as $t \geq 0$:

$$\|S_1(t+\tau)f - S_1(t)f\|^2 = \sum_{m=1}^{\infty} \left|\frac{e^{a_m(t+\tau)} - e^{a_m t}}{a_m}\right|^2 \cdot |f_m|^2$$

$$= \sum_{m=1}^{k} \left|\frac{e^{a_m(t+\tau)} - e^{a_m t}}{a_m}\right|^2 \cdot |f_m|^2 + \sum_{m=k+1}^{\infty} \left|\frac{e^{a_m(t+\tau)} - e^{a_m t}}{a_m}\right|^2 \cdot |f_m|^2.$$

The second term in the sum obtained can be made sufficiently small with fixed $t \geq 0$ and τ from a bounded subset of $\mathbb{R}$ since $\sum_{m=1}^{\infty} \left|\frac{e^{a_m(t+\tau)}}{a_m}\right|^2 \cdot |f_m|^2$ and $\sum_{m=1}^{\infty} \left|\frac{e^{a_m t}}{a_m}\right|^2 \cdot |f_m|^2$ converge, due to the choice of $k \in \mathbb{N}$. Then, with fixed $k \in \mathbb{N}$, one can make the first term sufficiently small as $\tau \to 0$ since this is a finite sum.

Thus, in this example we have shown that the defined family of operators $\{S_1(t),\, t \geq 0\}$ forms a one-time integrated semi-group which is not exponentially bounded.

Example 3.1.3 [110] *of an n-times local integrated semi-group which cannot be extended to the semi-axis $t \geq 0$.*

Let $\mathcal{X} = l_2$ and $T > 0$. Consider the operator A of multiplication by the sequence $a = (a_1, a_2, \ldots, a_m, \ldots)$ defined as follows:

$$Af := (a_1 f_1, a_2 f_2, \ldots, a_m f_m, \ldots),$$

where

$$a_m = \frac{m}{T} + i\sqrt{\frac{e^{2m}}{m^2} - \frac{m^2}{T^2}}, \qquad m \in \mathbb{N},$$

with $dom A = \{f \in l_2 : Af \in l_2\}$.

82   *3. Examples. Supplements*

As in Example 3.1.2, the operators $U(t)$, $t \geq 0$, defined formally as operators of multiplication by the sequence

$$e^{At} = (e^{a_1 t}, e^{a_2 t}, \ldots, e^{a_m t}, \ldots),$$

are not bounded as $t > 0$. However, similar to the previous example, the operator family $\{U(t), t \geq 0\}$ possesses the semi-group property on its domain.

Let us construct $\{S_n(t), t \geq 0\}$ as a primitive of order n of the family $\{U(t), t \geq 0\}$ by components. Then the family obtained satisfies the characteristic property $(S_n 1)$ of integrated semi-groups as a primitive of the family that possesses the semi-group property. Besides, if the operators $S_n(t)$ are proven to be bounded for t from a certain set, then by arguments of the previous example one can show that this family is strongly continuous wrt to t on the set.

Let $b_m(t)$ be an mth component of the vector corresponding to the operator $S_n(t)$. Then it equals the n-tuple integral of $e^{a_m s}$ on $[0, t]$, i.e.,

$$b_m(t) = \frac{e^{a_m t}}{a_m^n} - \sum_{k=1}^{n} \frac{t^{n-k}}{(n-k)! \cdot a_m^k} \cdot \tag{3.1.10}$$

Calculating the module of the first term,

$$\left| \frac{e^{a_m t}}{a_m^n} \right| = \frac{e^{\frac{mt}{T}}}{e^{mn}} \cdot m^n = m^n e^{m\left(\frac{t}{T} - n\right)}, \tag{3.1.11}$$

we obtain that for $t < nT$ this value tends to zero as $m \to \infty$, and for $t \geq nT$ it is unbounded as $m \to \infty$. The estimates of the rest of the terms in (3.1.10),

$$\left| \frac{t^{n-k}}{(n-k)! \cdot a_m^k} \right| = \frac{m^k}{e^{km}(n-k)!} |t|^{n-k} \leq \frac{m^k}{e^{km}} \frac{(nT)^{n-k}}{(n-k)!}, \qquad k = 1, 2, \ldots, n,$$

are true for $t \in [0, nT)$ and we see that these values tend to zero as $m \to \infty$ too. Hence, for each $t \in [0, nT)$, we have $|b_m(t)| \to 0$ as $m \to \infty$; this means that the operators $S_n(t)$ are bounded on the semi-interval $[0, nT)$.

Thus we have obtained a local n-times integrated semi-group. Moreover, it is seen from the equality (3.1.11) that it cannot be extended to the semi-axis $t \geq 0$. Furthermore, the same equality implies that the semi-group is defined only on a semi-interval and cannot be extended to its closure.

It should also be noted that the semi-interval $[0, nT)$, where the semi-group is defined, extends when the order of the primitive grows. This allows us to obtain solutions of the Cauchy problem for larger t; however, such a construction requires more and more regularization.

Example 3.1.4 [52] *of an operator generating an exponentially bounded R-semi-group and not generating a semi-group of classes C_0, C_1, A or a convoluted one.*

3.1. Examples of regularized semi-groups

Let $\mathcal{X} = L^p(\mathbb{R}) \times L^p(\mathbb{R})$, $1 \leq p < \infty$. Consider the operator A of multiplication by the matrix (3.1.1):

$$\begin{pmatrix} -h & 0 \\ -g & -h \end{pmatrix}, \quad \text{where} \quad h(x) = 1 + x^2, \quad g(x) = e^{(1+x^2)^\gamma}.$$

Define $Rf := (1 + |g|)^{-1}f$. This operator is injective and $\overline{ran\, R} = \mathcal{X}$. It is not difficult to see that the family of operators

$$S(t) := e^{-th}(1 + |g|)^{-1} \begin{pmatrix} 1 & 0 \\ -tg & 1 \end{pmatrix}, \quad t \geq 0,$$

forms an R-semi-group and that the family is exponentially bounded.

Let us find the resolvent set of A. The matrix

$$\frac{1}{(\lambda + 1 + x^2)^2} \begin{pmatrix} \lambda + 1 + x^2 & 0 \\ -e^{(1+x^2)^\gamma} & \lambda + 1 + x^2 \end{pmatrix}$$

corresponds to the operator $(\lambda I - A)^{-1}$. However, unlike Example 3.1.1, the operator of multiplication by the matrix is not bounded for all $\gamma > 0$. Therefore, $\rho(A) = \emptyset$ and A does not generate a semi-group of class C_0, C_1, $\mathcal{A}$, integrated or convoluted.

Example 3.1.5 *of an operator generating a global R-semi-group which is not exponentially bounded and does not generate a semi-group of classes C_0, C_1, $\mathcal{A}$ or a local convoluted one.*

Let $\mathcal{X} = L^2(\mathbb{R})$. Consider the operator A of multiplication by the variable x:

$$Af(x) := xf(x), \quad x \in \mathbb{R}, \quad dom\, A = \left\{ f \in \mathcal{X} : \int_{\mathbb{R}} |xf(x)|^2 \, dx < \infty \right\}.$$

The family

$$S(t)f(x) := e^{xt - |x|^2} f(x), \quad f \in \mathcal{X}, \quad x \in \mathbb{R},$$

as is easily seen, forms an R-semi-group with the generator A and with $Rf(x) = e^{-|x|^2} f(x)$, $x \in \mathbb{R}$, $f \in \mathcal{X}$. Moreover,

$$\|S(t)\| = \sup_{x \in \mathbb{R}} \left\{ e^{xt - |x|^2} \right\} = e^{\left(\frac{t^2}{4}\right)},$$

and the family $\{S(t), t \geq 0\}$ is not exponentially bounded. Hence A cannot generate an (exponentially bounded) semi-group of classes C_0, C_1, $\mathcal{A}$, or an exponentially bounded convoluted, in particular integrated, semi-group.

We show that this operator does not generate even a local integrated or convoluted semi-group. Let us find the resolvent of A:

$$(\lambda I - A)f(x) = \lambda f(x) - xf(x) = g(x), \quad x \in \mathbb{R}.$$

84   3. Examples. Supplements

The solution of this equation has the form

$$f(x) = (\lambda I - A)^{-1} g(x) = \frac{g(x)}{\lambda - x}, \qquad x \in \mathbb{R},$$

whence we obtain that $(\lambda I - A)^{-1}$ is unbounded for each $\lambda \in \mathbb{R}$ and that $Sp(A) = \mathbb{R}$. Therefore, the resolvent set of A does not contain any semi-interval of real axis of the kind $Re\lambda > \omega$. This means that A generates neither a local integrated nor a local convoluted semi-group.

3.2 Examples of solutions to Petrovsky correct, conditionally correct, and incorrect systems

Now we consider some examples of equations with differential operators $A = \mathbf{A}(i\partial/\partial x)$, where, in accordance with the results of Section 2.3, the behavior of semi-groups generated by such operators and the well-posedness of the related Cauchy problems are determined by estimates of $\|e^{t\mathbf{A}(s)}\|$, solution operators to the Fourier transformed systems. We will construct generalized solutions to the Cauchy problem for some differential systems with initial data from distribution spaces.

Example 3.2.1 *of a Petrovsky correct (parabolic) system.*

Consider the parabolic equation

$$\frac{\partial u(x;t)}{\partial t} = \frac{\partial^2 u(x;t)}{\partial x^2}, \qquad x \in \mathbb{R}, \quad t \geq 0, \tag{3.2.1}$$

with the initial data $u(x;0) = f(x)$ in a space of distributions Ψ' (it may be a space of the type $(\mathcal{S}_\alpha^\beta)'$ or $(W_M^\Omega)'$).

Let us apply the Fourier transform to the problem. We obtain the Cauchy problem for the ordinary differential equation

$$\frac{d\widetilde{u}(s;t)}{dt} = -s^2 \widetilde{u}(s;t), \quad t \geq 0, \qquad \widetilde{u}(s;0) = \widetilde{f}(s), \quad s \in \mathbb{C}, \tag{3.2.2}$$

in the corresponding space $\widetilde{\Phi}'$. The operator of this problem has the form $\mathbf{A}(s) = -s^2 I$; hence its characteristic root

$$\Lambda(s) = Re\lambda(s) = \tau^2 - \sigma^2, \qquad s = \sigma + i\tau \in \mathbb{C}, \qquad \Lambda(\sigma) = -\sigma^2.$$

Therefore, in accordance with Definition 2.3.1, Equation (3.2.1) is Petrovsky correct, in particular, it is parabolic.

3.2. Examples of solutions to systems 85

Solution operators of the problem (3.2.2) are equal to the operators of multiplication by the exponent e^{-ts^2}, $t \geq 0$. They are bounded in $L^2(\mathbb{R})$ as $s = \sigma \in \mathbb{R}$:

$$\|\widetilde{u}(t)\|^2_{L^2(\mathbb{R})} = \int_{\mathbb{R}} e^{-2t\sigma^2} |\widetilde{f}(\sigma)|^2 \, d\sigma \leq \|\widetilde{f}\|^2_{L^2(\mathbb{R})}. \tag{3.2.3}$$

Due to the properties of the Fourier transform, the solution of the Cauchy problem for (3.2.2) is defined as the convolution of f with the Green function:

$$u(x;t) = G_t(x) * f(x), \qquad x \in \mathbb{R}, \quad t \geq 0,$$

where $\widetilde{G}_t(s) = e^{-ts^2}$. Let us find the Green function. For this purpose here and in the follow-up examples we need some formulas of generalized Fourier transforms (see, e.g., [35, 111]):

$$\begin{aligned}
\mathcal{F}[s] &= -2\pi i \cdot \delta'(x) & \tag{3.2.4} \\
\mathcal{F}\left[s^k\right] &= 2\pi(-i)^k \cdot \delta^{(k)}(x) & \tag{3.2.5} \\
\mathcal{F}\left[\text{V.p.} \frac{1}{s}\right] &= \pi i \cdot \text{sgn}(x) & \tag{3.2.6} \\
\mathcal{F}[\cos(ts)] &= \pi \left(\delta(x-t) + \delta(x+t)\right) & \tag{3.2.7} \\
\mathcal{F}[\sin(ts)] &= \pi i \left(\delta(x-t) - \delta(x+t)\right) & \tag{3.2.8} \\
\mathcal{F}\left[e^{-ts^2}\right] &= \sqrt{\frac{\pi}{t}} \, e^{-\frac{x^2}{4t}}, \qquad t > 0, & \tag{3.2.9}
\end{aligned}$$

and the relation between the direct and inverse Fourier transform

$$\mathcal{F}^{-1}[f](s) = \frac{1}{2\pi} \mathcal{F} f(-s). \tag{3.2.10}$$

Using (3.2.9) and (3.2.10) we obtain

$$G_t(x) = \mathcal{F}^{-1}\left[e^{-ts^2}\right] = \frac{1}{2\pi} \mathcal{F}\left[e^{-ts^2}\right] = \frac{\sqrt{\pi}}{2\pi\sqrt{t}} e^{-\frac{x^2}{4t}} = \frac{1}{2\sqrt{\pi t}} e^{-\frac{x^2}{4t}}$$

and the well-known formula for the solution

$$u(x;t) = \frac{1}{2\sqrt{\pi t}} \int_{\mathbb{R}} e^{-\frac{(x-\xi)^2}{4t}} f(\xi) \, d\xi.$$

By the estimate (3.2.3) and the Parseval equality, we have

$$\|u(t)\|_{L^2(\mathbb{R})} = \|\widetilde{u}(t)\|_{L^2(\mathbb{R})} \leq \|\widetilde{f}\|_{L^2(\mathbb{R})} = \|f\|_{L^2(\mathbb{R})}.$$

Let us find the resolvent of $\mathbf{A}(s)$:

$$(\lambda I - \mathbf{A}(s))\widetilde{f}(s) = \widetilde{g}(s) \qquad \Longrightarrow \qquad \widetilde{f}(s) = \frac{\widetilde{g}(s)}{\lambda + s^2}.$$

86   3. Examples. Supplements

Hence,

$$\|\mathcal{R}_{\mathbf{A}(\sigma)}(\lambda)\widetilde{g}\|_{L^2(\mathbb{R})}^2 = \int_{\mathbb{R}} \frac{|\widetilde{g}(\sigma)|^2}{|\lambda + \sigma^2|^2}\, d\sigma = \frac{1}{|\lambda|^2}\int_{\mathbb{R}} |\widetilde{g}(\sigma)|^2\, d\sigma = \frac{1}{|\lambda|^2}\|\widetilde{g}\|_{L^2(\mathbb{R})}^2.$$

Therefore,

$$\|\mathcal{R}_{\mathbf{A}(\sigma)}(\lambda)\|_{L^2(\mathbb{R})} = \|\mathcal{R}_{\frac{d^2}{dx^2}}(\lambda)\|_{L^2(\mathbb{R})} \le \frac{1}{|\lambda|} \le \frac{1}{Re\lambda}, \qquad Re\lambda > 0.$$

It follows that the operator $\mathbf{A}(i\partial/\partial x) = d^2/dx^2$ generates a C_0-semi-group in the space $L^2(\mathbb{R})$ and the Cauchy problem for (3.2.1) is uniformly well-posed.

Note that in spaces $L^p(\mathbb{R}), p \ne 2$, this operator generates only semi-groups of growth order $\alpha = \alpha(p)$ and the Cauchy problem is not uniformly well-posed.

Example 3.2.2 *of a Petrovsky correct (hyperbolic) system.*

Consider the hyperbolic equation

$$\frac{\partial^2 u(x;t)}{\partial t^2} = \frac{\partial^2 u(x;t)}{\partial x^2}, \qquad x \in \mathbb{R}, \quad t \ge 0, \tag{3.2.11}$$

with the initial data

$$u(x;0) = f_1(x), \qquad \frac{\partial u(x;0)}{\partial t} = f_2(x),$$

in a space Ψ' as in Example 3.2.1. By the change

$$u_1(x;t) = u(x;t), \qquad u_2(x;t) = \frac{\partial u(x;t)}{\partial t}$$

we reduce Equation (3.2.11) to the system of differential equations of the first order wrt t:

$$\begin{cases} \dfrac{\partial u_1(x;t)}{\partial t} = u_2(x;t), \\[2mm] \dfrac{\partial u_2(x;t)}{\partial t} = \dfrac{\partial^2 u_1(x;t)}{\partial x^2}, \end{cases} \qquad x \in \mathbb{R}, \quad t \ge 0, \tag{3.2.12}$$

with the initial data

$$\begin{cases} u_1(x;0) = f_1(x), \\ u_2(x;0) = f_2(x), \end{cases} \qquad x \in \mathbb{R}. \tag{3.2.13}$$

We apply the Fourier transform to the problem (3.2.12)–(3.2.13). Then it transforms into the Cauchy problem for the system of ordinary differential equations

$$\begin{cases} \dfrac{d\widetilde{u}_1(s;t)}{dt} = \widetilde{u}_2(s;t), \\[2mm] \dfrac{d\widetilde{u}_2(s;t)}{dt} = -s^2\, \widetilde{u}_1(s;t), \end{cases} \qquad s \in \mathbb{C}, \quad t \ge 0, \tag{3.2.14}$$

3.2. Examples of solutions to systems

$$\begin{cases} \widetilde{u}_1(s;0) = \widetilde{f}_1(s), \\ \widetilde{u}_2(s;0) = \widetilde{f}_2(s), \end{cases} \quad s \in \mathbb{C}. \tag{3.2.15}$$

The matrix of the system (3.2.14) has the form

$$\mathbf{A}(s) = \begin{pmatrix} 0 & 1 \\ -s^2 & 0 \end{pmatrix}.$$

Characteristic roots of the system are the roots of the equation

$$\det(\mathbf{A}(s) - \lambda I) = \begin{vmatrix} -\lambda & 1 \\ -s^2 & -\lambda \end{vmatrix} = \lambda^2 + s^2 = 0, \quad \lambda_{1,2}(s) = \pm is = \pm i(\sigma + i\tau),$$

and

$$\Lambda(s) = \max\{Re\lambda_1(s), Re\lambda_2(s)\} = \max\{\pm\tau\} = |\tau| \quad \Longrightarrow \quad \Lambda(\sigma) = 0.$$

Therefore, the system (3.2.12) is Petrovsky correct. In addition, the growth order of the function $\Lambda(\cdot)$ in the complex plane is equal to one:

$$\Lambda(s) = |\tau| \leq |s|, \quad s = \sigma + i\tau \in \mathbb{C}.$$

Hence, due to Definition 2.3.1, the system (3.2.12) is hyperbolic.

We look for the solution of the Cauchy problem (3.2.14)–(3.2.15) in the form

$$\widetilde{u}(s;t) = e^{t\mathbf{A}(s)}\widetilde{f}(s), \quad s \in \mathbb{C}, \quad t \geq 0.$$

Let us find the matrix exponent using the Taylor series expansion

$$e^{t\mathbf{A}(s)} = \sum_{k=1}^{\infty} \frac{t^k \mathbf{A}^k(s)}{k!}.$$

In this example we obtain the equalities

$$\mathbf{A}^2(s) = \begin{pmatrix} -s^2 & 0 \\ 0 & -s^2 \end{pmatrix}, \quad \mathbf{A}^3(s) = \begin{pmatrix} 0 & -s^2 \\ s^4 & 0 \end{pmatrix},$$

$$\mathbf{A}^4(s) = \begin{pmatrix} s^4 & 0 \\ 0 & s^4 \end{pmatrix}, \quad \mathbf{A}^5(s) = \begin{pmatrix} 0 & s^4 \\ -s^6 & 0 \end{pmatrix}, \quad \dots$$

Thus the solving matrix-function of the system (3.2.14) has the form

$$e^{t\mathbf{A}(s)} = \begin{pmatrix} 1 & 0 \\ 0 & 1 \end{pmatrix} + t\begin{pmatrix} 0 & 1 \\ -s^2 & 0 \end{pmatrix} + \frac{t^2}{2}\begin{pmatrix} -s^2 & 0 \\ 0 & -s^2 \end{pmatrix}$$

$$+ \frac{t^3}{3!}\begin{pmatrix} 0 & -s^2 \\ s^4 & 0 \end{pmatrix} + \frac{t^4}{4!}\begin{pmatrix} s^4 & 0 \\ 0 & s^4 \end{pmatrix} + \frac{t^5}{5!}\begin{pmatrix} 0 & s^4 \\ -s^6 & 0 \end{pmatrix} + \dots$$

$$= \begin{pmatrix} 1 - \dfrac{t^2}{2}s^2 + \dfrac{t^4}{4!}s^4 - \cdots & t - \dfrac{t^3}{3!}s^2 + \dfrac{t^5}{5!}s^4 - \cdots \\[3mm] -ts^2 + \dfrac{t^3}{3!}s^4 - \dfrac{t^5}{5!}s^6 + \cdots & 1 - \dfrac{t^2}{2}s^2 + \dfrac{t^4}{4!}s^4 - \cdots \end{pmatrix}$$

$$= \begin{pmatrix} 1 - \dfrac{(ts)^2}{2} + \dfrac{(ts)^4}{4!} - \cdots & \left(ts - \dfrac{(ts)^3}{3!} + \dfrac{(ts)^5}{5!} - \cdots \right) \mathrm{V.p.}\dfrac{1}{s} \\[3mm] -s\left(ts - \dfrac{(ts)^3}{3!} + \dfrac{(ts)^5}{5!} - \cdots \right) & 1 - \dfrac{(ts)^2}{2} + \dfrac{(ts)^4}{4!} - \cdots \end{pmatrix}$$

$$= \begin{pmatrix} \cos(ts) & \sin(ts)\mathrm{V.p.}\dfrac{1}{s} \\[3mm] -s\,\sin(ts) & \cos(ts) \end{pmatrix}.$$

Then the solution of the Cauchy problem (3.2.14)–(3.2.15) can be written in the form

$$\widetilde{u}(s;t) = e^{t\mathbf{A}(s)}\widetilde{f}(s), \qquad t \geq 0, \quad s \in \mathbb{C},$$

and, due to the Fourier transform properties, the solution of the initial Cauchy problem is defined in the form of the convolution

$$u(x;t) = G_t(x) * f(x), \qquad t \geq 0, \quad x \in \mathbb{R},$$

where $\widetilde{G}_t(s) = e^{\mathbf{A}(s)t}$.

In order to obtain the elements of the matrix $G_t(x)$ we use the formulas (3.2.4)–(3.2.10). We have

$$\mathcal{F}^{-1}[\cos(ts)] = \frac{1}{2\pi}\,\mathcal{F}[\cos(ts)] = \frac{1}{2}\,\left(\delta(x - t) + \delta(x + t) \right), \qquad (3.2.16)$$

$$\mathcal{F}^{-1}[\sin(ts)] = -\frac{1}{2\pi}\,\mathcal{F}[\sin(ts)] = \frac{i}{2}\,\left(\delta(x + t) - \delta(x - t) \right), \qquad (3.2.17)$$

$$\mathcal{F}^{-1}[-s] = \frac{1}{2\pi}\,\mathcal{F}[s] = -i\delta'(x),$$

$$\mathcal{F}^{-1}\left[\mathrm{V.p.}\frac{1}{s}\right] = \frac{1}{2\pi}\,\mathcal{F}\left[\mathrm{V.p.}\left(-\frac{1}{s}\right)\right] = -\frac{i}{2}\,\mathrm{sgn}\,x.$$

3.2. Examples of solutions to systems

89

Let us write sgn x as $2H(x) - 1$, where H is the Heaviside function. Then

$$\mathcal{F}^{-1}\left[\mathrm{V.p.}\frac{1}{s}\right] = -\frac{i}{2}(2H(x) - 1),$$

$$\mathcal{F}^{-1}[-s\,\sin(ts)] = (-i\delta'(x)) * \left(\frac{i}{2}\,(\delta(x+t) - \delta(x-t))\right)$$

$$= \frac{1}{2}\,(\delta'(x+t) - \delta'(x-t)),$$

$$\mathcal{F}^{-1}\left[\mathrm{V.p.}\frac{1}{s}\sin(ts)\right] = \left(\frac{i}{2} - iH(x)\right) * \left(\frac{i}{2}\,(\delta(x+t) - \delta(x-t))\right)$$

$$= \frac{1}{2}\,(H(x+t) - H(x-t)).$$

Therefore, the Green matrix-function has the form

$$G_t(x) = \begin{pmatrix} \dfrac{\delta(x+t) + \delta(x-t)}{2} & \dfrac{H(x+t) - H(x-t)}{2} \\[3mm] \dfrac{\delta'(x+t) - \delta'(x-t)}{2} & \dfrac{\delta(x+t) + \delta(x-t)}{2} \end{pmatrix}$$

and again we obtain the well-known formulae

$$\begin{cases} u_1(x;t) = \dfrac{f_1(x+t) + f_1(x-t)}{2} + \dfrac{1}{2}\displaystyle\int_{x-t}^{x+t} f_2(\tau)d\tau, \\[4mm] u_2(x;t) = \dfrac{f_1'(x+t) - f_1'(x-t)}{2} + \dfrac{f_2(x+t) + f_2(x-t)}{2}. \end{cases} \tag{3.2.18}$$

Thus, if $f_1', f_2 \in L^2(\mathbb{R})$, then the generalized solution of the initial Cauchy problem is $u = \begin{pmatrix} u_1 \\ u_2 \end{pmatrix}$ in $L^2(\mathbb{R}) \times L^2(\mathbb{R})$. If, in addition, $f_1'', f_2' \in L^2(\mathbb{R})$, then u is a classical solution.

The abstract Cauchy problem corresponding to the one considered is (1.0.1), where

$$u = \begin{pmatrix} u_1 \\ u_2 \end{pmatrix}, \quad A = \mathbf{A}(i\partial/\partial x) = \begin{pmatrix} 0 & 1 \\ \dfrac{d^2}{dx^2} & 0 \end{pmatrix}, \quad f = \begin{pmatrix} f_1 \\ f_2 \end{pmatrix};$$

its solution

$$u(t) = U(t)\begin{pmatrix} f_1 \\ f_2 \end{pmatrix}, \qquad t \geq 0,$$

where the elements of the matrix $U(t)$, $t \geq 0$, are defined in accordance with the formula (3.2.18). Because of differential operators applied to the elements of the matrix, the family of operators $\{U(t), t \geq 0\}$ is unbounded in the space $L^2(\mathbb{R}) \times L^2(\mathbb{R})$. Nevertheless, the family of its primitives is bounded, i.e., the operator A generates a one-time integrated semi-group.

90   *3. Examples. Supplements*

Example 3.2.3 *of the system of differential equations of the first order generating a uniformly well-posed Cauchy problem.*

Consider the Cauchy problem for the system of equations

$$
\begin{cases}
\dfrac{\partial u_1(x;t)}{\partial t} = \dfrac{\partial u_2(x;t)}{\partial x}, \\[2mm]
\dfrac{\partial u_2(x;t)}{\partial t} = \dfrac{\partial u_1(x;t)}{\partial x},
\end{cases}
\qquad x \in \mathbb{R}, \quad t \ge 0,
$$

with initial data

$$
\begin{cases}
u_1(x;0) = f_1(x), \\
u_2(x;0) = f_2(x),
\end{cases}
\qquad x \in \mathbb{R},
$$

in spaces Ψ', Φ' as in the previous examples. Let us write the problem in the abstract form (1.0.1), where

$$
u = \begin{pmatrix} u_1 \\ u_2 \end{pmatrix}, \quad
A = \mathbf{A}(i\partial/\partial x) = \begin{pmatrix} 0 & \dfrac{d}{dx} \\[2mm] \dfrac{d}{dx} & 0 \end{pmatrix}, \quad
f = \begin{pmatrix} f_1 \\ f_2 \end{pmatrix}.
$$

Applying the Fourier transform to the problem, we obtain the system of ordinary differential equations

$$
\begin{cases}
\dfrac{d\tilde{u}_1(s;t)}{dt} = -is\,\tilde{u}_2(s;t), \\[3mm]
\dfrac{d\tilde{u}_2(s;t)}{dt} = -is\,\tilde{u}_1(s;t),
\end{cases}
\qquad s \in \mathbb{C}, \quad t \ge 0, \qquad (3.2.19)
$$

with initial data

$$
\begin{cases}
\tilde{u}_1(s;0) = \tilde{f}_1(s), \\
\tilde{u}_2(s;0) = \tilde{f}_2(s),
\end{cases}
\qquad s \in \mathbb{C}. \qquad (3.2.20)
$$

The matrix $\mathbf{A}(s)$ of this system is

$$
\mathbf{A}(s) = \begin{pmatrix} 0 & -is \\ -is & 0 \end{pmatrix}.
$$

The formal solution to the problem (3.2.19)–(3.2.20) can be written in the form

$$
\tilde{u}(s;t) = e^{t\mathbf{A}(s)}\tilde{u}(s) = \sum_{k=1}^{\infty} \frac{t^k \mathbf{A}^k(s)}{k!}.
$$

In this example

$$
\mathbf{A}^2(s) = \begin{pmatrix} -s^2 & 0 \\ 0 & -s^2 \end{pmatrix}, \qquad
\mathbf{A}^3(s) = \begin{pmatrix} 0 & is^3 \\ is^3 & 0 \end{pmatrix},
$$

3.2. Examples of solutions to systems 91

$$\mathbf{A}^4(s) = \begin{pmatrix} s^4 & 0 \\ 0 & s^4 \end{pmatrix}, \qquad \mathbf{A}^5(s) = \begin{pmatrix} 0 & -is^5 \\ -is^5 & 0 \end{pmatrix}, \ \dots$$

and the solving matrix-function of the problem is

$$e^{t\mathbf{A}(s)} = \begin{pmatrix} 1 - \dfrac{(ts)^2}{2} + \dfrac{(ts)^4}{4!} + \dots & -i\left(ts + \dfrac{(ts)^3}{3!} - \dfrac{(ts)^5}{5!} + \dots\right) \\[2ex] -i\left(ts + \dfrac{(ts)^3}{3!} - \dfrac{(ts)^5}{5!} + \dots\right) & 1 - \dfrac{(ts)^2}{2} + \dfrac{(ts)^4}{4!} + \dots \end{pmatrix}$$

$$= \begin{pmatrix} \cos(ts) & -i\,\sin(ts) \\ -i\,\sin(ts) & \cos(ts) \end{pmatrix}.$$

Hence we obtain the solution of the Fourier transformed Cauchy problem

$$\widetilde{u}(s;t) = e^{t\mathbf{A}(s)}\widetilde{f}(s) = \begin{pmatrix} \cos(ts) & -i\,\sin(ts) \\ -i\,\sin(ts) & \cos(ts) \end{pmatrix} \begin{pmatrix} \widetilde{f}_1(s) \\ \widetilde{f}_2(s) \end{pmatrix}$$

and the solution of the original problem is defined as the convolution

$$u(x;t) = G_t(x) * f(x), \qquad x \in \mathbb{R}, \qquad t \ge 0,$$

where $\widetilde{G}_t(s) = e^{t\mathbf{A}(s)}$. In order to find the elements of the matrix $G_t(x)$, we use the formulae (3.2.16)–(3.2.17). Then the Green matrix-function has the form

$$G_t(x) = \begin{pmatrix} \dfrac{\delta(x-t) + \delta(x+t)}{2} & \dfrac{\delta(x+t) - \delta(x-t)}{2} \\[2ex] \dfrac{\delta(x+t) - \delta(x-t)}{2} & \dfrac{\delta(x-t) + \delta(x+t)}{2} \end{pmatrix}$$

and

$$\begin{cases} u_1(x;t) = \dfrac{1}{2}\left(f_1(x-t) + f_1(x+t) + f_2(x+t) - f_2(x-t)\right), \\[3ex] u_2(x;t) = \dfrac{1}{2}\left(f_1(x+t) - f_1(x-t) + f_2(x-t) + f_2(x+t)\right). \end{cases}$$

Therefore, the initial Cauchy problem is uniformly well-posed in $L^2(\mathbb{R}) \times L^2(\mathbb{R})$.

Example 3.2.4 *of a differential system which in dependence on the parameter k generates semi-groups of different classes.*

92   *3. Examples. Supplements*

Consider the system of differential equations

$$\begin{cases} \dfrac{\partial u_1(x;t)}{\partial t} = \dfrac{\partial^2 u_1(x;t)}{\partial x^2}, \\[3mm] \dfrac{\partial u_2(x;t)}{\partial t} = i^k \dfrac{\partial^k u_1(x;t)}{\partial x^k} + \dfrac{\partial^2 u_2(x;t)}{\partial x^2}, \end{cases} \qquad x \in \mathbb{R}, \quad t \geq 0, \quad (3.2.21)$$

with initial data

$$\begin{cases} u_1(x;0) = f_1(x), \\ u_2(x;0) = f_2(x), \end{cases} \qquad x \in \mathbb{R},$$

and apply the Fourier transform to the system:

$$\begin{cases} \dfrac{d\tilde{u}_1(s;t)}{dt} = -s^2 \, \tilde{u}_1(s;t), \\[3mm] \dfrac{d\tilde{u}_2(s;t)}{dt} = s^k \, \tilde{u}_1(s;t) - s^2 \, \tilde{u}_2(s;t) \end{cases} \qquad s \in \mathbb{C}, \quad t \geq 0. \quad (3.2.22)$$

We look for a solution of the Cauchy problem for (3.2.22) in the form

$$\tilde{u}(s;t) = e^{t\mathbf{A}(s)} \, \tilde{f}(s).$$

The solving matrix-function of the Cauchy problem for (3.2.22) is

$$e^{t\mathbf{A}(s)} = \begin{pmatrix} 1 & 0 \\ 0 & 1 \end{pmatrix} + t \begin{pmatrix} -s^2 & 0 \\ s^k & -s^2 \end{pmatrix} + \frac{t^2}{2} \begin{pmatrix} s^4 & 0 \\ -2s^{k+2} & s^4 \end{pmatrix}$$

$$+ \frac{t^3}{3!} \begin{pmatrix} -s^6 & 0 \\ 3s^{k+4} & -s^6 \end{pmatrix} + \frac{t^4}{4!} \begin{pmatrix} s^8 & 0 \\ -4s^{k+6} & s^8 \end{pmatrix} + \frac{t^5}{5!} \begin{pmatrix} -s^{10} & 0 \\ 5s^{k+8} & -s^{10} \end{pmatrix} + \cdots$$

$$= \begin{pmatrix} 1 + (-ts^2) + \dfrac{(-ts^2)^2}{2} + \cdots & 0 \\[3mm] ts^k \left(1 + (-ts^2) + \dfrac{(-ts^2)^2}{2} + \cdots \right) & 1 + (-ts^2) + \dfrac{(-ts^2)^2}{2} + \cdots \end{pmatrix}$$

and

$$\tilde{u}(s;t) = e^{t\mathbf{A}(s)} \tilde{f}(s) = \begin{pmatrix} e^{-ts^2} & 0 \\ ts^k \, e^{-ts^2} & e^{-ts^2} \end{pmatrix} \begin{pmatrix} \tilde{f}_1(s) \\ \tilde{f}_2(s) \end{pmatrix} = e^{-ts^2} \begin{pmatrix} 1 & 0 \\ ts^k & 1 \end{pmatrix} \begin{pmatrix} \tilde{f}_1(s) \\ \tilde{f}_2(s) \end{pmatrix}$$

is a solution of the Cauchy problem.

The solution of the original Cauchy problem is defined in the form of convolution:

$$u(x;t) = G_t(x) * f(x), \qquad x \in \mathbb{R}, \quad t \geq 0,$$

3.3. Definitions and properties of spaces of test functions

where $G_t(s)$ is equal to the inverse Fourier transform of

$$e^{\mathbf{A}(s)t} = e^{-ts^2} \begin{pmatrix} 1 & 0 \\ ts^k & 1 \end{pmatrix}.$$

In order to find $G_t(x)$, we use the formulae (3.2.5) and (3.2.9) and the relation (3.2.10). We obtain

$$
\begin{aligned}
\mathcal{F}^{-1}\left[e^{-ts^2}\right] &= \frac{1}{2\pi}\mathcal{F}\left[e^{-ts^2}\right] = \frac{\sqrt{\pi}}{2\pi\sqrt{t}}\, e^{-\frac{x^2}{4t}} = \frac{1}{2\sqrt{\pi t}}\, e^{-\frac{x^2}{4t}}\,, \\
\mathcal{F}^{-1}\left[ts^k\right] &= \frac{1}{2\pi}\mathcal{F}\left[t(-s)^k\right] = \frac{t(-1)^k}{2\pi}\,\mathcal{F}\left[s^k\right] = t\, i^k\, \delta^{(k)}(x), \\
\mathcal{F}^{-1}\left[ts^k e^{-ts^2}\right] &= \left(t\, i^k\, \delta^{(k)}(x)\right) * \left(\frac{1}{2\sqrt{\pi t}}\, e^{-\frac{x^2}{4t}}\right) = \frac{i^k}{2}\sqrt{\frac{t}{\pi}}\,\frac{\partial^k}{\partial x^k}\left(e^{-\frac{x^2}{4t}}\right).
\end{aligned}
$$

Thus the Green matrix-function is defined as

$$
G_t(x) = \begin{pmatrix} \dfrac{1}{2\sqrt{\pi t}}\, e^{-\frac{x^2}{4t}} & 0 \\[4mm] \dfrac{i^k}{2}\sqrt{\dfrac{t}{\pi}}\,\dfrac{\partial^k}{\partial x^k}\left(e^{-\frac{x^2}{4t}}\right) & \dfrac{1}{2\sqrt{\pi t}}\, e^{-\frac{x^2}{4t}} \end{pmatrix}.
$$

Comparing this example with Example 3.1.1, where the operator A behaves similarly to the matrix $\mathbf{A}(s)$ of the system (3.2.22) with $2\gamma = k$, we can arrive at the following conclusion on generating various semi-groups in $L^2(\mathbb{R})$ by the operator of the system (3.2.21).

The operator $A = \mathbf{A}(i\partial/\partial x)$ of the system (3.2.21) generates a C_0-semi-group as $k = 0, 1, 2$ and a one-time integrated semi-group as $k = 3$; in addition, the operator generates a semi-group of growth order $\alpha > k/2 - 1$ and an R-semi-group with $R = (\lambda I - A)^{-n}$, $n = [\alpha] + 1$, as $k = 3, 4, \ldots$.

3.3 Definitions and properties of spaces of test functions $\mathcal{D}$, $\mathcal{S}$, $\mathcal{S}_\alpha$, $\mathcal{S}^\beta$, $\mathcal{S}_\alpha^\beta$, Z, W_M^Ω, $\mathcal{D}^{\{M_n\}}$, ...

For the convenience of readers, in this section we introduce spaces of test functions, which are used in construction of generalized solutions to deterministic problems in Part I and stochastic problems in Part II. We investigate their properties and define topology, convergence of sequences, and the notion of boundedness of sets in these spaces. In this connection we denote both linear spaces and corresponding topological spaces by the same symbols:

$$\mathcal{D},\ \mathcal{S}_\alpha^\beta,\ W_M^\Omega, \ldots.$$

94   *3. Examples. Supplements*

The space $\mathcal{D}$. The Schwartz space $\mathcal{D}(\mathbb{R})$, denoted simply as $\mathcal{D}$, consists of all infinitely differentiable functions $\varphi(x), x \in \mathbb{R}$ with compact supports.

Convergence of a sequence $\varphi_n \in \mathcal{D}$ to zero implies that the supports of all functions belong to a compact set K and both the sequence φ_n and the sequences of all its derivatives converge uniformly to zero on K. It is well known that in this space we cannot introduce a norm consistent with the convergence. However, $\mathcal{D}$ can be represented in the form of a countable union of normed spaces (the inductive limit) so that the convergence in the inductive limit topology coincides with the one introduced.

Recall that the topology in a countably normed space is introduced as follows [37, 53]. Let $p \in \mathbb{N}$ and $\varepsilon > 0$. The set of all functions φ satisfying

$$\|\varphi\|_1 < \varepsilon, \quad \|\varphi\|_2 < \varepsilon, \quad \ldots, \quad \|\varphi\|_p < \varepsilon$$

is called the neighborhood of zero and is denoted by $U_{p,\varepsilon}(0)$. The topology in a countably normed space is the topology of projective limit [105] and a set $\mathcal{B}$ is bounded in a countably normed space if it is bounded in each norm.

Consider the space $\mathcal{D}_A$ which consists of all infinitely differentiable functions with supports in $[-A, A]$.[1] Define a set of norms in the space as

$$\|\varphi\|_p = \sup_{q \leq p} \ \sup_{|x| \leq A} |\varphi^{(q)}(x)|, \qquad p \in \mathbb{N}_0.$$

The space $\mathcal{D}_A$ with such a system of norms is a complete countably normed space. Convergence of a sequence $\varphi_n \in \mathcal{D}_A$ to zero implies that both the sequence φ_n and the sequences of all its derivatives $\varphi_n^{(q)}(\cdot)$, $q \in \mathbb{N}$, converge uniformly to zero on $[-A, A]$.

The set $\mathcal{B}$ is bounded in $\mathcal{D}_A$ if for any $q \in \mathbb{N}_0$ there exists a constant C_q such that $|\varphi^{(q)}(x)| \leq C_q$, $x \in [-A, A]$, $\varphi \in \mathcal{B}$. The space $\mathcal{D}_A$ with the considered system of norms is a perfect countably normed space; it follows that in the space and in the conjugate one the strong and weak convergences coincide.

It follows from the definition of $\mathcal{D}_A$ that, if $A_1 < A_2$, then $\mathcal{D}_{A_1} \subset \mathcal{D}_{A_2}$ and any sequence convergent in $\mathcal{D}_{A_1}$ is convergent in $\mathcal{D}_{A_2}$. If A takes its values in the set of positive integers, then we have the sequence of embedded expanding spaces

$$\mathcal{D}_1 \subset \mathcal{D}_2 \subset \ldots \subset \mathcal{D}_j \subset \mathcal{D}_{j+1} \subset \ldots$$

Consider the space equal to the union of all $\mathcal{D}_j$ where $j \in \mathbb{N}$: $\bigcup_j \mathcal{D}_j$. This space coincides with the space $\mathcal{D}$ as a set. Convergence to zero of a sequence φ_n in $\mathcal{D} = \bigcup_j \mathcal{D}_j$ with the inductive limit topology implies that all functions $\varphi_n(\cdot)$ belong to a space $\mathcal{D}_j$ and converge to zero in the $\mathcal{D}_j$. Hence the convergence in $\mathcal{D}$ defined in the beginning of the section is wrt the inductive limit topology.

---

[1]In this section again we use the notations accepted in the theory of Gelfand–Shilov spaces denoting constants by letters A, B. In order not to confuse these constants with the differential operator in Sections 2.3 and 3.2 and further in Part II, we denote the operator by $\mathbf{A}(i\partial/\partial x)$.

3.3. Definitions and properties of spaces of test functions

According to the definition of $\mathcal{D}$ as the inductive limit of $\mathcal{D}_j$, a set $\mathcal{B}$ is *bounded in $\mathcal{D}$* if and only if it is bounded in some $\mathcal{D}_j$.

The space $\mathcal{S}_\alpha$. The space $\mathcal{S}_\alpha$ ($\alpha \geq 0$) consists of all infinitely differentiable functions $\varphi(x), x \in \mathbb{R}$, satisfying the inequalities

$$|x^k \varphi^{(q)}(x)| \leq C_q A^k k^{k\alpha}, \qquad k, q \in \mathbb{N}_0, \quad x \in \mathbb{R}, \tag{3.3.1}$$

with constants $A = A(\varphi)$ and $C_q = C_q(\varphi)$ ($k^{k\alpha} = 1$ as $k = 0$).

This definition imposes a limitation on growth (more precisely, on decrease) of functions and their derivatives as $|x| \to \infty$. Indeed, if we write (3.3.1) in the form

$$|\varphi^{(q)}(x)| \leq C_q \left(\frac{A}{|x|}\right)^k k^{k\alpha}, \qquad k, q \in \mathbb{N}_0, \quad x \in \mathbb{R}, \tag{3.3.2}$$

one can note that the smaller is α, the faster the functions $\varphi(\cdot)$ tend to zero as $|x| \to \infty$. It turns out that if $\alpha = 0$, then any function $\varphi(\cdot)$ satisfying (3.3.1) is equal to zero as $|x| > A$. Indeed, as $\alpha = 0$ and $q = 0$, we obtain from (3.3.2)

$$|\varphi(x)| \leq C_0 \left(\frac{A}{|x|}\right)^k, \qquad k \in \mathbb{N}_0, \quad x \in \mathbb{R}.$$

Passing to infimum in the inequality with respect to all $k \in \mathbb{N}_0$, we conclude that $\varphi(x) = 0$ as $|x| > A$. The converse is evident: for any infinitely differentiable function equal to zero at $|x| > A$, the estimate

$$|x^k \varphi^{(q)}(x)| \leq C_q A^k, \qquad |x| \leq A,$$

holds providing (3.3.1) with $\alpha = 0$. Thus $\mathcal{S}_0$ coincides with the space $\mathcal{D}$. Since the structure of this space and the convergence and the notion of a bounded set were described above, we will further consider the spaces $\mathcal{S}_\alpha$ in the case $\alpha > 0$.

The elements of the space $\mathcal{S}_\alpha, \alpha > 0$, can be equivalently defined as the functions satisfying the estimates

$$|\varphi^{(q)}(x)| \leq C_q' e^{-a|x|^{1/\alpha}}, \qquad q \in \mathbb{N}_0, \quad x \in \mathbb{R}, \tag{3.3.3}$$

where $C_q' = C_q'(\varphi)$ and $a = a(\varphi) = \frac{\alpha}{eA^{1/\alpha}}$. In order to clarify the idea of the proof of this important property, we consider the function

$$\mu_\alpha(y) = \inf_{k \in \mathbb{N}_0} \frac{k^{k\alpha}}{|y|^k}.$$

Then from (3.3.2) the estimate follows:

$$|\varphi^{(q)}(x)| \leq C_q \mu_\alpha\left(\frac{|x|}{A}\right), \qquad q \in \mathbb{N}_0, \quad x \in \mathbb{R}. \tag{3.3.4}$$

96   3. *Examples. Supplements*

For the infimum of the function $f(k) = \frac{k^{k\alpha}}{|y|^k}$ wrt the continuous parameter k, we obtain

$$\inf_{k \in \mathbb{R}_+} f(k) = f(k^*) = e^{-\frac{\alpha}{e}|y|^{1/\alpha}} \quad \text{as} \quad k^* = \frac{1}{e}|y|^{1/\alpha}.$$

Then $\mu_\alpha(y)$, as the infimum with respect to natural values of k, satisfies the inequality

$$e^{-\frac{\alpha}{e}|y|^{1/\alpha}} \leq \mu_\alpha(y). \tag{3.3.5}$$

Applying the Tailor expansion of f in a neighborhood of the point k^*, it is easy to show the converse estimate:

$$\mu_\alpha(y) \leq C e^{-\frac{\alpha}{e}|y|^{1/\alpha}}. \tag{3.3.6}$$

Now, if $\varphi \in S_\alpha$, then (3.3.4) is true and, due to (3.3.6), the inequality (3.3.3) follows. Conversely, if (3.3.3) holds, then from (3.3.5) we obtain

$$|\varphi^{(q)}(x)| \leq C'_q e^{-a|x|^{1/\alpha}} \leq C'_q \mu_\alpha\left(\frac{|x|}{A}\right) = C'_q \inf_{k \in \mathbb{N}_0} \frac{A^k k^{k\alpha}}{|x|^k} \leq C'_q \frac{A^k k^{k\alpha}}{|x|^k}.$$

Hence (3.3.1) holds and φ belongs to the space S_α. Thus the space $S_\alpha, \alpha > 0$, can be equivalently defined as the set of all infinitely differentiable functions satisfying (3.3.3).

Similar to the particular case $S_0 = \mathcal{D}$, the space S_α for $\alpha > 0$ can be introduced in the form of a countable union of normed spaces. Denote by $S_{\alpha,A}$ the set of all infinitely differentiable functions satisfying the condition

$$|x^k \varphi^{(q)}(x)| \leq C_{q,\varepsilon}(A + \varepsilon)^k k^{k\alpha}, \qquad k, q \in \mathbb{N}_0, \quad x \in \mathbb{R}, \tag{3.3.7}$$

for any $\varepsilon > 0$ with some constant $C_{q,\varepsilon} = C_{q,\varepsilon}(\varphi)$. Let us introduce the system of norms

$$\|\varphi\|_{q,p} = \sup_{k \in \mathbb{N}_0} \sup_{x \in \mathbb{R}} \frac{|x^k \varphi^{(q)}(x)|}{\left(A + \frac{1}{p}\right)^k k^{k\alpha}}, \qquad p \in \mathbb{N}, \quad q \in \mathbb{N}_0, \tag{3.3.8}$$

in the space $S_{\alpha,A}$. Due to (3.3.7), the supremum is well defined. The space $S_{\alpha,A}$ with such a system of norms is a perfect countably normed space.

The set $\mathcal{B}$ is bounded in $S_{\alpha,A}$ if for any $p \in \mathbb{N}$, $q \in \mathbb{N}_0$ there exists a constant $C_{q,p}$ such that $\|\varphi\|_{q,p} \leq C_{q,p}$ for all φ from $\mathcal{B}$.

Convergence to zero of a sequence $\varphi_n \in S_{\alpha,A}$ means that it is bounded in the space and for any $q \in \mathbb{N}_0$ the sequence $\varphi_n^{(q)}$ tends to zero uniformly on any segment $|x| \leq x_0 < \infty$.

At first sight, this convergence is weaker than the convergence wrt the topology defined by the system of norms (3.3.8); however, it is shown in [37] that for $S_{\alpha,A}$ as well as for the spaces considered below, the convergence of

3.3. Definitions and properties of spaces of test functions 97

a sequence in the space topology is equivalent to the uniform convergence on bounded sets in $\mathbb{R}$.

By analogy with $\mathcal{S}_\alpha$, the space $\mathcal{S}_{\alpha,A}$ can be equivalently defined as the set of all infinitely differentiable functions satisfying the condition

$$|\varphi^{(q)}(x)| \leq C'_{q,\rho} e^{-(a-\rho)|x|^{1/\alpha}}, \qquad q \in \mathbb{N}_0, \quad x \in \mathbb{R},$$

for any $\rho > 0$, with $C'_{q,\rho} = C'_{q,\rho}(\varphi)$, $a = \frac{\alpha}{eA^{1/\alpha}}$. If we introduce the system of norms in $\mathcal{S}_{\alpha,A}$ by

$$\|\varphi\|_p = \sup_{q \leq p} \sup_{x \in \mathbb{R}} e^{a(1-\frac{1}{p})|x|^{1/\alpha}} |\varphi^{(q)}(x)|, \qquad p \in \mathbb{N},$$

then it is equivalent to the system (3.3.8) in the sense of convergence or divergence of sequences $\varphi_n \in \mathcal{S}_{\alpha,A}$ with respect to both norms.

It follows from (3.3.7) that, if $A_1 < A_2$, then $\mathcal{S}_{\alpha,A_1}$ is a subspace of $\mathcal{S}_{\alpha,A_2}$. Hence the union of the spaces $\mathcal{S}_{\alpha,A}$ with respect to the indices $A \in \mathbb{N}$ coincides with the space $\mathcal{S}_\alpha$. Thus $\mathcal{S}_\alpha$ is presented in the form of the countable union of normed spaces

$$S_\alpha = \bigcup_A S_{\alpha,A},$$

and the topology in this space is the topology of the strong inductive limit. This provides the following definition of convergence in $\mathcal{S}_\alpha$: $\varphi_n \in \mathcal{S}_\alpha$ converges to zero if all the functions $\varphi_n(\cdot)$ belong to some space $\mathcal{S}_{\alpha,A}$ and the sequence converges to zero in it.

A set $\mathcal{B}$ is bounded in $\mathcal{S}_\alpha$ if there exists such $A > 0$ that $\mathcal{B}$ is bounded in $\mathcal{S}_{\alpha,A}$.

The space Z. The space Z consists of all entire functions $\varphi(z)$, $z \in \mathbb{C}$, satisfying the condition

$$|z^k \varphi(z)| \leq C_k e^{b|y|}, \qquad k \in \mathbb{N}_0, \quad z = x + iy \in \mathbb{C}, \tag{3.3.9}$$

with some constants $b = b(\varphi)$, $C_k = C_k(\varphi)$.

Denote by Z^b the set of all entire functions satisfying (3.3.9) with fixed $b > 0$ and define norms in Z^b as

$$\|\varphi\|_r = \sup_{k \leq r} \sup_{z \in \mathbb{C}} |z^k \varphi(z)| e^{-b|y|}, \qquad r \in \mathbb{N}.$$

The space Z^b with the system of norms is a perfect countably normed space. Convergence to zero of a sequence φ_n in Z^b means that $\varphi_n(\cdot)$ converges uniformly on every bounded segment of the real axis $|x| \leq x_0 < \infty$ and the sequence of norms $\|\varphi_n\|_r$ is bounded for any $r \in \mathbb{N}$.

A set $\mathcal{B}$ is bounded in Z^b if for any $k \in \mathbb{N}_0$ the inequality (3.3.9) holds for all elements of $\mathcal{B}$.

The embedding $Z^{b_1} \subset Z^{b_2}$ holds if $b_1 < b_2$ and

$$Z = \bigcup_b Z^b,$$

98   3. Examples. Supplements

which provides the definition of convergence and of bounded sets in Z.

The space $\mathcal{S}^\beta$. The space $\mathcal{S}^\beta$ ($\beta \geq 0$) consists of all infinitely differentiable functions $\varphi(x)$, $x \in \mathbb{R}$ satisfying the inequalities

$$|x^k \varphi^{(q)}(x)| \leq C_k B^q q^{q\beta}, \qquad k, q \in \mathbb{N}_0, \quad x \in \mathbb{R},$$

with some constants $B = B(\varphi)$, $C_k = C_k(\varphi)$. (For $q = 0$ we suppose $q^{q\beta} = 1$.)

This definition imposes certain restrictions on the growth of derivatives: the smaller β is, the stronger they are. Every function from $\mathcal{S}^\beta$ with $\beta < 1$ can be expanded analytically to the complex plane as an entire function satisfying the condition

$$|x^k \varphi(x + iy)| \leq C_{k,\rho} e^{(b+\rho)|y|^{\frac{1}{1-\beta}}}, \qquad k \in \mathbb{N}_0, \quad x + iy \in \mathbb{C}, \qquad (3.3.10)$$

for any $\rho > 0$, where $C_k' = C_k'(\varphi)$, $b = b(\varphi) = \frac{1-\beta}{e}(Be)^{\frac{1}{1-\beta}}$ [37]. The converse is true: if an entire function φ satisfies the inequality

$$|x^k \varphi(x + iy)| \leq C_k e^{b|y|^\gamma}, \qquad \gamma > 1, \quad k \in \mathbb{N}_0, \quad x + iy \in \mathbb{C},$$

then for any $\rho > 0$

$$|x^k \varphi^{(q)}(x)| \leq C_{k,\rho} B^q q^{q\beta}, \quad x \in \mathbb{R}, \quad \beta = 1 - 1/\gamma, \quad B = \frac{1}{e}\left((b+\rho)e\gamma\right)^{\frac{1}{\gamma}}.$$

It is easy to derive from (3.3.10) that the space $\mathcal{S}^\beta$ with $\beta = 0$ coincides with Z described above; therefore, further consideration of $\mathcal{S}^\beta$ will be carried out for $\beta > 0$. We present the space $\mathcal{S}^\beta$ as the countable union of normed spaces $\mathcal{S}^{\beta,B}$, which are defined as spaces of all infinitely differentiable functions satisfying the condition

$$|x^k \varphi^{(q)}(x)| \leq C_{k,\delta}(B + \delta)^q q^{q\beta}, \qquad k, q \in \mathbb{N}_0, \quad x \in \mathbb{R}, \qquad (3.3.11)$$

for any $\delta > 0$ with some constant $C_{k,\delta} = C_{k,\delta}(\varphi)$ and the norms

$$\|\varphi\|_{k,m} = \sup_{q \in \mathbb{N}_0} \sup_{x \in \mathbb{R}} \frac{|x^k \varphi^{(q)}(x)|}{\left(B + \frac{1}{m}\right)^q q^{q\beta}}, \qquad k \in \mathbb{N}_0, \quad m \in \mathbb{N}.$$

The space $\mathcal{S}^{\beta,B}$ with the system of norms is a perfect countably normed space. Convergence to zero of a sequence $\varphi_n \in \mathcal{S}^{\beta,B}$ means that for any $q \in \mathbb{N}_0$ the sequence of functions $\varphi_n^{(q)}$ converges to zero uniformly on any bounded segment and for any $k \in \mathbb{N}_0$, $m \in \mathbb{N}$ the collection of norms $\|\varphi_n\|_{k,m}$ is bounded (with the constant $C_{k,m}$).

A set $\mathcal{B}$ is bounded in $\mathcal{S}^{\beta,B}$ if, for any $k \in \mathbb{N}_0$, $m \in \mathbb{N}$ the inequality (3.3.11), where $\delta = 1/m$, holds with some constant $C_{k,m}$ independent of the elements of $\mathcal{B}$. It follows from the definition of $\mathcal{S}^{\beta,B}$ that

$$\mathcal{S}^{\beta,B_1} \subset \mathcal{S}^{\beta,B_2} \quad \text{for} \quad B_1 < B_2, \quad \text{and} \quad \mathcal{S}^\beta = \bigcup_B \mathcal{S}^{\beta,B},$$

3.3. Definitions and properties of spaces of test functions

where the topology in S^β is defined as the inductive limit topology. This allows us to introduce convergence in S^β as follows: $\varphi_n \to 0$ in S^β if the sequence $\varphi_n(\cdot)$ and all the sequences of its derivatives $\varphi_n^{(q)}(\cdot)$ converge uniformly to zero on any bounded segment.

A set $\mathcal{B}$ is bounded in S^β if it is bounded in some $S^{\beta,B}$.

The space S_α^β. The space S_α^β ($\alpha \geq 0$, $\beta \geq 0$) consists of all infinitely differentiable functions $\varphi(x)$, $x \in \mathbb{R}$, satisfying the inequalities

$$|x^k \varphi^{(q)}(x)| \leq C A^k B^q k^{k\alpha} q^{q\beta}, \qquad k, q \in \mathbb{N}_0, \quad x \in \mathbb{R},$$

with constants $A = A(\varphi)$, $B = B(\varphi)$, $C = C(\varphi)$. This definition imposes restrictions on both the rate of decrease of the functions as $|x| \to \infty$ and on the rate of growth of their derivatives. Therefore, the question of the existence of such functions arises. The answer is the following [37]: The spaces S_α^β are non-trivial only in the three cases below.

$$
\begin{array}{lll}
1) & \alpha + \beta \geq 1, & \alpha > 0, \ \beta > 0; \\
2) & \beta > 1, & \alpha = 0; \\
3) & \alpha > 1, & \beta = 0.
\end{array}
$$

We denote by $S_{\alpha,A}^{\beta,B}$ the space of all infinitely differentiable functions satisfying the condition

$$|x^k \varphi^{(q)}(x)| \leq C_{\varepsilon,\delta}(A+\varepsilon)^k (B+\delta)^q k^{k\alpha} q^{q\beta}, \qquad k, q \in \mathbb{N}_0, \quad x \in \mathbb{R}, \quad (3.3.12)$$

for any $\varepsilon > 0, \delta > 0$ with a constant $C_{\varepsilon,\delta} = C_{\varepsilon,\delta}(\varphi)$ and the norms

$$\|\varphi\|_{p,m} = \sup_{k \in \mathbb{N}_0} \ \sup_{q \in \mathbb{N}_0} \ \sup_{x \in \mathbb{R}} \ \frac{|x^k \varphi^{(q)}(x)|}{\left(A+\frac{1}{p}\right)^k \left(B+\frac{1}{m}\right)^q k^{k\alpha} q^{q\beta}}, \qquad p, m \in \mathbb{N}.$$

The space $S_{\alpha,A}^{\beta,B}$ with this system of norms is a perfect countably normed space.

A set $\mathcal{B}$ is bounded in $S_{\alpha,A}^{\beta,B}$ if, for any $p, m \in \mathbb{N}$, the condition (3.3.12) holds with a certain constant $C_{p,m}$ independent of the elements of $\mathcal{B}$. A set $\mathcal{B}$ is bounded in S_α^β if it is bounded in some $S_{\alpha,A}^{\beta,B}$.

Convergence to zero of a sequence $\varphi_n \in S_{\alpha,A}^{\beta,B}$ means that it is bounded in $S_{\alpha,A}^{\beta,B}$ and for any $q \in \mathbb{N}_0$ the sequence $\varphi_n^{(q)}(\cdot)$ tends uniformly to zero on every bounded segment of the real axis.

For $A_1 < A_2$, $B_1 < B_2$ the embedding $S_{\alpha,A_1}^{\beta,B_1} \subset S_{\alpha,A_2}^{\beta,B_2}$ holds and the space S_α^β is the union of perfect countably normed spaces $S_{\alpha,A}^{\beta,B}$. A sequence $\varphi_n \in S_\alpha^\beta$ converges to zero if all functions $\varphi_n(\cdot)$ belong to a common space $S_{\alpha,A}^{\beta,B}$ and converge to zero in this space.

The space S. The space S consists of all infinitely differentiable functions $\varphi(x)$, $x \in \mathbb{R}$, satisfying the inequalities

$$|x^k \varphi^{(q)}(x)| \leq C_{k,q}, \qquad k, q \in \mathbb{N}_0, \quad x \in \mathbb{R}, \quad (3.3.13)$$

100   3. Examples. Supplements

with some constant $C_{k,q} = C_{k,q}(\varphi)$, that is, $\mathcal{S}$ is the space of infinitely differentiable functions decreasing faster than any degree of $\frac{1}{|x|}$ does as $|x| \to \infty$.

The space $\mathcal{S}$ is a perfect countably normed space with the system of norms

$$\|\varphi\|_p = \sup_{k,q \leq p} \sup_{x \in \mathbb{R}} |x^k \varphi^{(q)}(x)|, \quad p \in \mathbb{N}.$$

Convergence to zero of a sequence $\varphi_n \in S$ means that for any $q, k \in \mathbb{N}_0$ the inequality (3.3.13) holds with some constants $C_{k,q}$ independent of the numbers n and φ_n and the sequence of all its derivatives $\varphi_n^{(q)}$ converges uniformly to zero on any bounded segment.

A set $\mathcal{B}$ is bounded in $\mathcal{S}$ if for any $q, k \in \mathbb{N}_0$ there exists a constant $C_{k,q}$ independent of the elements of $\mathcal{B}$ such that (3.3.13) holds. The space $\mathcal{S}$ is the widest of all the spaces considered above. Due to its properties the space $\mathcal{S}$ can be regarded as the limit of spaces $\mathcal{S}_\alpha^\beta$, i.e., $\mathcal{S} = \mathcal{S}_\infty^\infty$.

Note that in [37] some extensions of spaces $\mathcal{S}_\alpha^\beta$ were introduced by replacing the sequences $k^{k\alpha}$ and $q^{q\beta}$ with the more general a_k, b_q. The Roumieu and Beurling spaces, where b_q are denoted by M_q, provide such an extension of the spaces $\mathcal{S}_0^\beta$. Conditions imposed on the sequences $\{M_q\}$ guarantee non-triviality of the spaces, boundedness of infinite order differential operators, and other important properties.

The Roumieu and Beurling spaces of ultra-differentiable functions. According to [54], we consider a sequence of positive numbers M_q, $q \in \mathbb{N}_0$ satisfying the conditions

(M.1) $M_q^2 \leq M_{q-1} M_{q+1}, \quad q - 1 \in \mathbb{N};$

(M.2) $M_q \leq a\, b^q \min_{0 \leq p \leq q} M_p M_{q-p}, \quad q \in \mathbb{N}_0; \quad (\,(\mathbf{M.2})'\ M_{q+1} \leq a\, b^q M_q\,);$

(M.3) $\displaystyle\sum_{p=q+1}^{\infty} \frac{M_{p-1}}{M_p} \leq q\, c\, \frac{M_q}{M_{q+1}}, \quad q \in \mathbb{N}; \quad (\,(\mathbf{M.3})'\ \displaystyle\sum_{q=1}^{\infty} \frac{M_{q-1}}{M_q} < \infty\,);$

with some positive constants a, b, and c.

The Gevrey sequences

$$(q!)^\beta, \quad q^{q\beta}, \quad \Gamma(1 + q\beta) \quad \text{with } \beta > 1$$

provide the example of sequences M_q satisfying all these conditions.

The function

$$M(x) := \sup_{q \in \mathbb{N}_0} \ln \frac{x^q M_0}{M_q}, \quad x > 0,$$

is called *a function associated with* M_q. The following equality gives an equivalent definition of the function associated with M_q:

$$e^{-M(x)} = \inf_{q \in \mathbb{N}_0} \frac{M_q}{M_0 x^q}, \quad x > 0.$$

3.3. Definitions and properties of spaces of test functions   101

Compared with (3.3.5)–(3.3.6), where $M_q = q^{q\alpha}$, $M_0 = 1$, we have the equality $e^{-M(x)} = \mu_\alpha(x)$, which shows the connection of the functions $M(\cdot)$ and $\mu_\alpha(\cdot)$.

Associated functions play an important role in the theory of ultra-differentiable functions; in particular, they specify the behavior of the Fourier transform in these spaces.

If the ratio $(M_q/M_0)^{1/q}$ is bounded from below, then M is an increasing, logarithmic convex function equal to zero in a certain neighborhood of zero and growing faster than $\ln x^p$ with $p > 0$ as $x \to \infty$.

The function $M(x) \sim \frac{\beta}{e}|x|^{1/\beta}$ is the associated function for the sequence $M_q = q^{q\beta}$ and $M(x) \sim |x|^{1/\beta}$ for the sequence $M_q = (q!)^\beta$ [14].

The Roumieu space $\mathcal{D}^{\{M_q\}}$ consists of all infinitely differentiable functions $\varphi(x)$, $x \in \mathbb{R}$, with compact supports satisfying the inequality

$$|\varphi^{(q)}(x)| \leq CB^q M_q, \qquad q \in \mathbb{N}_0, \quad |x| \leq A,$$

with some constants $A = A(\varphi)$, $B = B(\varphi)$, $C = C(\varphi)$.

According to the Denjoy–Carleman–Mandelbrojt theorem (see the proof, e.g., in [54, 65]), this space is nontrivial if M_q satisfies (M.1) and (M.3)$'$.

One can see that the Roumieu space $\mathcal{D}^{\{M_q\}}$ provides the extension of the spaces $\mathcal{S}_0^\beta$. However, being the space of functions with compact supports, it is a particular case of the space $\mathcal{S}_{a_k}^{b_q}$ [37]. Therefore, we will not consider the structure of this space in detail here. We just note that it is the strong inductive limit of perfect countably normed spaces $\mathcal{D}_A^{\{M_q\},B}$ with the system of norms

$$\|\varphi\|_m = \sup_{q \in \mathbb{N}_0} \sup_{|x| \leq A} \frac{|\varphi^{(q)}(x)|}{\left(B + \frac{1}{m}\right)^q M_q}, \qquad m \in \mathbb{N}, \tag{3.3.14}$$

that is,

$$\mathcal{D}^{\{M_q\}} = \bigcup_{A,B} \mathcal{D}_A^{\{M_q\},B}$$

with the corresponding definition of convergence and notion of a bounded set.

The space of ultra-differentiable functions of Beurling class defined by $\{M_q\}$ is the space

$$\mathcal{D}_{\{M_q\}} = ind \lim_{A \in \mathbb{R}} proj \lim_{h \to 0} \mathcal{D}_{\{M_q\},h,A},$$

where $\mathcal{D}_{\{M_q\},h,A}$ is a normed space of functions $\varphi \in C^\infty(\mathbb{R})$ with compact supports A satisfying the inequalities $\|\varphi^{(q)}\|_{C(A)} \leq CM_q h^q$ with the norm

$$\|\varphi\|_{\{M_q\},h,A} = \sup_q \left(\frac{\|\varphi^{(q)}\|_{C(A)}}{M_q h^q} \right)$$

and the corresponding space of abstract ultra-distributions

$$\mathcal{D}'_{\{M_q\}}(H) := \mathcal{L}(\mathcal{D}_{\{M_q\}}, H).$$

102   *3. Examples. Supplements*

The space W_M. Let $\mu(\cdot)$ be an increasing continuous function on $[0, \infty)$ and $\mu(0) = 0$, $\lim_{\xi \to \infty} \mu(\xi) = \infty$. Let

$$M(x) = \int_0^x \mu(\xi)\, d\xi, \qquad x \geq 0. \tag{3.3.15}$$

Then $M(\cdot)$ increases at infinity faster than any linear function and is convex:

$$M(x_1) + M(x_2) \leq M(x_1 + x_2), \qquad x_1, x_2 \geq 0.$$

Set $M(x) := M(-x)$, $x < 0$.

The space W_M consists of all infinitely differentiable functions $\varphi(x)$, $x \in \mathbb{R}$, satisfying the condition

$$|\varphi^{(q)}(x)| \leq C_q e^{-M(ax)}, \qquad x \in \mathbb{R},$$

with some constants $C_q = C_q(\varphi)$, $a = a(\varphi)$. Since $M(x)$ increases as $|x| \to \infty$ faster than any linear function, the functions from W_M decrease at infinity faster than any exponent of type $e^{-a|x|}$.

The space W_M can be defined as the union of countably normed spaces. Denote by $W_{M,a}$, $a > 0$, the space of all infinitely differentiable functions $\varphi(\cdot)$ in $\mathbb{R}$ which satisfy the inequalities

$$\left|\varphi^{(q)}(x)\right| \leq C_{q,\delta}\, e^{-M((a-\delta)x)}, \qquad x \in \mathbb{R}, \quad q \in \mathbb{N}_0,$$

for any $\delta > 0$. The space $W_{M,a}$ with the system of norms

$$\|\varphi\|_p = \sup_{q \leq p} \sup_{x \in \mathbb{R}} \left|\varphi^{(q)}(x)\right| e^{M\left(\left(a - \frac{1}{p}\right)x\right)}, \qquad p \in \mathbb{N},$$

is complete countably normed. Similar to the spaces introduced above, this defines bounded sets and convergence to zero of a sequence φ_n in $W_{M,a}$. This also applies to W_M, which is defined as the union of countably normed spaces:

$$W_M = \bigcup_a W_{M,a},$$

where the topology is the strong inductive limit topology.

The space W_M has a non-empty intersection with S_α: the space S_α is W_M with $M(x) = |x|^{1/\alpha}$, $\alpha < 1$.

The space W^Ω. Let ω be an increasing continuous function on $[0, \infty)$ and let $\omega(0) = 0$, $\lim_{\eta \to \infty} \omega(\eta) = \infty$. Suppose

$$\Omega(y) = \int_0^y \omega(\eta)\, d\eta, \quad y \geq 0, \qquad \Omega(y) := \Omega(-y), \qquad y < 0. \tag{3.3.16}$$

The properties of the function $\Omega(\cdot)$ are similar to those of the function $M(\cdot)$

3.3. Definitions and properties of spaces of test functions 103

introduced in the previous subsection. It increases at infinity faster than any linear function and is convex.

The space W^Ω consists of all entire functions $\varphi(z)$, $z \in \mathbb{C}$, satisfying the inequality

$$|z^k \varphi(z)| \le C_k e^{\Omega(by)}, \qquad z = x + iy \in \mathbb{C},$$

with some constants $C_k = C_k(\varphi)$, $b = b(\varphi)$.

The space W^Ω can be defined as a union of countably normed spaces. Denote by $W^{\Omega, b}$, $b > 0$, the set of all entire functions φ satisfying the inequalities

$$|z^k \varphi(z)| \le C_{k,\rho} e^{\Omega((b+\rho)y)}, \qquad z = x + iy \in \mathbb{C}, \quad k \in \mathbb{N}_0,$$

for any $\rho > 0$ endowed with the norms

$$\|\varphi\|_{k,m} = \sup_{z \in \mathbb{C}} |z^k \varphi(z)| \, e^{-\Omega((b+\frac{1}{m})y)}, \qquad k \in \mathbb{N}_0, \quad m \in \mathbb{N}.$$

The space is perfect countably normed.

The definition implies that the imbedding $W^{\Omega, b_1} \subset W^{\Omega, b_2}$ holds true if $b_1 < b_2$ and

$$W^\Omega = \bigcup_b W^{\Omega, b},$$

where the topology is the strong inductive limit topology. This allows us to introduce bounded sets and convergence as follows. A sequence $\varphi_n \in W^\Omega$ is convergent to zero if all functions $\varphi_n(\cdot)$ belong to some space $W^{\Omega, b}$ and the sequence converges in this space. A set $\mathcal{B}$ is bounded in W^Ω if it is bounded in some $W^{\Omega, b}$.

The space W_M^Ω. The space consists of all entire functions φ of variable $z \in \mathbb{C}$ satisfying the condition

$$|z^k \varphi^{(q)}(z)| \le C e^{-M(ax)+\Omega(by)}, \qquad z = x + iy \in \mathbb{C},$$

with some constants $C = C_{k,q}(\varphi)$, $a = a_{k,q}(\varphi)$, $b = b_{k,q}(\varphi)$ and can be defined as the union of countably normed perfect spaces $W_{M,a}^{\Omega, b}$, $a > 0$, $b > 0$:

$$W_M^\Omega = \bigcup_{a,b} W_{M,a}^{\Omega, b},$$

where the topology is the strong inductive limit topology. This allows one to introduce the convergence and bounded sets in the space.

Recall bounded operators of multiplication in some of the spaces.

- *In the space* $\mathcal{S}_0 = \mathcal{D}$ any infinitely differentiable function f_0 defines the bounded operator of multiplication by f_0.

- *In the space* Z any entire function $f_0(\cdot)$ satisfying the inequality for some $C > 0$, $b > 0$, $h \ge 0$

$$|f_0(z)| \le C(1 + |z|)^h e^{b|y|}, \qquad q \in \mathbb{N}_0,$$

  defines the bounded multiplication operator.

- *In the space $\mathcal{S}_\alpha^\beta$, $\alpha > 0$, every infinitely differentiable function f_0 satisfying inequality for any $\varepsilon > 0$*

$$|f_0^{(q)}(x)| \le C_\varepsilon \varepsilon^q q^{q\beta} e^{\varepsilon |x|^{1/\alpha}}, \qquad q \in \mathbb{N}_0,$$

  defines the bounded multiplication operator; it transforms the space $\mathcal{S}_{\alpha,A}^{\beta,B}$ to itself.

- $\mathcal{S}_{\alpha,A}^{\beta,B} \to \mathcal{S}_{\alpha,A_1}^{\beta,B+B_0}$. Let a function f_0 satisfy the inequalities

$$|f_0^{(q)}(x)| \le CB_0^q q^{q\beta} e^{a_0 |x|^{1/\alpha}}, \qquad \alpha > 0, \quad q \in \mathbb{N}_0. \tag{3.3.17}$$

  For arbitrary $a > a_0$ put $A = \left(\frac{\alpha}{ea}\right)^\alpha$. Then, denote $a_1 = a - a_0$ and set $A_1 = \left(\frac{\alpha}{ea_1}\right)^\alpha$. Let $B > 0$. The function f_0 defines a multiplier on the space $\mathcal{S}_{\alpha,A}^{\beta,B}$ to $\mathcal{S}_{\alpha,A_1}^{\beta,B+B_0} \supset \mathcal{S}_{\alpha,A}^{\beta,B}$.

- *In the space $\mathcal{S}$ every infinitely differentiable function f_0 satisfying the inequalities*

$$|f_0^{(q)}(x)| \le C_q (1 + |x|)^{h_q}, \qquad q \in \mathbb{N}_0, \tag{3.3.18}$$

  defines a bounded multiplication operator.

- $W_{M,a}^{\Omega,b} \to W_{M,a-a_0}^{\Omega,b+b_0}$.

  Every entire function f_0 satisfying the inequality

$$|f_0(z)| \le C e^{M(a_0 x) + \Omega(b_0 y)}, \qquad z = x + iy \in \mathbb{C}, \tag{3.3.19}$$

  defines a bounded multiplication operator on $W_{M,a}^{\Omega,b}$ to $W_{M,a-a_0}^{\Omega,b+b_0}$ for $a > a_0$, $b > 0$.

3.4 Generalized Fourier and Laplace transforms. Structure theorems

Let f be an absolutely integrable function and $\widetilde{f}$ (or $\mathcal{F}f$, $\mathcal{F}[f]$) be its classical Fourier transform:

$$\widetilde{f}(\sigma) = \int_{-\infty}^{+\infty} e^{ix\sigma} f(x) \, dx, \qquad \sigma \in \mathbb{R}.$$

The basis of the definition of the generalized Fourier transform is the Parseval equality for the Fourier transform in $L^2(\mathbb{R})$ (see, e.g., [53, 89]), where it has the form of equality for scalar products

$$\langle \mathcal{F}[\varphi], \mathcal{F}[f] \rangle = 2\pi \langle \varphi, f \rangle, \qquad f, \varphi \in L^2(\mathbb{R}).$$

3.4. Generalized Fourier and Laplace transforms. Structure theorems   105

The equality may be used for functions φ from a space Φ of test functions and $f \in \Phi'$:

$$\langle \mathcal{F}[\varphi], \mathcal{F}[f] \rangle = 2\pi \langle \varphi, f \rangle, \qquad \varphi \in \Phi,$$

and it serves as the definition of the generalized Fourier transform and the inverse Fourier transform:

$$\langle \psi, \mathcal{F}f \rangle = 2\pi \langle \mathcal{F}^{-1}\psi, f \rangle, \qquad \psi \in \widetilde{\Phi}, \ \ f \in \Phi'.$$
$$\langle \varphi, \mathcal{F}^{-1}g \rangle := \frac{1}{2\pi} \langle \mathcal{F}\varphi, g \rangle, \qquad \varphi \in \Phi, \ \ g \in \widetilde{\Phi}'. \tag{3.4.1}$$

Thus, in the case when the Fourier transform and the inverse Fourier transform perform one-to-one mappings between Φ and $\widetilde{\Phi}$, we have the following embeddings:

$$\widetilde{\Phi'} \subseteq \left(\widetilde{\Phi} \right)' \qquad \text{and} \qquad \left(\widetilde{\Phi} \right)' \subseteq \widetilde{\Phi'}.$$

Therefore, since the classical Fourier transform performs a one-to-one mapping of Φ onto $\widetilde{\Phi}$, then $\left(\widetilde{\Phi} \right)' = \widetilde{\Phi'}$.

The definition (3.4.1) and properties of the classical Fourier transform provide the following rules of differentiation for the generalized Fourier transform, which are valid in the classical theory:

$$\frac{d}{d\sigma} \widetilde{f}(\sigma) = \mathcal{F}[ixf(x)](\sigma), \qquad \mathcal{F}\left[i\frac{d}{dx}f(x) \right](\sigma) = \sigma \widetilde{f}(\sigma),$$

and the formulas for the Fourier transform of convolution of distributions:

$$\mathcal{F}[f * g] = \mathcal{F}f \cdot \mathcal{F}g.$$

The Fourier transformed spaces of test functions. For any $\alpha \geq 0, \beta \geq 0$ we have [37]:

$$\widetilde{S_\alpha} = S^\alpha, \quad \widetilde{S^\beta} = S_\beta, \quad \widetilde{S^\beta_\alpha} = S^\alpha_\beta, \quad \widetilde{S_{\alpha,A}} = S^{\alpha,A}, \quad \widetilde{S^{\beta,B}} = S_{\beta,B};$$

for $\alpha, \ \beta > 0, \ \alpha + \beta > 1$

$$\widetilde{S^{\beta,B}_{\alpha,A}} = S^{\alpha,A}_{\beta,B};$$

for $\alpha = 0, \ \beta = 0$

$$\widetilde{S_0} = \widetilde{D} = Z, \qquad \widetilde{S^0} = \widetilde{Z} = \mathcal{D}.$$

In order to describe the Fourier transform in the spaces W^Ω_M we recall the definition of dual by Young functions. The functions $M(\cdot)$ and $\Omega(\cdot)$ given by (3.3.15)–(3.3.16) are called *dual by Young* if the functions $\mu(\cdot)$ and $\omega(\cdot)$ are reciprocal, i.e., $\mu(\omega(\eta)) = \eta$ and $\omega(\mu(\xi)) = \xi$ for any $\xi, \eta \in \mathbb{R}$. In this case the Young inequality holds:

$$\nu \vartheta \leqslant M(\nu) + \Omega(\vartheta), \qquad \nu, \vartheta \in \mathbb{R}.$$

106   *3. Examples. Supplements*

The following duality result for the Fourier transform in the spaces $W_{M,a}$ and $W^{\Omega,b}$ follows from it. If $M(\cdot)$ and $\Omega(\cdot)$ are functions dual by Young, then

$$\widetilde{W}_{M,a} = W^{\Omega,1/a}, \qquad \widetilde{W}^{\Omega,b} = W_{M,1/b}.$$

If $\Omega_2(\cdot)$ and $M_2(\cdot)$ are functions dual by Young with functions $M_1(\cdot)$ and $\Omega_1(\cdot)$, respectively, then

$$\widetilde{W}_{M_1}^{\Omega_1} = W_{M_2}^{\Omega_2}.$$

Generalized Laplace transform. We define the Laplace transform of distributions with supports in the positive semiaxis. As usual, an equality obtained for classical functions and correct in the generalized sense will serve as the basis of this definition.

Let f be an exponentially bounded function ($|f(t)| \leq Ce^{\omega t}$) equal to zero at $t < 0$ and let $\mathcal{L}f$ be its classical Laplace transform:

$$\mathcal{L}f(\lambda) = \int_0^{+\infty} e^{-\lambda t} f(t)\, dt, \qquad Re\lambda > \omega.$$

Using the substitution $-\lambda = is = i(\sigma + ib)$, we obtain the relation between the Laplace and the Fourier transforms:

$$g(\lambda) = \mathcal{L}f(\lambda) = \int_{-\infty}^{+\infty} e^{i\sigma t} e^{-bt} f(t)\, dt = \mathcal{F}[e^{-bt} f(t)](\sigma).$$

Hence, according to the definition of the inverse Fourier transform, we obtain

$$\mathcal{L}^{-1}g(t) = \frac{i}{2\pi} \int_{b+i\infty}^{b-i\infty} e^{(b-i\sigma)t} g(\lambda)\, d\lambda = \frac{1}{2\pi i} \int_{b-i\infty}^{b+i\infty} e^{\lambda t} g(\lambda)\, d\lambda, \quad b > \omega.$$

Let φ be an element of a space Φ of test functions which are exponentially bounded. We denote the classical Laplace transform of φ by $\psi = \mathcal{L}\varphi$. Using the properties of scalar products and the definition of the Fourier transform for real-valued functions φ and f, we obtain

$$
\begin{aligned}
\langle \varphi, f \rangle = \langle f, \varphi \rangle &= \int_0^{+\infty} f(t)\overline{\varphi(t)}\, dt \\
&= \frac{1}{2\pi i} \int_0^{+\infty} \int_{b-i\infty}^{b+i\infty} e^{\lambda t} \mathcal{L}f(\lambda)\overline{\varphi(t)}\, d\lambda\, dt \\
&= \frac{1}{2\pi i} \int_{b-i\infty}^{b+i\infty} \mathcal{L}f(\lambda) \left(\int_0^{+\infty} e^{\overline{\lambda} t}\varphi(t)\, dt \right) d\lambda \\
&= \frac{1}{2\pi i} \int_{b-i\infty}^{b+i\infty} \mathcal{L}f(\lambda)\overline{\mathcal{L}\varphi(-\overline{\lambda})}\, d\lambda = \frac{1}{2\pi} \langle \psi(-\overline{\lambda}), \mathcal{L}f(\lambda) \rangle.
\end{aligned}
$$

Let now f be an element of a distribution space Φ'. If Φ is such that the set $\mathcal{L}\Phi$ forms a space Ψ of test functions, then for any $\psi \in \Psi$ there exists the

3.4. Generalized Fourier and Laplace transforms. Structure theorems

inverse Laplace transform. Therefore, the Laplace transform defines one-to-one correspondence between Φ and Ψ and the equality obtained provides the definition of the generalized Laplace transform as a functional on Ψ:

$$\langle \psi(-\overline{\lambda}), \mathcal{L}f(\lambda) \rangle := 2\pi \langle \mathcal{L}^{-1}\psi, f \rangle, \quad \overline{\lambda} = b + i\sigma, \quad b > \omega, \quad \psi \in \Psi.$$

The definition obtained is consistent with that introduced by Fattorini in [30] for the spaces $\mathcal{S}'_\omega(\mathcal{X})$. The generalized Laplace transform of a distribution f in this space is defined by applying f to the test function $\{e^{-\lambda t}\}$ (see Section 2.1):

$$(\mathcal{L}f)(\lambda) := \langle \{e^{-\lambda t}\}, f(t) \rangle, \quad Re\lambda > \omega.$$

The generalized Laplace transform, similarly to the classical one, possesses the following properties for $Re\lambda > \omega$:

$$\frac{d}{d\lambda}\mathcal{L}u(\lambda) = -\mathcal{L}(tu(t));$$

$$\left(\mathcal{L}\frac{d}{dt}u(t)\right)(\lambda) = \lambda\mathcal{L}u(\lambda) - u(0);$$

$$\mathcal{L}[u * v](\lambda) = \mathcal{L}u\mathcal{L}v.$$

Structure theorems for distributions and analytical functions. Traditionally, the structure theorems represent a distribution as the result of applying a differentiation operator (of finite or infinite order) to some continuous function. For example, the structure theorem in $\mathcal{D}'$ states that locally for any $f \in \mathcal{D}'$ there exists a continuous primitive of a finite order, and the structure theorem in $\mathcal{S}'$ states that for any $f \in \mathcal{S}'$ there exists a continuous primitive of a finite order at the whole axis (see, for example, [12]).

In the present section we give structure theorems for distributions and ultra-distributions. In addition, we give a few theorems which describe the behavior in the complex plane of analytical functions defined on the real axis. We call them "the structure theorems" as well.

The structure theorem for the Schwartz space of distributions is the following.

Theorem 3.4.1 (The structure theorem in $\mathcal{D}'$) *Let $u \in \mathcal{D}'$ and let G be an open bounded set in $\mathbb{R}$. Then there exist a continuous function $g : \mathbb{R} \to \mathbb{R}$ and an integer $m > 0$ such that*

$$\langle \varphi, u \rangle = \langle \varphi, g^{(m)} \rangle \tag{3.4.2}$$

for all $\varphi \in \mathcal{D}$ with supp $\varphi \subset G$. If $u = 0$ on $(-\infty, a)$, then $g(t) = 0$ for $t < a$.

For each bounded set $\mathcal{B} \subset \mathcal{D}'$ and each open bounded set $G \subset \mathbb{R}$ there exists $m \in \mathbb{N}$ independent of the elements of $\mathcal{B}$ that provides the equality (3.4.2) for all $u \in \mathcal{B}$ and uniform boundedness of functions g on compact sets in $\mathbb{R}$.

The structure theorem in $\mathcal{S}'$ and in the space of exponentially bounded distributions $\mathcal{S}'_\omega$ has global character.

108   *3. Examples. Supplements*

Theorem 3.4.2 (The structure theorem in $\mathcal{S}'$ (in $\mathcal{S}'_\omega$)) *For any distribution $u \in \mathcal{S}'$ ($u \in \mathcal{S}'_\omega$) there exist $m \in \mathbb{N}$, $r > 0$ and a continuous function $g : \mathbb{R} \to \mathbb{R}$ such that*

$$\langle \varphi, u \rangle = \langle \varphi, g^{(m)} \rangle, \qquad \varphi \in \mathcal{S}, \tag{3.4.3}$$

and

$$\|g(t)\| \le C|t|^r \quad (\|g(t)e^{-\omega t}\| \le C|t|^r) \quad as \quad |t| \to \infty. \tag{3.4.4}$$

If $u = 0$ on $(-\infty, a)$, then $g(t) = 0$ for $t < a$.

For each bounded set $\mathcal{B} \subset \mathcal{S}'$ there exists $m \in \mathbb{N}$ independent of the elements of $\mathcal{B}$ that provides the equality (3.4.3) for all elements $u \in \mathcal{B}$; moreover, all the corresponding functions g ($g(t)e^{-\omega t}$) have the same order of growth $r > 0$ in (3.4.4).

The structure of distributions with compact supports is given in the following theorem.

Theorem 3.4.3 (Paley–Winner–Shwartz) [37] *If an entire function f satisfies the conditions*

$$|f(z)| \le C_1 e^{b|z|}, \quad z \in \mathbb{C}, \qquad |f(x)| \le C_2(1 + |x|^h), \quad x \in \mathbb{R},$$

then its Fourier transform $\widetilde{f}$ is the element of $\mathcal{D}'$ with support in $[-b, b]$. Moreover, for any $\varepsilon > 0$ there exist such an integrable function $g(\cdot)$ with support in $[-b - \varepsilon, b + \varepsilon]$ and such a polynomial $P_k(\cdot)$ of power $k \le h + 1$ that $\widetilde{f}(\sigma) = P_k(d/d\sigma)\, g(\sigma)$.

The converse result is also true.

Now we present structure theorems for spaces of ultra-distributions.

Theorem 3.4.4 (The first structure theorem for $\left(\mathcal{D}^{\{M_q\}}\right)'$) [54] *Let a sequence M_q satisfy (M.1) and (M.3)'. For each $u \in \left(\mathcal{D}^{\{M_q\}}\right)'$ there exists such a sequence $u_n \in C'(\mathbb{R})$ of integrable functions such that for any open bounded set $G \subset \mathbb{R}$ and any $B > 0$ there exists $C > 0$ such that*

$$\|u_n\|_{C'(\overline{G})} \le C \frac{B^n}{M_n}, \qquad n \in \mathbb{N}_0. \tag{3.4.5}$$

In this case the equality $u|_G = \sum_{n=0}^{\infty} D^n u_n$ holds in $\left(\mathcal{D}^{\{M_q\}}\right)'$.

The inverse is true: if the sequence of measures $u_n \in C'(\overline{G})$ satisfies (3.4.5), then the series in the definition of u converges absolutely in $\left(\mathcal{D}^{\{M_q\}}\right)'$.

Note that all the structure theorems listed above hold for functions from corresponding spaces of abstract distributions $\mathcal{D}'(\mathcal{X})$, $\mathcal{S}'(\mathcal{X})$, $\mathcal{S}'_\omega(\mathcal{X})$, $\left(\mathcal{D}^{\{M_q\}}\right)'(\mathcal{X})$ [30].

In Part II we apply the structure theorems to constructing generalized wrt t solutions of the stochastic Cauchy problems with A generating integrated and convoluted semi-groups and also to constructing generalized wrt x solutions with A generating an R-semi-group.

3.4. Generalized Fourier and Laplace transforms. Structure theorems   109

Theorem 3.4.5 *If an entire function f satisfies the conditions*

$$|f(z)| \leq C_1 e^{b|z|^p}, \quad z \in \mathbb{C}^n, \qquad |f(x)| \leq C_2 e^{a|x|^h}, \quad x \in \mathbb{R}^n,$$

where $0 < h \leq p$, $a \neq 0$, then for each $a' > a$ there exists a region

$$H_\mu = \{z = x + iy : |y| \leq c(1 + |x|)^\mu\}, \qquad \mu \geq 1 - (p - h), \quad c = c(b; a; a'),$$

where

$$|f(z)| \leq C_3 e^{a'|x|^h}, \qquad z \in H_\mu,$$

with $C_3 = \max\{C_1; C_2\}$. If $h < p$ and p is the exact growth order of f or if $h = p$ and $a < 0$, then $\mu \leq 1$.

Theorem 3.4.6 *If an entire function f satisfies the conditions*

$$|f(z)| \leq C_1 e^{b|z|^p}, \quad z \in \mathbb{C}^n, \qquad |f(x)| \leq C_2 (1 + |x|)^h, \quad x \in \mathbb{R}^n,$$

then there exists a region

$$H_\mu = \{z = x + iy : |y| \leq c(1 + |x|)^\mu\}, \qquad 1 - p \leq \mu \leq 1, \quad c = c(b; h),$$

where

$$|f(z)| \leq C_3 (1 + |x|)^h, \qquad z \in H_\mu,$$

with $C_3 = \max\{C_1; C_2\}$.

Theorem 3.4.7 *If an entire function f satisfies the condition*

$$|f(z)| \leq C_1 e^{b|z|^p}, \qquad z \in \mathbb{C}^n,$$

and the estimate

$$|f(z)| \leq C_2 e^{a|x|^h}, \qquad z = x + iy,$$

holds in the region $H_\mu = \{z = x + iy : |y| \leq c(1 + |x|)^\mu\}$, $0 < \mu \leq 1$, where $h < p$ or $h \leq p$ for $a < 0$. Then for all $z = x + iy$ the estimate holds

$$|f(z)| \leq C_3 e^{a|x|^h + b'|y|^{p/\mu}},$$

with $C_3 = \max\{C_1; C_2\}$, $b' = b'(b; a; K)$.

Theorem 3.4.8 *If an entire function f satisfies the condition*

$$|f(z)| \leq C_1 e^{b|z|^p}, \qquad z \in \mathbb{C}^n,$$

and the estimate

$$|f(z)| \leq C_2 (1 + |x|)^h,$$

holds in the region $H_\mu = \{z = x + iy : |y| \leq c(1 + |x|)^\mu\}$, $0 < \mu \leq 1$, then for all $z \in \mathbb{C}^n$

$$|f(z)| \leq C_3 (1 + |x|)^h e^{b'|y|^{p/\mu}},$$

with $C_3 = \max\{C_1; C_2\}$, $b' = b'(b; h; K)$.

110   *3. Examples. Supplements*

Theorem 3.4.9 *If an entire function f satisfies the condition*

$$|f(z)| \leq Ce^{a|x|^h + b'|y|^\gamma}, \qquad h \leq \gamma, \quad z = x + iy \in \mathbb{C}^n,$$

then the estimates for its derivatives

$$|f^{(q)}(x)| \leq C_1 B^{|q|} |q|^{|q|(1-\frac{1}{\gamma})} e^{a_1|x|^h}, \qquad q = (q_1, \ldots, q_n), \quad q_j \in \mathbb{N}_0,$$

hold on the real axis with $a_1 = a + \varepsilon$, $\varepsilon > 0$.

Theorem 3.4.10 *If an entire function f satisfies the condition*

$$|f(z)| \leq Ce^{a|x|^h}$$

in the region $H_\mu = \{z = x + iy : |y| \leq c(1 + |x|)^\mu\}$, $\mu \leq 0$, then there exists such $B > 0$, $B = B(a, \mu)$ that the estimates for its derivatives

$$|f^{(q)}(x)| \leq C_1 B^q q^{q(1-\frac{\mu}{h})} e^{a_1|x|^h}, \qquad q = (q_1, \ldots, q_n), \quad q_j \in \mathbb{N}_0,$$

hold on the real axis. Here $a_1 = a + \varepsilon$, $\varepsilon > 0$ and it is of the same sign as a.

Theorem 3.4.11 *If an entire function f satisfies the condition*

$$|f(z)| \leq C(1 + |x|)^h$$

in the region $H_\mu = \{z = x + iy : |y| \leq c(1 + |x|)^\mu\}$, $\mu \leq 0$, then there exists such $B > 0$, $B = B(\mu)$ that the estimates for its derivatives

$$|f^{(q)}(x)| \leq C_1 B^{|q|} |q|^{|q|} (1 + |x|)^{h - \mu q}, \qquad q = (q_1, \ldots, q_n), \quad q_j \in \mathbb{N}_0,$$

hold on the real axis.

Part II

Infinite-Dimensional Stochastic Cauchy Problems

Chapter 4

Weak, regularized, and mild solutions to Itô integrated stochastic Cauchy problems in Hilbert spaces

After the "deterministic" Part I containing necessary results from the semi-group, abstract distribution, and regularization theory, we proceed to the main theme of the book, infinite-dimensional stochastic problems (P.1)

$$X'(t) = AX(t) + F(t, X) + B(t, X)\mathbb{W}(t), \quad t \geq 0, \quad X(0) = \zeta.$$

with white noise processes and generators of regularized semi-groups.

As mentioned in the Introduction, one can mark out different approaches to overcoming obstacles caused by irregularity of the white noise in stochastic equations. In this chapter we address the first one, which uses infinite-dimensional extensions of the Itô calculus as a framework for solving (P.1). Under this approach one deals with Wiener processes considered as "primitives" of the white noise. The second approach considers the stochastic problems as differential equations with the white noise itself, but in a generalized statement. It will be addressed in Chapters 5 and 6.

On the basis of the extension of the Itô approach to Hilbert spaces, we study the integrated stochastic Cauchy problem (I.1):

$$dX(t) = AX(t)\,dt + F(t, X) + B(t, X)\,dW(t),$$
$$t \in [0, T], \quad X(0) = \zeta, \quad (4.0.1)$$

with a stochastic integral wrt a Wiener process W and A generating a regularized semi-group. For the linear and semi-linear case of (4.0.1) we explore weak, weak regularized, and mild solutions and their interrelations.

We begin this chapter with the preliminary section presenting necessary information on stochastic integrals and their properties.

113

4.1 Hilbert space-valued variables, processes, and stochastic integrals. Main properties and results

This preliminary section presents the necessary stochastic techniques for solving stochastic Cauchy problems in Hilbert spaces. First, random variables with values in Hilbert spaces are defined and their properties are presented. Then the notion of a Hilbert space valued stochastic process is introduced and cylindrical and Q-Wiener processes are studied. In this connection we recall the notions of nuclear, trace class, and Hilbert–Schmidt operators, which are widely used in studying Hilbert space valued stochastic processes and their probability characteristics. In conclusion, we define Hilbert space valued stochastic Itô integrals and present basic results on stochastic integrals necessary for the next sections.

4.1.1 Random variables in Hilbert spaces. Properties of operators related to probability characteristics

Along with the notion of a Hilbert space valued random variable, we introduce here such characteristics of random variables as expectation, correlation and covariance operators and study their properties. Particular attention is paid to special properties connected with infinite-dimensions. For a better understanding of these properties we recall necessary definitions of compact, Hilbert–Schmidt, trace class, and nuclear operators, as well as properties of these operators and relations between them. A connection of Hilbert space valued random variables with their distribution laws is shown via the comparison with the finite-dimensional case.

Let Ω be a set, $\mathcal{F}$ be a σ-algebra of subsets of Ω, and P be a probability measure on $\mathcal{F}$. Then the pair $(\Omega, \mathcal{F})$ is called a *measurable space* and the triplet $(\Omega, \mathcal{F}, P)$ is called a *probability space*.

Let $(\Omega, \mathcal{F})$ and $(\mathcal{X}, \mathcal{G})$ be measurable spaces. A mapping $u : \Omega \to \mathcal{X}$ such that

$$\{\omega \in \Omega : u(\omega) \in G\} \in \mathcal{F} \quad \text{for any} \quad G \in \mathcal{G}$$

is called a *random variable* on $(\Omega, \mathcal{F})$ with values in $(\mathcal{X}, \mathcal{G})$. A random variable u that can take only a finite number of values is referred to as a *simple random variable*.

Let $\mathcal{X} = H$ be a Hilbert space; then the smallest σ-algebra $\mathcal{B}(H)$ containing all open subsets of H is called the *Borel σ-algebra on H* and a $(\mathcal{F}, \mathcal{B}(H))$-measurable mapping $u : \Omega \to H$ is called an *H-valued random variable*.

Similar to the case of real-valued random variables, where the collection of sets

$$\{x \in \mathbb{R} : x \leq \alpha\}, \quad \alpha \in \mathbb{R},$$

generates the Borel σ-algebra $\mathcal{B}(\mathbb{R})$, the Borel σ-algebra $\mathcal{B}(H)$ is generated by

4.1. Hilbert space valued variables, processes, and stochastic integrals   115

the collection of sets

$$\{g \in H : \langle g, h \rangle_{\mathrm{H}} \leq \alpha\}, \quad h \in H, \ \alpha \in \mathbb{R}.$$

This implies that u is an H-valued random variable if and only if for each $h \in H$ the mapping $\langle u(\cdot), h \rangle_{\mathrm{H}} : \Omega \to \mathbb{R}$ is a real-valued random variable. Thus

$$\|u(\cdot)\|_{\mathrm{H}} = \langle u(\cdot), u(\cdot) \rangle_{\mathrm{H}}^{1/2} : \Omega \to \mathbb{R}$$

is a real-valued random variable for any H-valued random variable u.

To introduce the Bochner integral of an H-valued random variable we first define an integral for a simple H-valued random variable

$$u(\omega) = \sum_{i=1}^{N} x_i \chi_{G_i}(\omega), \qquad \omega \in \Omega, \ x_i \in H, \ G_i \in \mathcal{F},$$

as

$$\int_F u(\omega) P(d\omega) := \sum_{i=1}^{N} x_i P(G_i \cap F), \qquad F \in \mathcal{F},$$

where χ_G is the characteristic function of set G. In general, an H-valued random variable u is said to be *Bochner integrable* if $\int_\Omega \|u(\omega)\|_{\mathrm{H}} P(d\omega) < \infty$ and its *Bochner integral* is defined as

$$\int_\Omega u(\omega) P(d\omega) := \lim_{n \to \infty} \int_\Omega u_n(\omega) P(d\omega),$$

where $\{u_n\}$ is a sequence of simple random variables such that

$$\|u(\omega) - u_n(\omega)\|_{\mathrm{H}} \to 0 \quad \text{as} \quad n \to \infty \quad \text{for} \quad \omega \in \Omega \text{ a.s.}$$

and

$$\lim_{n \to \infty} \int_\Omega \|u(\omega) - u_n(\omega)\|_{\mathrm{H}} P(d\omega) \to 0.$$

Similar to the case of real-valued random variables (i.e., of real-valued measurable functions) it is not difficult to show that this definition is independent of the choice of the sequence (see, e.g., [53, 103]). We also note the straightforward inequality

$$\left\| \int_F u(\omega) P(d\omega) \right\|_{\mathrm{H}} \leq \int_F \|u(\omega)\| P(d\omega).$$

Any H-valued random variable u induces a measure on $(H, \mathcal{B}(H))$,

$$\mu(G) := P\{\omega : x = u(\omega) \in G\} \equiv \mathcal{L}_u(G), \qquad G \in \mathcal{B}(H),$$

which is called the *distribution law* (or simply the *distribution*) of u; if an

116   *4. Itô integrated stochastic Cauchy problems in Hilbert spaces*

H-valued random variable $u(\omega)$, $\omega \in \Omega$, is integrable, then (using different notations) we have the equalities

$$\int_\Omega u(\omega)P(d\omega) = \int_\Omega u(\omega)dP(\omega) = \int_H x\mu(dx) = \int_H x\mathcal{L}_u(dx) = \int_H x\,d\mathcal{L}_u(x).$$

Conversely, if μ is a measure on $(H, \mathcal{B}(H))$, then we can construct a probability space $(\Omega, \mathcal{F}, P)$ and an H-valued random variable u on this space with the distribution law equal to μ.

If μ is a probability measure on $(H, \mathcal{B}(H))$, then its *characteristic function* $\hat{\mu}(\lambda) : H \to \mathbb{C}$ is defined as the Fourier transform

$$\hat{\mu}(\lambda) := \int_H e^{i\langle\lambda,x\rangle_{\text{H}}}\mu(dx), \quad \lambda \in H.$$

Note that if μ and μ_1 are two probability measures on $(H, \mathcal{B}(H))$ such that $\hat{\mu} = \hat{\mu}_1$, then $\mu = \mu_1$ on $(H, \mathcal{B}(H))$.

Let $L^p(\Omega, \mathcal{F}, P; H) \equiv L^p(\Omega; H)$, $p \geq 1$, be the space of all equivalence classes of p-order-integrable H-valued random variables on $(\Omega, \mathcal{F}, P)$ with the norm

$$\|u\|_p := \left(\int_\Omega \|u(\omega)\|_{\text{H}}^p P(d\omega)\right)^{1/p}.$$

Here random variables u and v belong to the same equivalence class if and only if $u = v$ $P_{\text{a.s.}}$ (almost surely wrt P). The space $L^2(\Omega; H)$ is Hilbert with the scalar product

$$\langle u, v\rangle_{L^2(\Omega;H)} := \int_\Omega \langle u(\omega), v(\omega)\rangle_{\text{H}}P(d\omega).$$

Definition 4.1.1 *The expectation of a random variable $u \in L^1(\Omega; H)$ is*

$$\mathbf{E}(u) := \int_\Omega u(\omega)\,P(d\omega),$$

the correlation operator of random variables $u, v \in L^2(\Omega; H)$ is

$$\mathbf{Cor}(u, v) := \mathbf{E}[(u - \mathbf{E}[u]) \otimes (v - \mathbf{E}[v])]^1$$

and the covariance operator of $u \in L^2(\Omega; H)$ is

$$\mathbf{Cov}(u) := \mathbf{E}[(u - \mathbf{E}[u]) \otimes (u - \mathbf{E}[u])]. \tag{4.1.1}$$

Note that the expectation of an H-valued random variable is an element of H. In particular, the expectation of an $\mathbb{R}^n$-valued random variable is an n-dimensional vector. The covariance operator of an H-valued random variable is

---

[1]For the given $h_1 \in H_1$, $h_2 \in H_2$ the linear operator $h_1 \otimes h_2 : H_2 \to H_1$ is defined by $(h_1 \otimes h_2)h := h_1\langle h_2, h\rangle_{H_2}$, $h \in H_2$.

4.1. Hilbert space valued variables, processes, and stochastic integrals 117

a generalization of the covariance matrix in the $\mathbb{R}^n$-valued case. Recall that an $n \times n$ matrix is a covariance matrix of an $\mathbb{R}^n$-valued random variable if and only if it is non-negative and symmetric. The covariance operator of an H-valued random variable inherits these properties. To describe specific properties of the covariance operator of an H-valued random variable, we introduce some useful subclasses of bounded linear operators in Hilbert spaces (for more detail see, e.g., [28, 57, 102, 103, 104, 114]).

Let H_1 and H_2 be Hilbert spaces and let $\mathcal{L}(H_1, H_2)$ be the space of all bounded linear operators from H_1 to H_2 equipped with the norm

$$\|A\| := \inf\{C : \|Ax\|_{H_2} \le C\|x\|_{H_1}, \ x \in H_1\}.$$

If $H_1 = H_2 = H$, then we write $\mathcal{L}(H) := \mathcal{L}(H, H)$. An operator $A \in \mathcal{L}(H_1, H_2)$ is called *compact* (or *completely continuous*) if the image of any bounded subset from H_1 has compact closure in H_2.

We note some basic properties of compact operators:

1) if A is compact, then for any bounded linear operators B_1 and B_2 the products AB_1 and B_2A are compact;

2) if $A \in \mathcal{L}(H_1, H_2)$ is compact, then its adjoint $A^* \in \mathcal{L}(H_2, H_1)$ is compact (recall that by the Riesz theorem, Hilbert space H_i is identified with H_i^* and due to this we write $A^* \in \mathcal{L}(H_2, H_1)$ instead of $A^* \in \mathcal{L}(H_2^*, H_1^*)$);

3) if $A \in \mathcal{L}(H)$ is a self-adjoint compact operator and $\ker A = \{0\}$, then there exists an orthonormal basis $\{e_j\}_{j=1}^{\infty}$ in H consisting of eigenvectors of A: $Ae_j = \lambda_j e_j$, where corresponding eigenvalues λ_j are real and $\lim_{j \to \infty} \lambda_j = 0$;

4) if $A \in \mathcal{L}(H_1, H_2)$ is compact, then it admits the polar decomposition $A = = UT$, where $T = \sqrt{A^*A}$ is a non-negative self-adjoint compact operator on H_1 and U is an isometric operator from $ran\,T$ into H_2.

To summarize the above properties, we observe that an operator $A \in \mathcal{L}(H_1, H_2)$ is compact if and only if it admits the representation

$$Ax = \sum_{j=1}^{\infty} \lambda_j \langle x, e_j \rangle_{H_1} h_j, \quad x \in H_1, \tag{4.1.2}$$

where $\{e_j\}_{j=1}^{\infty}$ and $\{h_j\}_{j=1}^{\infty}$ are orthonormal systems in H_1 and H_2, respectively, $\lambda_j > 0$ and $\lim_{j \to \infty} \lambda_j = 0$. In this case λ_j are eigenvalues of the non-negative self-adjoint compact operator T from the polar decomposition of A; e_j are eigenvectors of T and $h_j = Ue_j$.

By introducing certain requirements on the rate of convergence of sequence $\{\lambda_j\}_{j=1}^{\infty}$ in representation (4.1.2), we arrive at the class of Hilbert–Schmidt operators, namely, an operator $A \in \mathcal{L}(H_1, H_2)$ is said to be of *Hilbert–Schmidt type* if it admits representation (4.1.2), where $\sum_{j=1}^{\infty} \lambda_j^2 < \infty$.

Among the useful properties of Hilbert–Schmidt operators are the following:

118   *4. Itô integrated stochastic Cauchy problems in Hilbert spaces*

1) $A \in \mathcal{L}(H_1, H_2)$ is of Hilbert–Schmidt type if and only if $\sum_{j=1}^{\infty} \|Ae_j\|^2 < \infty$ for at least one orthonormal basis $\{e_j\}_{j=1}^{\infty}$ of H_1;

2) the value

$$\|A\|_{\text{HS}} := \left(\sum_{j=1}^{\infty} \|Ae_j\|_{\text{H}_2}^2 < \infty \right)^{1/2} \tag{4.1.3}$$

does not depend on the choice of basis $\{e_j\}$ of H_1 and is referred to as the *Hilbert–Schmidt* norm of operator A;

3) the set of all Hilbert–Schmidt operators $\mathcal{L}_{\text{HS}}(H_1, H_2)$ is a normed space with the norm $\|A\|_{\text{HS}}$; it is also a separable Hilbert space with the scalar product $\langle A, B \rangle_{\text{HS}} = \sum_{j=1}^{\infty} \langle Ae_j, Be_j \rangle_{\text{H}_2}$;

4) if $A : H_1 \to H_2$ is a Hilbert–Schmidt operator, then its adjoint operator $A^* : H_2 \to H_1$ is also Hilbert–Schmidt;

5) if A is a Hilbert–Schmidt operator and B is a bounded linear operator, then their products AB and BA are also Hilbert–Schmidt operators.

By introducing a more restrictive requirement on the rate of convergence of the sequence $\{\lambda_j\}_{j=1}^{\infty}$ in representation (4.1.2), we arrive at the class of nuclear operators, namely, an operator $A \in \mathcal{L}(H_1, H_2)$ is called *nuclear* if it admits representation (4.1.2), where $\sum_{j=1}^{\infty} \lambda_j < \infty$.

We summarize the basic properties of nuclear operators:

1) the product of two Hilbert–Schmidt operators is a nuclear operator; conversely, any nuclear operator is the product of two Hilbert–Schmidt operators;

2) if A is a nuclear operator, then its adjoint A^* is also a nuclear operator;

3) if A is a nuclear operator and B is a bounded linear operator, then their products AB and BA are also nuclear;

4) a non-negative operator $A \in \mathcal{L}(H)$ is nuclear if and only if it is an *operator with finite trace*, that is, if and only if the series

$$\sum_{j=1}^{\infty} \langle Ae_j, e_j \rangle_{\text{H}} =: TrA,$$

converges for any orthonormal basis $\{e_j\}_{j=1}^{\infty}$ of H and does not depend on the choice of basis. The sum is referred to as the *trace of* A;

5) if $A \in \mathcal{L}(H_1, H_2)$ is a nuclear operator, then it admits the polar decomposition $A = UT$, where $T = \sqrt{A^*A} : H_1 \to H_1$ is a non-negative self-adjoint compact operator with finite trace and $U : H_1 \to H_2$ is an isometric operator;

4.1. Hilbert space valued variables, processes, and stochastic integrals 119

6) an operator $A \in \mathcal{L}(H_1, H_2)$ is nuclear if and only if $\sum_{j=1}^{\infty} \|A e_j\|_{H_2} < \infty$ for at least one orthonormal basis $\{e_j\}_{j=1}^{\infty}$ of H_1;

7) if $A \in \mathcal{L}(H_1, H_2)$ is a nuclear operator, then

$$\sup \sum_{j=1}^{\infty} \langle A e_j, h_j \rangle_{H_2} =: \|A\|_N,$$

where the supremum is taken over all orthonormal bases of H_1 and H_2;

8) the set of all nuclear operators $\mathcal{L}_N(H_1, H_2)$ is a normed space with the norm $\|A\|_N$ and for any $A \in \mathcal{L}_N(H_1, H_2)$ we have

$$\|A\| \le \|A\|_{HS} \le \|A\|_N.$$

Now, after the survey of the properties of compact, Hilbert–Schmidt, nuclear, and trace class operators, we can continue studying H-valued random variables and summarize the main properties of the covariance operator $\mathbf{Cov}(u)$ of an H-valued random variable u defined by (4.1.1).

Proposition 4.1.1 *Let H be a separable Hilbert space over $\mathbb{R}$ and u be an H-valued random variable. The operator $\mathbf{Cov}(u)$ is a non-negative self-adjoint nuclear operator on H with the finite trace*

$$Tr\mathbf{Cov}(u) = \mathbf{E}[\,\|u - E(u)\|_H^2\,].$$

Proof. Let $u \in L^2(\Omega; H)$; then for any $h_1, h_2 \in H$ we have

$$
\begin{aligned}
\langle \mathbf{Cov}(u) h_1, h_2 \rangle &= \langle \mathbf{E}[\langle u - \mathbf{E}(u), h_1 \rangle_H \langle u - \mathbf{E}(u) \rangle], h_2 \rangle_H \\
&= \mathbf{E}[\langle u - \mathbf{E}(u), h_1 \rangle_H \langle u - \mathbf{E}(u), h_2 \rangle_H] \\
&= \langle h_1, \mathbf{Cov}(u) h_2 \rangle_H.
\end{aligned}
$$

Thus the operator $\mathbf{Cov}(u)$ is self-adjoint. It is non-negative since for any $h \in H$

$$\langle \mathbf{Cov}(u) h, h \rangle_H = \int_\Omega \langle u(\omega) - \mathbf{E}(u), h \rangle_H^2 \, P(d\omega) \ge 0.$$

Finally, for any orthogonal basis in H we have

$$
\begin{aligned}
Tr\mathbf{Cov}(u) &= \sum_{j=1}^{\infty} \langle \mathbf{Cov}(u) e_j, e_j \rangle_H = \sum_{j=1}^{\infty} \langle \mathbf{E}\left[(u - \mathbf{E}(u))\langle u - \mathbf{E}(u), e_j \rangle\right], e_j \rangle_H \\
&= \sum_{j=1}^{\infty} \int_\Omega |\langle u(\omega) - \mathbf{E}(u), e_j \rangle_H|^2 \, P(d\omega) = \mathbf{E}[\,\|u - \mathbf{E}(u)\|_H^2\,] < \infty.
\end{aligned}
$$

In particular, if $\mathbf{E}(u) = 0$, then

$$Tr\mathbf{Cov}(u) = \sum_{j=1}^{\infty} \int_\Omega |\langle u(\omega), e_j \rangle_H|^2 \, P(d\omega) = \mathbf{E}\|u\|_H^2. \qquad (4.1.4)$$

120   4. Itô integrated stochastic Cauchy problems in Hilbert spaces

$\square$

We say that H-valued random variables u and v are *independent* if

$$P\{\omega \in \Omega : u(\omega) \in B_1 \text{ and } v(\omega) \in B_2\} = P\{\omega : u(\omega) \in B_1\}P\{\omega : v(\omega) \in B_2\}$$

for any $B_1, B_2 \in \mathcal{B}(H)$. If $u, v \in L^1(\Omega; \mathbb{R})$, this implies $\mathbf{E}(uv) = \mathbf{E}(u)\mathbf{E}(v)$.

Let $\mathcal{G} \subseteq \mathcal{F}$ be a σ-subalgebra of $\mathcal{F}$; then for any H-valued random variable $u : (\Omega, \mathcal{F}) \to (H, \mathcal{B}(H))$ with $u \in L^1(\Omega; H)$, there exists a random variable $v : (\Omega, \mathcal{G}) \to (H, \mathcal{B}(H))$ defined by

$$\int_G v \, dP = \int_G u \, dP \quad \text{for any} \quad G \in \mathcal{G}.$$

It is unique up to sets of P-measure zero and is referred to as the *conditional expectation* $\mathbf{E}(u|\mathcal{G}) = v$ of random variable u with respect to the σ-algebra $\mathcal{G}$.

Let us introduce Hilbert space valued *Gaussian random variables*. Recall that a real-valued random variable v is called Gaussian if its distribution law $\mathcal{L}_v$ has Gaussian density:

$$\mathcal{L}_v(a, b) = \int_a^b \frac{1}{\sqrt{2\pi q}} e^{-\frac{(y-m)^2}{2q}} \, dy, \qquad q > 0, \quad m \in \mathbb{R}.$$

The distribution function of Gaussian random variable v is therefore

$$\mathcal{L}_v(x) = \mathcal{L}_v(-\infty, x) = \int_{-\infty}^x \frac{1}{\sqrt{2\pi q}} e^{-\frac{(y-m)^2}{2q}} \, dy, \qquad x \in \mathbb{R},$$

and its expectation and covariance are

$$\mathbf{E}(v) = \int_G v(\omega) \, P(d\omega) = \int_{\mathbb{R}} x \, d\mathcal{L}_v(x) = m$$

and

$$\mathbf{Cov}(v) = \mathbf{E}[(v - m)^2] = \int_{\mathbb{R}} (x - m)^2 \, d\mathcal{L}_v(x) = q,$$

respectively. The distribution law of real-valued Gaussian random variable v defines a probability measure on $(\mathbb{R}, \mathcal{B}(\mathbb{R}))$, which is referred to as a *Gaussian measure* and is denoted by $\mathcal{N}(m, q)$.

Likewise, an $\mathbb{R}^n$-valued random variable v is called Gaussian if its distribution law has the form

$$\mathcal{L}_v(F) = \int_F \frac{1}{(2\pi)^{n/2}(\det Q)^{1/2}} e^{-\frac{1}{2}\langle Q^{-1}(y-m), y-m \rangle} \, dy, \qquad F \subset \mathbb{R}^n,$$

where $m \in \mathbb{R}^n$ and Q is an $n \times n$ non-negative symmetric matrix. This

4.1. Hilbert space valued variables, processes, and stochastic integrals 121

distribution law defines a probability measure on $(\mathbb{R}^n, \mathcal{B}(\mathbb{R}^n))$, which is referred to as (non-degenerate) *Gaussian measure* $\mathcal{N}(m, Q)$. In this case vector $m = (m_i, \ldots, m_n)$ is the expectation of v:

$$\mathbf{E}(v) = \int_\Omega v(\omega)\, P(d\omega) = \int_{\mathbb{R}^n} x_1 \ldots x_n\, d\mathcal{L}_v(x_1, \ldots, x_n) = m_1 \ldots m_n.$$

The matrix $Q = [\sigma_{ij}]_{i,j=1}^n$ is the *covariance* of v since

$$\sigma_{ij} = \int_{\mathbb{R}^n} (x_i - m_i)(x_j - m_j)\, d\mathcal{L}_v(x_1, \ldots, x_n).$$

Thus it defines a covariance operator on $\mathbb{R}^n$ by

$$\mathbf{Cov}(v)y = \int_{\mathbb{R}^n} (x_1 - m_1) \ldots (x_n - m_n) \sum_{i=1}^n (x_i - m_i)y_i\, d\mathcal{L}_v(x_1, \ldots, x_n) = Qy.$$

The characteristic function of an $\mathbb{R}^n$-valued Gaussian random variable v is the Fourier transform of its distribution law $\mathcal{L}_v = \mathcal{N}(m, Q)$:

$$\widehat{\mathcal{N}}(m, Q)(\lambda) = \int_{\mathbb{R}^n} e^{i\langle \lambda, x \rangle}\, d\mathcal{L}_v(x) = e^{i\langle \lambda, m \rangle} e^{-\frac{1}{2}\langle Q\lambda, \lambda \rangle}, \quad \lambda \in \mathbb{C}. \tag{4.1.5}$$

Therefore, an $\mathbb{R}^n$-valued random variable v is Gaussian if and only if its characteristic function has the above form (4.1.5). This relation reduces to the following definition of Gaussian random variables with values in a separable Hilbert space H.

Definition 4.1.2 *An H-valued random variable $v : \Omega \to H$ is called Gaussian if $v(h) = \langle v, h \rangle_H : \Omega \to \mathbb{R}$ for any $h \in H$ is a real-valued Gaussian random variable.*

It follows from the definition that an H-valued random variable v is Gaussian if and only if the random variables $v(e_j) = \langle e_j, v \rangle$ are real-valued Gaussian for an orthogonal basis $\{e_j\}_{j=1}^\infty$ in H. It follows that for an H-valued random variable v the expectations $\mathbf{E}[v(e_j)]$ define a linear functional m_h on H:

$$m_h := \mathbf{E}[v(h)] := \sum_{j=1}^\infty \mathbf{E}[v(e_j)\langle e_j\, h \rangle], \qquad h \in H,$$

and the correlation of random H-valued random variables v_1, v_2 defines a positive bilinear functional $\mathcal{Q}$ on H by

$$\mathcal{Q}(h_1, h_2) := \mathbf{E}[(v(h_1) - m_{h_1})(v(h_2) - m_{h_2})], \qquad h_1, h_2 \in H.$$

Continuity of the inner product in H implies that these functionals are continuous and therefore there exists an element $m \in H$ and a non-negative self-adjoint operator Q on H such that

$$m_h = (h, m)_H, \qquad \mathcal{Q}(h_1, h_2) = (Qh_1, h_2)_H, \qquad h, h_1, h_2 \in H.$$

122   *4. Itô integrated stochastic Cauchy problems in Hilbert spaces*

The element m is referred to as the *mean* of the H-valued random variable v and the operator $Q : H \to H$ as its *covariance* operator. It is not difficult to show that operator Q is nuclear with a finite trace.

Moreover, the converse is true: an element $m \in H$ and a non-negative self-adjoint operator Q with a finite trace on H induce an H-valued random variable v with the distribution law $\mathcal{L}_v = \mu$, where the measure μ on $(H, \mathcal{B}(H))$ is defined via its characteristic function as

$$\hat{\mu}(h) = \int_H e^{i\langle h, x\rangle}\, d\mathcal{L}_v(x) = e^{i\langle h, m\rangle} e^{-\frac{1}{2}\langle Qh, h\rangle}, \qquad h \in H. \qquad (4.1.6)$$

It is referred to as a *Gaussian measure* with mean m and covariance operator Q. Thus the following result holds.

Proposition 4.1.2 *For any positive symmetric trace class operator $Q : H \to H$ and $m \in H$ there exists an H-valued Gaussian random variable with expectation m and covariance operator Q. Its distribution law is a Gaussian measure on $(H, \mathcal{B}(H))$ with mean m and covariance operator Q.*

Proof. We show how to construct H-valued Gaussian random variable u with given expectation m and covariance operator Q. Let $(\Omega, \mathcal{F}, P)$ be a probability space and $\{\beta_j\}$ be a sequence of independent real-valued random variables with distribution law $\mathcal{N}(0, 1)$. (The existence of such a sequence follows from the Kolmogorov extension theorem and we construct such a sequence in Section 6.1.) Consider

$$u = m + \sum_{j=1}^{\infty} \sigma_j \beta_j e_j, \qquad (4.1.7)$$

where $\{e_j\}$ is the orthonormal basis in H consisting of eigenvectors of Q:

$$Qe_j = \sigma_j^2 e_j, \qquad \sum_{j=1}^{\infty} \sigma_j^2 = \sum_{j=1}^{\infty} \langle Qe_j, e_j\rangle_\text{H} < \infty.$$

The series (4.1.7) is convergent in $L^2(\Omega; H)$ due to the following equalities:

$$\mathbf{E}\left(\sum_{j=1}^{\infty} (\sigma_j \beta_j)^2\right) = \lim_{n\to\infty} \sum_{j=1}^{n} \sigma_j^2\, \mathbf{E}(\beta_j)^2 = \sum_{j=1}^{\infty} \sigma_j^2 = Tr\, Q.$$

Let $h \in H$. Consider the characteristic functional of u:

$$\begin{aligned}
\mathbf{E}(e^{i\langle h, u\rangle}) &= e^{i\langle h, m\rangle} \lim_{n\to\infty} \mathbf{E}\left(e^{i\sum_{j=1}^n \sigma_j \beta_j \langle h, e_j\rangle}\right) \\
&= e^{i\langle h, m\rangle} \lim_{n\to\infty} e^{-\frac{1}{2}\sum_{j=1}^n \sigma_j^2 \langle h, e_j\rangle^2} = e^{i\langle h, m\rangle - \frac{1}{2}\langle Qh, h\rangle}.
\end{aligned}$$

In addition, we have $\mathbf{E}(u) = m$ and for any $h_1, h_2 \in H$

$$\langle \mathbf{Cov}(u)h_1, h_2\rangle_H = \lim_{n\to\infty} \mathbf{E}\Big(\sum_{j=1}^{n} \sigma_j \beta_j \langle e_j, h_1\rangle_\text{H} \sum_{k=1}^{n} \sigma_k \beta_k \langle e_k, h_2\rangle_\text{H}\Big) = \langle Qh_1, h_2\rangle_\text{H}.$$

4.1. Hilbert space valued variables, processes, and stochastic integrals 123

Thus Q is the covariance operator of u. It follows that the H-valued random variable u is Gaussian with distribution law $\mathcal{N}(m, Q)$. $\qquad\square$

Now we show that an H-valued Gaussian random variable cannot have a characteristic function with $Q = I$ (or with a bounded, not a trace class operator Q). Let $\{e_j\}_{j=1}^{\infty}$ be an orthonormal basis of H. If $Q = I$, formula (4.1.6) takes the form

$$\widehat{\mu}(e_j) = \int_H e^{i\langle e_j, x\rangle} \, d\mathcal{L}_v(x) = e^{i\langle e_j, m\rangle} e^{-\frac{1}{2}\langle Ie_j, e_j\rangle}.$$

Since $\langle e_j, m\rangle \to 0$ as $j \to \infty$, we have $\widehat{\mu}(e_j) \to 1$ for the left-hand side; however, for the right-hand side we have $e^{i\langle e_j, m\rangle} e^{-\frac{1}{2}\langle Ie_j, e_j\rangle} \to e^{-1/2}$ as $j \to \infty$. Thus the requirement that $\langle Qe_j, e_j\rangle_{\mathrm{H}} \to 0$ as $j \to \infty$ is crucial here.

4.1.2 Stochastic processes in Hilbert spaces. Definitions, properties, and comparison with the finite-dimensional case. Cylindrical and Q-Wiener processes

Here we introduce the notion of a stochastic process with values in a Hilbert space H and investigate properties of H-valued stochastic processes. Special attention is given to Gaussian and operator-valued processes. We consider Brownian motion, which is an important example of a Gaussian process in the particular case $H = \mathbb{R}^n$. Further on we consider two types of generalization of Brownian motion to infinite-dimensional Hilbert spaces: a Q-Wiener process with a trace class covariance operator Q and a cylindrical Wiener process, which can be regarded as a process in a certain weak sense. We define stochastic Itô integrals for operator-valued stochastic processes $\{\Phi(t), t \geq 0\}$ wrt both types of Wiener processes. In conclusion, we present theorems which are most important and useful for the study of stochastic problems addressed in this book.

Let $(\Omega, \mathcal{F}, P)$ be a probability space and H be a Hilbert space.

Definition 4.1.3 *A parameterized family of H-valued random variables $\{u(t), t \in \mathcal{T}\}$ defined on $(\Omega, \mathcal{F}, P)$ is called an H-valued stochastic process.*

It follows from the definition that for each fixed $t \in \mathcal{T}$ we have an H-valued random variable

$$u(t, \cdot): \ \Omega \to H.$$

On the other hand, fixing $\omega \in \Omega$, we obtain the function $u(\cdot, \omega): \ \mathcal{T} \to H$ called a *trajectory (or path)* of $u(t)$. Sometimes it is convenient to identify each ω with the corresponding path $u(\cdot, \omega)$ and Ω with a subset of the space of all functions from $\mathcal{T}$ into H.

In the next sections, while studying stochastic Cauchy problems in Hilbert spaces, we will use $[0, T]$, or $[0, T)$ with $T < \infty$, or $[0, \infty)$ as the parameter set

124 4. *Itô integrated stochastic Cauchy problems in Hilbert spaces*

$\mathcal{T}$. In either case a solution of the stochastic problem is a stochastic process $\{X(t), t \in \mathcal{T}\}$.

Now we recall definitions of measurability and certain types of continuity of stochastic processes, then introduce the notion of a version of a process and give an instructive example of a continuous stochastic process with a discontinuous version.

Let $\{u(t), t \in \mathcal{T}\}$ be a stochastic process. We say that

- $u(t)$ is *measurable* if the map $u(\cdot, \cdot) : \mathcal{T} \times \Omega \to H$ is $\mathcal{B}(\mathcal{T}) \times \mathcal{F}$ measurable;

- $u(t)$ is *stochastically continuous* at $t_0 \in \mathcal{T}$ if for any positive numbers ε, δ there exists a positive number ρ such that

$$P\left(\|u(t) - u(t_0)\|_H \geq \varepsilon\right) \leq \delta \quad \text{for any} \quad t \in [t_0 - \rho, t_0 + \rho] \cap \mathcal{T};$$

- $u(t)$ is *mean square continuous* at $t_0 \in \mathcal{T}$ if

$$\lim_{t \to t_0} \mathbf{E}\left[\|u(t) - u(t_0)\|_H^2\right] = 0.$$

The process is called stochastically continuous or mean square continuous on $\mathcal{T}$ if it has the corresponding property at each point of $\mathcal{T}$.

The question of continuity of trajectories of a stochastic process arises in different aspects of the research. In particular, the question of coincidence (in a sense) of continuous and discontinuous processes is closely related to the question of uniqueness of solutions to stochastic problems. In the next chapter we will see that uniqueness of solutions of stochastic Cauchy problems is understood in the sense of coincidence of versions. In this connection let us consider the example mentioned above of two processes $\{u(t), t \in \mathcal{T}\}$ and $\{v(t), t \in \mathcal{T}\}$ with different continuity properties which, nevertheless, coincide in the following sense:

$$\text{for each } t \in \mathcal{T}, \quad P\{\omega : u(t, \omega) \neq v(t, \omega)\} = 0. \tag{4.1.8}$$

The stochastic process $\{v(t), t \in \mathcal{T}\}$ satisfying the condition (4.1.8) is called a *version* or *modification* of $\{u(t), t \in \mathcal{T}\}$.

Example 4.1.1 *Consider the probability space $(\Omega, \mathcal{F}, P)$, where $\Omega = [0, 1]$, $\mathcal{F}$ is the σ-field of Lebesgue measurable subsets of $[0, 1]$ and P is the Lebesgue measure on $[0, 1]$. Let $\mathcal{T} = \mathbb{R}$. For all $t \in \mathbb{R}, \omega \in [0, 1]$ consider*

$$u(t, \omega) = 0 \quad and \quad v(t, \omega) = \begin{cases} 0, & t \neq \omega, \\ h, & t = \omega, \end{cases} \quad h \in H.$$

Then for any $t \in \mathbb{R}, P\{\omega : u(\omega, t) = 0\} = 1$ and $P\{\omega : v(\omega, t) = 0\} = 1$.

Obviously we have

$$P\{\omega : u(\omega, t) = v(\omega, t)\} = 1,$$

4.1. Hilbert space valued variables, processes, and stochastic integrals 125

i.e., both processes are versions of one another. It is easy to check that both processes have the same finite-dimensional distributions. From this point of view they can be regarded as different versions of one process. However,

$$P\{\omega:\ u(\cdot,\omega)\text{ is continuous}\} = 1, \quad \text{but} \quad P\{\omega:\ v(\cdot,\omega)\text{ is continuous}\} = 0,$$

because for any $\omega \in [0, 1]$ the function $v(\cdot, \omega)$ has a discontinuity at $t = \omega$.

This example suggests the idea of replacing the question of continuity of trajectories of a stochastic process by the question of existence of a continuous version of this process. In the next section we will see that uniqueness of solutions of stochastic Cauchy problems is understood up to a version.

The important result on the existence of a continuous version is

Theorem 4.1.1 (The Kolmogorov theorem) *If an H-valued stochastic process $\{u(t),\ t \geq 0\}$ has the property: for any $t, s \in [0, \infty)$ there exist $C > 0$, $\varepsilon > 0$, $\delta > 0$ such that*

$$\boldsymbol{E}\|u(t) - u(s)\|_H^{\delta} \leq C|t - s|^{1+\varepsilon},$$

then there exists a continuous version of the process.

As shown in the previous subsection, the properties of random variables (including H-valued ones) are closely connected with the properties of their distribution laws. A similar role in the theory of stochastic processes is played by their finite-dimensional distributions.

Definition 4.1.4 *Probability measures $\mu_{t_1,\ldots,t_k}$ defined on Hilbert spaces $H^k = H \times \ldots \times H$, $k \in \mathbb{N}$, by*

$$\mu_{t_1,\ldots,t_k}(G_1 \times \ldots \times G_k) = P\{\omega : u(t_1,\omega) \in G_1, \ldots, u(t_k,\omega) \in G_k\}, \quad (4.1.9)$$

where $G_i \in \mathcal{B}(H)$, are called (finite-dimensional) distributions of the process $\{u(t),\ t \in \mathcal{T}\}$.

Similar to the distribution law of a random variable, the family of all finite-dimensional distributions determines many, but not all, properties of a process. In the particular case $H = \mathbb{R}^n$ the Kolmogorov extension theorem [61] states that under two natural consistency conditions imposed on the set of measures $\{\mu_{t_1,\ldots,t_k} :\ k \in \mathbb{N},\ t_1,\ldots,t_k \in \mathcal{T}\}$ on $\mathbb{R}^{nk}$, they define an $\mathbb{R}^n$-valued stochastic process with $\mu_{t_1,\ldots,t_k}$ as the finite-dimensional distributions.

Consider an important example of a stochastic process defined by the following family of measures on the spaces $\mathbb{R}^{nk}$:

$$\mu_{t_1,\ldots,t_k}(G_1 \times \ldots \times G_k) =$$

$$\int_{G_1 \times \ldots \times G_k} p(t_1, x, x_1)p(t_2 - t_1, x_1, x_2) \cdots p(t_k - t_{k-1}, x_{k-1}, x_k)\, dx_1 \cdots dx_k.$$

$$(4.1.10)$$

126   *4. Itô integrated stochastic Cauchy problems in Hilbert spaces*

Here the function

$$p(t, x, y) = \frac{1}{(2\pi t)^{n/2}} e^{-\frac{|x-y|^2}{2t}}, \qquad x, y \in \mathbb{R}^n, \quad t > 0,$$

for any fixed t and x is the probability density of an $\mathbb{R}^n$-valued Gaussian random variable with distribution law

$$\mathcal{N}(x, t)(G) = \frac{1}{(2\pi t)^{n/2}} \int_G e^{-\frac{|x-y|^2}{2t}} \, dy, \qquad G \in \mathcal{B}(\mathbb{R}^n).$$

The family of measures defined by (4.1.10) satisfies the Kolmogorov extension theorem, hence there exist a probability space $(\Omega, \mathcal{F}, P = P(x))$ and a stochastic process $\{\beta(t), \, t \geq 0\}$ with values in $\mathbb{R}^n$ (*n-dimensional Brownian motion*) such that its finite-dimensional distributions are defined by measures (4.1.9):

$$P\{\omega \in \Omega : \; \beta(t_1, \omega) \in G_1, \dots, \beta(t_k, \omega) \in G_k\} = \mu_{t_1, \dots, t_k}(G_1 \times \dots \times G_k).$$

In particular the distribution law of the $\mathbb{R}$-valued Brownian motion starting at point x is

$$P\{\omega : \; \beta(t, \omega) \in G\} = \mu_t(G) = \mathcal{N}(x, t)(G).$$

Now we list the main properties of the Brownian motion defined above.

Proposition 4.1.3 [95] *Let $\{\beta(t), \, t \geq 0\}$ be the process defined by measures (4.1.10). It has the following properties.*

(B1) $P\{\omega : \beta(0, \omega) = x\} = 1$, *i.e., the process $\beta(t)$ starts at the point x $P_{\text{a.s.}}$.*

(B2) *$\beta(t)$ has independent increments, i.e., for any set $0 \leq t_1 < \cdots < t_k$ the random variables*

$$\beta(t_1), \; \beta(t_2) - \beta(t_1), \; \dots, \; \beta(t_k) - \beta(t_{k-1})$$

are independent wrt P.

(B3) *$\beta(t)$ is a Gaussian process, which means that $(\beta(t_1), \dots, \beta(t_k))$ is an $\mathbb{R}^{nk}$-valued Gaussian random variable for any $0 \leq t_1 \leq \cdots \leq t_k$. The expectation of this vector is equal to $m = (x, \dots, x) \in \mathbb{R}^{nk}$ and the covariance matrix is equal to*

$$Q_{t_1, \dots, t_k} = \begin{bmatrix} t_1 I_n & t_1 I_n & t_1 I_n & \cdots & t_1 I_n \\ t_1 I_n & t_2 I_n & t_2 I_n & \cdots & t_2 I_n \\ t_1 I_n & t_2 I_n & t_3 I_n & \cdots & t_3 I_n \\ \vdots & \vdots & \vdots & & \vdots \\ t_1 I_n & t_2 I_n & t_3 I_n & \cdots & t_k I_n \end{bmatrix}$$

where I_n is the $n \times n$ unit matrix.

4.1. Hilbert space valued variables, processes, and stochastic integrals **127**

(B4) $\beta(t)$ *has continuous trajectories* $P_{\text{a.s.}}$, *i.e., the mapping* $t \to \beta(t, \omega)$, $t \geq 0$, *is continuous for almost all* $\omega \in \Omega$.

The process defined by the family of measures (4.1.10) is one of the class of processes with properties (B1)–(B4). This gives rise to the following definition.

Definition 4.1.5 *An* $\mathbb{R}^n$*-valued stochastic process* $\{\beta(t), t \geq 0\}$ *satisfying conditions* (B1)–(B4) *is called Brownian motion (starting at point* x*).*

Brownian motion is a very important Gaussian process in the theory of finite-dimensional stochastic equations and serves as a basis for its infinite-dimensional extensions, which are H-valued Wiener processes.

Definition 4.1.6 *Let* H *be a Hilbert space. An* H*-valued stochastic process* $\{u(t), t \geq 0\}$ *is said to be a Gaussian process if for any* $k \in \mathbb{N}$ *and arbitrary positive numbers* $t_1, \ldots, t_k$, *the measure* $\mu_{t_1, \ldots, t_k}$ *defined by (4.1.9) is Gaussian.*

It follows from the definition that an H-valued stochastic process $\{u(t), t \geq 0\}$ is Gaussian if and only if the H^k-valued random variable $(u(t_1), \ldots, u(t_k))$ is Gaussian for any choice of positive numbers $t_1, \ldots, t_k$.

As follows from the properties obtained of an H-valued Gaussian random variable, it cannot have the covariance operator equal to the unity operator on H: by (4.1.4) it must be a trace class operator. Motivated by Definition 4.1.5 we introduce an important class of stochastic processes, called *Q-Wiener processes*, where Q are trace class operators. The notion of a Q-Wiener process can be regarded as a generalization of the notion of Brownian motion to the case of infinite-dimensional Hilbert spaces.

Definition 4.1.7 *Let* Q *be a symmetric non-negative trace class operator in a Hilbert space* H. *An* H*-valued stochastic process* $W = \{W(t), t \geq 0\}$ *is called a Q-Wiener process (starting at zero) if*

(W1) $W(0) = 0\ P_{\text{a.s.}}$.

(W2) W *has independent increments.*

(W3) $\mathcal{L}_{[W(t)-W(s)]} = \mathcal{N}(0, (t-s)Q), \quad 0 \leq s \leq t$.

(W4) W *has continuous trajectories.*

As shown in the beginning of this section, for any non-negative symmetric trace class operator Q in a separable Hilbert space H there exists an orthonormal basis of eigenvectors $\{e_j\}$ with $Qe_j = \sigma_j^2 e_j$ such that

$$TrQ = \sum_{j=1}^{\infty} \langle Qe_j, e_j \rangle_{\text{H}} = \sum_{j=1}^{\infty} \sigma_j^2.$$

It follows from Definition 4.1.7 and properties of Gaussian random variables that a Q-Wiener process has the following probability characteristics.

128   4. Itô integrated stochastic Cauchy problems in Hilbert spaces

Proposition 4.1.4 *Let W be a Q-Wiener process in a Hilbert space H. Then the process is Gaussian and*

$$\boldsymbol{E}(W(t)) = 0, \quad \boldsymbol{Cov}(W(t)) = tQ, \quad t \geq 0. \tag{4.1.11}$$

Proof. In addition to the properties (4.1.11), let us show that for any $t \geq 0$ the random variable $W(t)$ has the following expansion in $L^2(\Omega, P; H) = L^2(\Omega; H)$:

$$W(t) = \sum_{j=1}^{\infty} \sigma_j \beta_j(t) e_j, \tag{4.1.12}$$

where $\beta_j(t) = \frac{1}{\sigma_j}\langle W(t), e_j \rangle_{\mathrm{H}}$, $t \geq 0$, are independent real-valued Brownian motions on $(\Omega, \mathcal{F}, P)$.

To show the representation (4.1.12) we prove that the series is convergent in $L^2(\Omega; H)$. For this it is enough to prove that partial sums form a fundamental sequence in the complete space $L^2(\Omega; H)$. For $1 \leq n < m$ we have

$$\boldsymbol{E} \left\| \sum_{j=n}^{m} \sigma_j \beta_j(t) e_j \right\|_H^2 = t \sum_{j=n}^{m} \sigma_j^2 \|e_j\|_H^2 = t \sum_{j=n}^{m} \sigma_j^2. \tag{4.1.13}$$

Since Q is a trace class operator and $\sum_{j=1}^{\infty} \sigma_j^2 = TrQ < \infty$, the series in (4.1.12) is convergent in $L^2(\Omega; H)$. $\square$

Thus we have proved that for any Q-Wiener process W there exists a sequence of independent real-valued Brownian motions $\beta_j(t) = \frac{1}{\sigma_j}\langle W(t), e_j \rangle_H$, $t \geq 0$, $j = 1, 2, \ldots$. The converse result follows from the proof of the next proposition.

Proposition 4.1.5 *For a Hilbert space H and arbitrary trace class symmetric non-negative operator Q in H there exists an H-valued Q-Wiener process.*

Proof. Let $\{\beta_j(t), t \geq 0\}$, $j \in \mathbb{N}$, be a sequence of independent real-valued Brownian motions on $(\Omega, \mathcal{F}, P)$. Consider the series (4.1.12) with σ_j^2 being the eigenvalues of Q. Since for any $1 \leq n < m$ we have (4.1.13), the series (4.1.12) is convergent in $L^2(\Omega; H)$.

Now we prove that the process W defined by (4.1.12) is a Q-Wiener process. Without loss of generality we may assume that $\beta_j(0) = 0$, $j \in \mathbb{N}$; then $W(0) = 0$ and the property (W1) holds. Further, for any $t \geq s \geq 0$, the increment $W(t) - W(s)$ is a Gaussian random variable as it is the mean square limit of the sequence of Gaussian random variables

$$\sum_{j=1}^{n} \sigma_j \beta_j(t) e_j - \sum_{j=1}^{n} \sigma_j \beta_j(s) e_j.$$

4.1. Hilbert space valued variables, processes, and stochastic integrals 129

Since the increments are Gaussian random variables, they are independent if and only if they are uncorrelated. Consider $0 \leq t_1 < t_2 < \ldots < t_n$. Denote $\Delta W_i = W(t_{i+1}) - W(t_i)$, $i = 1, \ldots, n-1$. Let us show that they are uncorrelated. The independence of increments of any Brownian motion and the independence of Brownian motions $\beta_j(t)$ and $\beta_l(t)$ with $j \neq l$ imply for any $h, g \in H$:

$$\langle \mathbf{Cor}(\Delta W_i, \Delta W_k)h_1, h_2 \rangle_{\mathrm{H}} = \mathbf{E}\langle (\Delta W_i \otimes \Delta W_k)h_1, h_2 \rangle_{\mathrm{H}}$$

$$= \mathbf{E}\langle \Delta W_i, h_1 \rangle_{\mathrm{H}} \langle \Delta W_k, h_2 \rangle_{\mathrm{H}} = 0, \qquad i \neq k.$$

Thus (W2) holds. To verify (W3) consider the covariance operator of the difference $W(t) - W(s)$. We have

$$\langle \mathbf{Cov}(W(t) - W(s))h, g \rangle_{\mathrm{H}}$$

$$= \mathbf{E}\left(\langle h, \sum_{j=1}^{\infty} \sigma_j(\beta_j(t) - \beta_j(s))e_j \rangle_{\mathrm{H}} \langle g, \sum_{j=1}^{\infty} \sigma_j(\beta_j(t) - \beta_j(s))e_j \rangle_{\mathrm{H}} \right)$$

$$- (t-s)\sum_{j=1}^{\infty} \sigma_j^2 \langle h, e_j \rangle_{\mathrm{H}} \langle g, e_j \rangle_{\mathrm{H}} = (t-s)\langle Qh, g \rangle_{\mathrm{H}}.$$

So, (W3) holds. Since $\beta_j(t)$, $j \in \mathbb{N}$, are continuous, (W4) is also fulfilled. $\quad\square$

In Section 6.1 devoted to Hilbert space valued generalized random variables we present a construction of a sequence of independent real-valued Brownian motions $\{\beta_j(t), t \geq 0\}$ within the framework of the theory of generalized random variables.

Now we consider the case of a bounded operator Q in a separable Hilbert space H with $TrQ = \infty$ and define weak (or cylindrical) Wiener processes. If Q is not a trace class operator, in particular $Q = I$, then the series (4.1.12) is divergent in $L^2(\Omega; H)$. There are two ways of avoiding the problems that arise. The first one is to consider the cylindrical Wiener process as a process in H in a weak sense. The second one is to construct a Q_1-Wiener process on an appropriate space $H_1 \supset H$.

We begin with the first approach. We will see that instead of the factors σ_j ensuring the convergence of the series (4.1.12), some other regularizing multipliers appear in the case $TrQ = \infty$, in particular, $Q = I$.

For any $h \in H$ consider the scalar product

$$\langle h, W(t) \rangle_{\mathrm{H}} := \sum_{j=1}^{\infty} \beta_j(t) \langle h, e_j \rangle_{\mathrm{H}}, \qquad t \geq 0, \tag{4.1.14}$$

where $\{\beta_j(t), t \geq 0\}$ is a family of independent Brownian motions and $\{e_j\}$ is an orthogonal basis in H. Then for a bounded self-adjoint positive operator

130   *4. Itô integrated stochastic Cauchy problems in Hilbert spaces*

Q in H with $TrQ = \infty$, the process defined by (4.1.14) is a real-valued Q_h-Wiener process with zero expectation and covariance $tQ_h = t\langle Qh, h\rangle_H$.

We can see that the role of the regularizing multipliers $\langle h, e_j\rangle$ with $\sum_{j=1}^{\infty}\langle h, e_j\rangle^2 < \infty$ in this case is similar to the role of the multipliers σ_j in (4.1.12).

Summarizing the properties of the stochastic process defined by (4.1.14), we arrive at the following definition.

Definition 4.1.8 *Let Q be a bounded self-adjoint positive operator in a Hilbert space H with $TrQ = \infty$. Let, for any $h \in H$, the real-valued process $\{\langle h, W(t)\rangle =: W_h, t \geq 0\}$ be Gaussian with independent increments and continuous version and*

$$\boldsymbol{E}(W_h) = 0 \quad and \quad \boldsymbol{Cov}(W_h) = t\langle Qh, h\rangle_H. \tag{4.1.15}$$

Then $\{W(t), t \geq 0\}$ is called a weak (or cylindrical) Wiener process in H.

If a process $\{W(t), t \geq 0\}$ satisfies (4.1.15), it is also referred to as a (weak) Q-Wiener process with $TrQ \leq \infty$.

The second way to define a *cylindrical Wiener process* W is to consider W as a Q_1-Wiener process with a trace class operator Q_1 in a certain Hilbert space H_1. This is a slightly more complicated way. At the beginning we present the definition for the general case of a Q-Wiener process with $TrQ \leq \infty$ and then for the particular case $Q = I$. To do this we need to introduce a space H_Q, which later will be useful in the definition of stochastic integrals wrt Q-Wiener and cylindrical Wiener processes.

Let Q be a bounded self-adjoint positive operator in H with $TrQ \leq \infty$. Define the Hilbert space

$$H_Q := Q^{1/2}(H), \quad \langle u, v\rangle_{H_Q} := \langle Q^{-1/2}u, Q^{-1/2}v\rangle_H, \tag{4.1.16}$$

and take a Hilbert space H_1 in such a way that the embedding J of H into H_1 is a continuous operator and embedding J_Q of H_Q into H_1 is a Hilbert–Schmidt operator.

Obviously, H_Q is a subspace of H and H_1 is a somewhat wider space than H. We construct cylindrical Wiener process $\{W(t), t \geq 0\}$ as a Q_1-Wiener process on H_1.

Proposition 4.1.6 *Let Q be a bounded self-adjoint positive operator in a Hilbert space H with $TrQ = \infty$. Let $\{\beta_j(t), t \geq 0\}$ be a family of independent real-valued Brownian motions. Then the series*

$$W(t) = \sum_{j=1}^{\infty} \beta_j(t)g_j, \quad t \geq 0, \tag{4.1.17}$$

where $\{g_j\}$ is an orthonormal basis in H_Q, defines an H_1-valued Q_1-Wiener process with the zero expectation and covariance operator $Q_1 := J_Q J_Q^*$ in the space H_1 defined by the equality $Q_1^{1/2}(H_1) = H_Q$ and equipped with the norm $\|h\|_{H_Q} = \|Q_1^{-1/2}h\|_{H_1}$.

4.1. Hilbert space valued variables, processes, and stochastic integrals 131

Proof. To show that the series (4.1.17) is convergent in $L^2(\Omega; H_1)$, similarly to (4.1.13), we obtain

$$\mathbf{E}\left(\left\|\sum_{j=n}^{m}\beta_j(t)g_j\right\|_{H_1}^2\right) = t\sum_{j=n}^{m}\|g_j\|_{H_1}^2 = t\sum_{j=n}^{m}\|J_Q g_j\|_{H_1}^2.$$

Here $\|g_j\|_{H_1} = \|J_Q g_j\|_{H_1}$ since J_Q is the imbedding operator of H_Q into H_1 and $g_j \in H_Q \subset H_1$. Since the embedding is a Hilbert–Schmidt operator, we have $\sum_{j=1}^{\infty}\|J_Q g_j\|_{H_1}^2 < \infty$.

Now we define the covariance operator Q_1 of $\{W(t),\, t \geq 0\}$ in H_1. Consider $0 \leq s \leq t$. Similar to equalities for $\mathbf{Cov}(W(t) - W(s))$ obtained in Proposition 4.1.5 we have

$$\begin{aligned}
\langle \mathbf{Cov}(W(t) - W(s))h, g\rangle_{H_1} &= (t-s)\sum_{j=1}^{\infty}\langle J_Q^* h, g_j\rangle_{H_Q}\langle J_Q^* g, g_j\rangle_{H_Q}\\
&= (t-s)\langle J_Q^* h, J_Q^* g\rangle_{H_Q} = (t-s)\langle J_Q J_Q^* h, g\rangle_{H_1}
\end{aligned}$$

for any $h, g \in H_1$. Hence it is natural to define

$$Q_1 := J_Q J_Q^* \quad : \quad H_1^* \simeq H_1 \to H_1.$$

Being a covariance operator, Q_1 is self-adjoint and positive. Moreover, being a product of Hilbert–Schmidt operators, Q_1 is a trace class operator on H_1.

The fact that the process obtained has all the properties of a Q_1-Wiener process in H_1 can be proved by the similar arguments as in Proposition 4.1.5. $\square$

Remark 4.1.1 Note that in the particular case $Tr\,Q < \infty$ we can take $H_1 = H$ and $Q_1 = Q$, while in the case $Q = I$ we have $H_Q = H$ and take H_1 in such a way that projection from H_1 to H is a trace class operator Q_1. Due to Proposition 4.1.6, any cylindrical Wiener process is a Q_1-Wiener process in suitable H_1.

4.1.3 Stochastic integrals wrt Wiener processes

Now we define stochastic integrals of operator-valued stochastic processes $\{\Phi(t),\, t \geq 0\}$ wrt a Q-Wiener or a cylindrical Wiener process $\{W(t),\, t \geq 0\}$ in the sense of Itô:

$$\int_0^t \Phi(s)\, dW(s), \quad t \in [0, \infty). \tag{4.1.18}$$

We will pay special attention to the particular form of such an integral

$$\int_0^t U(t-s)\, B dW(s) \quad t \in [0, \infty),$$

132   *4. Itô integrated stochastic Cauchy problems in Hilbert spaces*

called a *stochastic convolution* since it plays an important role in studying stochastic equations. We will show that the integral (4.1.18) wrt either a Q-Wiener or a cylindrical Wiener process (with $Q = I$) is defined under the condition

$$\mathbf{E}\left[\int_0^t \|\Phi(s)\|_{\mathrm{HS}}^2 \, ds\right] < \infty, \qquad 0 \le t < \infty, \qquad (4.1.19)$$

where $\|\Phi(s)\|_{\mathrm{HS}}^2$ is the Hilbert–Schmidt operator norm of $\Phi(s) : \mathbb{H}_Q \to H$. The condition (4.1.19) formally looks the same for cylindrical and Q-Wiener processes. However, the condition is different for each type of Wiener process since the operator Q is different in each case.

Before going further, let us explain the notation of spaces $\mathbb{H}$ and $\mathbb{H}_Q$. Since in the next sections stochastic integrals will be used for constructing solutions to stochastic Cauchy problems (P.1) and (I.1), we will use notations in accordance with the problems, where A is the generator of a semi-group defined in a Hilbert space H and Wiener process $\{W(t), t \ge 0\}$ or white noise $\{\mathbb{W}(t), t \ge 0\}$ are, generally, processes with values in another Hilbert space $\mathbb{H}$. In compliance with these notations we will consider the operator-valued stochastic process $\{\Phi(t), t \ge 0\}$ as $\mathcal{L}(\mathbb{H}_Q, H)$ valued, $\mathcal{L}(\mathbb{H}, H)$ valued, or $\mathcal{L}_{\mathrm{HS}}(\mathbb{H}, H)$ valued.

Thus let $\mathbb{H}$, H be separable Hilbert spaces, $(\Omega, \mathcal{F}, P)$ be a probability space, and $\mathcal{T}$ be equal to $[0, \tau), \tau \le \infty)$, or $[0, T]$.

An increasing family of σ-fields $\{\mathcal{F}_t, t \in \mathcal{T}\}$ on Ω is called a *filtration*. A filtration $\{\mathcal{F}_t, t \in \mathcal{T}\}$ is said to be *normal* if $\{G \in \mathcal{F} : P(G) = 0\} \subset \mathcal{F}_0$ and $\mathcal{F}_t = \cap_{s>t}\mathcal{F}_s$ for all $t \in \mathcal{T}$. If, for any $t \in \mathcal{T}$, the random variable $u(t)$ is $\mathcal{F}_t$-measurable, then the process $\{u(t), t \in \mathcal{T}\}$ is called *adapted* to the filtration $\{\mathcal{F}_t\}$.

To define the concept of a *predictable* Hilbert space valued random process, we denote $\Omega_\infty = [0, \infty) \times \Omega$ $(\Omega_T = [0, T] \times \Omega)$ and introduce the σ-field $\mathcal{B}_\infty$ $(\mathcal{B}_T)$ on Ω_∞ (Ω_T) generated by sets of the form

$$\begin{aligned} (s, t] \times G, \qquad & G \in \mathcal{F}_s, \quad 0 \le s < t < \infty \ (0 \le s < t \le T), \\ \{0\} \times G, \qquad & G \in \mathcal{F}_0. \end{aligned} \qquad (4.1.20)$$

Denote by $\mathcal{P}_\infty$ the product of the Lebesgue measure on $[0, \infty)$ with the probability measure P on Ω and by $\mathcal{P}_T$ its restriction on $[0, T] \times \Omega$.

A measurable mapping from $(\Omega_\infty, \mathcal{P}_\infty)$ or $(\Omega_T, \mathcal{P}_T)$ to $(\mathbb{H}, \mathcal{B}(\mathbb{H}))$ is called a *predictable process*. It is known that an adapted and stochastically continuous process on $[0, T]$ has a predictable version [20, 81].

Now we arrive at the definition of a stochastic integral and, as usual in definitions of integrals, we start with the definition for elementary processes.

Let $(\Omega, \mathcal{F}, P)$ be a probability space with a normal filtration $\{\mathcal{F}_t, t \ge 0\}$ generated by a Wiener process W, i.e., $W(t)$ is $\mathcal{F}_t$-measurable, and by (W3) the increments $W(t + h) - W(t)$ are independent of $\mathcal{F}_t$.

An $\mathcal{L}(\mathbb{H}, H)$-valued process $\{\Phi(t), t \in [0, T]\}$ is said to be *elementary* if there exist a set $0 = t_0 < t_1 < \ldots < t_k = T$ and $\mathcal{L}(\mathbb{H}, H)$-valued random

4.1. Hilbert space valued variables, processes, and stochastic integrals 133

variables $\Phi_0, \Phi_1, \ldots, \Phi_{k-1}$ such that Φ_m is $\mathcal{F}_{t_m}$-measurable and $\Phi(t) = \Phi_m$, $t \in (t_m, t_{m+1}]$, for each $m = 0, \ldots k - 1$.

For an elementary process $\Phi \in \mathcal{L}(\mathbb{H}, H)$ a *stochastic integral* wrt a Wiener process W is defined as usual for elementary functions:

$$\int_0^t \Phi(s) dW(s) := \sum_{m=0}^{k-1} \Phi_m \left(W(t_{m+1} \wedge t) - W(t_m \wedge t) \right), \ t \in [0, T].^2 \quad (4.1.21)$$

Further, we indicate the class of $\mathcal{L}_{HS}(\mathbb{H}_Q, H)$-valued processes Φ for which the stochastic integral with respect to a Wiener process can be defined as the mean square limit of the sums (4.1.21) for elementary processes approximating Φ. Moreover, we will show that for both types of Wiener processes the condition on Φ can be formally written in the same form as (4.1.19).

If an operator $Q : \mathbb{H} \to \mathbb{H}$ is the covariance operator of a Q-Wiener process W, it is symmetric, non-negative, and trace class and there exists an orthonormal basis $\{e_j\}$ of eigenvectors of Q such that

$$Q e_j = \sigma_j^2 e_j, \qquad \sum_{j=1}^{\infty} \sigma_j^2 < \infty.$$

While defining a cylindrical Wiener process, we introduced the Hilbert space $\mathbb{H}_Q$ with $Tr\, Q \leq \infty$. Note that for such a Q the definition (4.1.16) implies that $\{g_j = \sigma_j e_j\}$ is an orthonormal basis in $\mathbb{H}_Q$.

Now consider the space $\mathcal{L}_{HS}(\mathbb{H}_Q, H)$ of Hilbert–Schmidt operators from $\mathbb{H}_Q$ into H. The following two propositions are concerned with relations between spaces $\mathcal{L}(\mathbb{H}, H)$ and $\mathcal{L}_{HS}(\mathbb{H}_Q, H)$. The relation will be crucial for the definition of stochastically integrable processes $\{\Phi(t), t \in [0, T]\}$.

Proposition 4.1.7 *Let $\mathbb{H}, H$ be separable Hilbert spaces and Q be a trace class operator on $\mathbb{H}$. For any $\Phi \in \mathcal{L}_{HS}(\mathbb{H}_Q, H)$, the operators $\Phi\Phi^*$ and $\Phi Q^{1/2}(\Phi Q^{1/2})^*$ act in H and*

$$Tr[\Phi\Phi^*] = \|\Phi\|_{HS}^2 = Tr[\Phi Q \Phi^*].$$

Proof. Let $\{f_k\}_{k=1}^{\infty}$ be an orthonormal basis in H and let $\{g_k\}_{k=1}^{\infty}$ be an orthonormal basis in $\mathbb{H}_Q$. Using the definition of the space $\mathbb{H}_Q$ and by the definition of adjoint operators, we obtain

$$
\begin{aligned}
Tr[\Phi\Phi^*] &= \sum_{k=1}^{\infty} \langle \Phi\Phi^* f_k, f_k \rangle_H = \sum_{k=1}^{\infty} \langle \Phi^* f_k, \Phi^* f_k \rangle_{\mathbb{H}_Q} \\
&= \sum_{k=1}^{\infty} \left\langle \sum_{j=1}^{\infty} \langle \Phi^* f_k, g_j \rangle_{\mathbb{H}_Q} g_j, \sum_{j=1}^{\infty} \langle \Phi^* f_k, g_j \rangle_{\mathbb{H}_Q} g_j \right\rangle_{\mathbb{H}_Q} \\
&= \sum_{k=1}^{\infty} \sum_{j=1}^{\infty} \langle \Phi^* f_k, g_j \rangle_{\mathbb{H}_Q} \langle \Phi^* f_k, g_j \rangle_{\mathbb{H}_Q} = \sum_{k=1}^{\infty} \sum_{j=1}^{\infty} \langle f_k, \Phi g_j \rangle_H^2
\end{aligned}
$$

$^2 t_{m+1} \wedge t := \min\{t_{m+1}; t\}.$

134   4. Itô integrated stochastic Cauchy problems in Hilbert spaces

$$= \sum_{j=1}^{\infty} \sum_{k=1}^{\infty} \langle f_k, \Phi g_j \rangle_{\mathbb{H}}^2 = \sum_{j=1}^{\infty} \|\Phi g_j\|_{\mathbb{H}}^2 = \|\Phi\|_{\mathrm{HS}}^2. \qquad (4.1.22)$$

By the definition of the scalar product in $\mathbb{H}_Q$, for any $h_1, h_2 \in \mathbb{H}_Q$, we have

$$\langle h_1, h_2 \rangle_{\mathbb{H}_Q} = \langle Q^{1/2} h_1, Q^{1/2} h_2 \rangle_{\mathbb{H}}.$$

Taking into account that $Q^{1/2}$ is self-adjoint in $\mathbb{H}_Q$, we obtain

$$\|\Phi\|_{\mathrm{HS}}^2 = \sum_{k=1}^{\infty} \langle \Phi^* f_k, \Phi^* f_k \rangle_{\mathbb{H}_Q} = \sum_{k=1}^{\infty} \langle \Phi Q \Phi^* f_k, f_k \rangle_{\mathbb{H}} = Tr[\Phi Q \Phi^*]. \qquad \square$$

Now we show that any operator from $\mathcal{L}(\mathbb{H}, H)$ can be regarded as an element of $\mathcal{L}_{\mathrm{HS}}(\mathbb{H}_Q, H)$ for trace class operators $Q : \mathbb{H} \to \mathbb{H}$.

Proposition 4.1.8 *If* $\Phi \in \mathcal{L}(\mathbb{H}, H)$, *then* $\Phi_Q := \Phi|_{\mathbb{H}_Q}$ *belongs to the space* $\mathcal{L}_{HS}(\mathbb{H}_Q, H)$.

Proof. First we show that $\Phi_Q \in \mathcal{L}_{\mathrm{HS}}(\mathbb{H}_Q, H)$. Let $\{e_j\}_{j=1}^{\infty}$ be an orthonormal basis in $\mathbb{H}$ consisting of eigenvectors of Q with $Q e_j = \sigma_j^2 e_j$. Then $\{g_j = \sigma_j e_j\}_{j=1}^{\infty}$ is an orthonormal basis in $\mathbb{H}_Q$ and the following equalities hold:

$$\|\Phi_Q\|_{\mathrm{HS}}^2 = \sum_{j=1}^{\infty} \|\Phi_Q g_j\|_{\mathbb{H}}^2 = \sum_{j=1}^{\infty} \sigma_j^2 \|\Phi e_j\|_{\mathbb{H}}^2 \le \|\Phi\|^2 \sum_{j=1}^{\infty} \sigma_j^2 < \infty.$$

In addition, similar to equalities (4.1.22), we have $Tr[\Phi Q \Phi^*] = \|\Phi_Q\|_{\mathrm{HS}}^2$. $\quad \square$

The following theorem states the fundamental equality, called Itô isometry. The equality will be used in the definition of stochastic integrals in the general case. Here again at the beginning we formulate and prove the Itô isometry for elementary processes. Then, by passing to the mean square limit, we will extend it to the general case.

Theorem 4.1.2 (The Itô isometry) *If* $\{\Phi(t), t \ge 0\}$ *is an elementary process with values in* $\mathcal{L}_{\mathrm{HS}}(\mathbb{H}_Q, H)$ *satisfying the condition (4.1.19) and* W *is a* Q-*Wiener process or a cylindrical Wiener process (with* $Q = I$ *for simplicity), then*

$$\mathbf{E}\left[\left\|\int_0^t \Phi(s) dW(s)\right\|_H^2\right] = \mathbf{E}\left[\int_0^t \|\Phi(s)\|_{\mathrm{HS}}^2 \, ds\right]. \qquad (4.1.23)$$

Proof. Let $t \in [0, T]$. Denote $\Delta W_j(t) = W(t_{j+1} \wedge t) - W(t_j \wedge t)$, $j = 1, \ldots, m$. Then

$$\mathbf{E}\left[\left\|\int_0^t \Phi(s) dW(s)\right\|_H^2\right] = \mathbf{E}\left[\sum_{j=0}^m \|\Phi_j \Delta W_j(t)\|^2\right]$$

$$+ 2\mathbf{E}\left[\sum_{j=1}^m \sum_{i=0}^{j-1} \langle \Phi_i \Delta W_i(t), \Phi_j \Delta W_j(t) \rangle_{\mathbb{H}}\right],$$

4.1. Hilbert space valued variables, processes, and stochastic integrals 135

where

$$\mathbf{E}\left[\|\Phi_j \Delta W_j(t)\|^2\right] = \mathbf{E}\left[\sum_{k=1}^{\infty} \langle \Phi_j \Delta W_j(t), f_k \rangle_{\mathrm{H}}^2\right] = \sum_{k=1}^{\infty} \mathbf{E}\left[\langle \Delta W_j(t), \Phi_j^* f_k \rangle_{\mathrm{H}}^2\right].$$

By the definition of an elementary process, for any $u \in \mathbb{H}$, the random variable $\Phi_j u$ is $\mathcal{F}_{t_j}$-measurable with respect to $\mathcal{B}(H)$. Hence $\langle \Phi_j u, h \rangle_{\mathrm{H}} = \langle u, \Phi_j^* h \rangle_{\mathrm{H}}$ is $\mathcal{F}_{t_j}$-measurable with respect to $\mathcal{B}(\mathbb{R})$. By this and since $\Delta W_j(t)$ is independent of $\mathcal{F}_{t_j}$, we obtain

$$\mathbf{E}\left[\|\Phi_j \Delta W_j(t)\|_H^2\right] = \sum_{k=1}^{\infty} \mathbf{E}\left[\mathbf{E}\left[\langle \Delta W_j(t), \Phi_j^* f_k \rangle_{\mathrm{H}}^2 | \mathcal{F}_{t_j}\right]\right],$$

where, by the definition of covariance operator and by the properties of conditional expectations, we have

$$\mathbf{E}\left[\mathbf{E}\left[\langle \Delta W_j(t), \Phi_j^* f_k \rangle_{\mathrm{H}}^2 | \mathcal{F}_{t_j}\right]\right] = \mathbf{E}[\langle \Delta W_j(t), \Phi_j^* f_k \rangle_{\mathrm{H}}^2]$$

$$= \langle \mathbf{Cov}[\Delta W_j(t)] \Phi_j^* f_k, \Phi_j^* f_k \rangle_{\mathrm{H}}^2 = (t_{j+1} - t_j)\langle Q \Phi_j^* f_k, \Phi_j^* f_k \rangle_{\mathrm{H}}.$$

Hence

$$\mathbf{E}\left[\|\Phi_j \Delta W_j(t)\|_H^2\right] = (t_{j+1} - t_j)\mathbf{E}\left[Tr[\Phi_j Q \Phi_j^*]\right].$$

Since for $j = m$ the increment $\Delta W_m(t)$ is independent of $\mathcal{F}_{t_m}$ as $t > t_m$, by the same arguments, we have

$$\mathbf{E}\left[\|\Phi_m \Delta W_m(t)\|_{\mathrm{H}}^2\right] = (t - t_m)\mathbf{E}\left[Tr[\Phi_m Q \Phi_m^*]\right].$$

Thus

$$\mathbf{E}\left[\sum_{j=0}^{m} \|\Phi_j \Delta W_j(t)\|_{\mathrm{H}}^2\right] = \sum_{j=0}^{m} (t_{j+1} - t_j)\mathbf{E}\left[\|\Phi_j\|_{\mathrm{HS}}^2\right] \mathbf{E}\left[\int_0^t \|\Phi(s)\|_{\mathrm{HS}}^2 \, ds\right].$$

Since for $j \neq i$ we have $\mathbf{E}\left[\langle \Phi_i \Delta W_i(t), \Phi_j \Delta W_j(t) \rangle_{\mathrm{H}}\right] = 0$, finally we obtain (4.1.23). $\qquad\square$

Now, after clarifying the relation between the spaces $\mathcal{L}(\mathbb{H}, H)$ and $\mathcal{L}_{\mathrm{HS}}(\mathbb{H}_Q, H)$ and proving the Itô isometry for elementary processes, we can prove the following result on the mean square approximation of a predictable $\mathcal{L}_{\mathrm{HS}}(\mathbb{H}_Q, H)$ valued (in particular, $\mathcal{L}(\mathbb{H}, H)$ valued) process Φ by elementary processes. As a consequence, we define a stochastic integral of such Φ wrt a Q-Wiener process as the mean square limit of the stochastic integrals defined for elementary processes.

Proposition 4.1.9 *If $\{\Phi(t), t \in [0, T]\}$ is an $\mathcal{L}_{\mathrm{HS}}(\mathbb{H}_Q, H)$-valued predictable process satisfying (4.1.19), then there exists a sequence of elementary processes $\{\Phi_n(t), t \in [0, T]\}$ such that*

$$\mathbf{E}\left[\int_0^T \|\Phi(s) - \Phi_n(s)\|_{\mathrm{HS}}^2 \, ds\right] \to 0 \quad as \quad n \to \infty. \tag{4.1.24}$$

136   4. Itô integrated stochastic Cauchy problems in Hilbert spaces

Proof. By Proposition 4.1.8, the space $\mathcal{L}(\mathbb{H}, H)$ is embedded into $\mathcal{L}_{HS}(\mathbb{H}_Q, H)$. This implies that $\{\Phi(t),\ t \in [0, T]\}$ is $\mathcal{L}_{HS}(\mathbb{H}_Q, H)$-predictable.

Next, similar to the case of $\mathbb{R}$-valued random variables, there exists a sequence $\{\Phi_n(t),\ t \in [0, T]\}$ of elementary $\mathcal{L}(\mathbb{H}, H)$ valued predictable processes (hence $\mathcal{L}_{HS}(\mathbb{H}_Q, H)$-predictable) such that

$$\|\Phi(\omega, t) - \Phi_n(\omega, t)\|_{HS} \to 0 \quad \text{as} \quad n \to \infty$$

for all $(\omega, t) \in \Omega_T$. Consequently, (4.1.24) holds true. $\qquad\square$

Thus we can introduce the class of stochastically integrable processes as the set of $\mathcal{L}_{HS}(\mathbb{H}_Q, H)$ valued predictable processes satisfying the condition (4.1.19).

Definition 4.1.9 *Let $\mathbb{H}$ and H be separable Hilbert spaces. Let W be an $\mathbb{H}$-valued Q-Wiener or cylindrical process (considered as a (weak) Q-Wiener process with $Tr\, Q = \infty$) and $\{\Phi(t),\ t \in [0, T]\}$ be an $\mathcal{L}_{HS}(\mathbb{H}_Q, H)$ valued predictable process satisfying (4.1.19). Let $\{\Phi_n\}$ be a sequence of elementary processes mean square convergent to Φ. The stochastic integral of Φ wrt W is defined as the mean square limit of stochastic integrals:*

$$\int_0^t \Phi(s)\, dW(s) := l.i.m._{n \to \infty} \int_0^t \Phi_n(s)\, dW(s), \quad t \in [0, T]. \qquad (4.1.25)$$

The existence of such a sequence of elementary processes Φ_n is proved by Proposition 4.1.9. The stochastic integral for $\{\Phi_n\}$ is well defined by (4.1.21), namely,

$$\int_0^t \Phi_n(s) dW(s) := \sum_{k=0}^{n-1} (\Phi_n)_k \Delta W(s_k), \qquad t \in [0, T].$$

The existence of the limit in (4.1.25) follows from the Itô isometry.

Remark 4.1.2 Despite the fact that the equality (4.1.25) is taken as the definition for a stochastic integral of Φ wrt a cylindrical Wiener process under formally the same condition (4.1.19) as wrt a Q-Wiener process, it turns out to be much more restrictive in the former case. This is because in the case of a cylindrical process, the condition (4.1.19) really means $\Phi \in \mathcal{L}_{HS}((\mathbb{H}, H))$. In addition, as proved in Theorem 4.1.6, an $\mathbb{H}$-valued cylindrical process can also be regarded as a Q_1-Wiener process in a Hilbert space $\mathbb{H}_1$ defined so that the inclusion of $\mathbb{H}$ into $\mathbb{H}_1$ is a Hilbert–Schmidt operator. In this case the requirement for Φ to be $\mathcal{L}_{HS}((\mathbb{H}_1)_{Q_1}, H)$ valued is the same as to be $\mathcal{L}_{HS}(\mathbb{H}_Q, H)$ valued, in particular for $Q = I$, is the same as to be $\mathcal{L}_{HS}(\mathbb{H}, H)$ valued.

We conclude the section with two important theorems on stochastic integrals. They are Theorem 4.1.3 on change of order of integration, called the stochastic Tonelli–Fubini theorem, and Theorem 4.1.4 on change of variables in stochastic integrals, called the Itô formula. We present the theorems without proofs (for the proofs, see, e.g., [34]), but we consider in detail the terms in the equalities obtained.

4.1. Hilbert space valued variables, processes, and stochastic integrals 137

Theorem 4.1.3 *Let $(E, \mathcal{B}(E), \mu)$ be a probability space, where μ is a finite positive measure and*

$$\Phi = \Phi(t, \omega, x) \; : \; (\Omega_T \times E, \, \mathcal{B}_T \times \mathcal{B}(E)) \; \rightarrow \; (\mathcal{L}_{HS}(\mathbb{H}_Q, H), \, \mathcal{B}(\mathcal{L}_{HS}(\mathbb{H}_Q, H))$$

is integrable with respect to μ. Then

$$\int_E \left[\int_0^T \Phi(t, \cdot, x) \, dW(t) \right] d\mu(x) = \int_0^T \left[\int_E \Phi(t, \cdot, x) \, d\mu(x) \right] dW(t).$$

Theorem 4.1.4 *Let W be an $\mathbb{H}$-valued Q-Wiener or cylindrical Wiener process and $\{\Phi(t), \, t \geq 0\}$ be a predictable $\mathcal{L}_{HS}(\mathbb{H}_Q, H)$-valued process under the condition (4.1.19). Let φ be a predictable Bochner integrable H-valued process and $X(0)$ be a measurable H-valued random variable. Let $F : [0, \infty) \times H \to \mathbb{R}$ have continuous derivatives $F_t, F_x,$ and F_{xx}. Then for the process*

$$\{X(t) = X(0) + \int_0^t \varphi(s)ds + \int_0^t \Phi(s) \, dW(s), \quad t \geq 0\}$$

the Itô formula holds:

$$
\begin{aligned}
F(t, X(t)) \; = \; & F(0, X(0)) + \int_0^t \langle F_x(s, X(s)), \Phi(s) \, dW(s) \rangle \\
& + \int_0^t \Big\{ F_t(s, X(s)) + \langle F_x(s, X(s)), \varphi(s) \rangle \\
& \qquad + \frac{1}{2} Tr \left[F_{xx}(s, X(s)) \Phi(s) Q \Phi^*(s) \right] \Big\} ds. \quad (4.1.26)
\end{aligned}
$$

Now we give an interpretation for the terms of the infinite-dimensional Itô formula (4.1.26), especially the last term, and explain the formula, comparing it with the Itô formula in the finite-dimensional case.

The derivatives F_x and F_{xx} in (4.1.26) are understood in the Frechet sense (see, e.g., [53, 57]). It means that

$$F_x : [0, T] \times H \to H^* \quad \text{and} \quad F_{xx} : [0, T] \times H \to \mathcal{L}(H, H^*).$$

More precisely, for each fixed $t \in [0, T]$ and any $x \in H$,

$$F_x(t, x)(\cdot) : H \to \mathbb{R}, \qquad F_x(t, x) \in \mathcal{L}(H, \mathbb{R}) = H^*,$$

$$F_{xx}(t, x)(\cdot) : H \to H^*, \qquad F_{xx}(t, x) \in \mathcal{L}(H, H^*).$$

In the case of a Q-Wiener process, i.e., when $Q : \mathbb{H} \to \mathbb{H}$ is trace class, the term $Tr \left[F_{xx}(s, X(s)) \Phi(s) Q \Phi^*(s) \right]$ is well defined. This is indeed the case since

$$F_{xx}(s, X(s)) \Phi(s) Q \Phi^*(s) \; : \; H^* \simeq H \; \rightarrow \; H^* \simeq H$$

is a trace class operator. This is due to the fact that

$$F_{xx} : H^* \to H, \qquad \Phi : \mathbb{H} \to H, \qquad \Phi^* : H^* \to \mathbb{H}^*$$

138   *4. Itô integrated stochastic Cauchy problems in Hilbert spaces*

are bounded operators and since we identify H^* with H and $\mathbb{H}^*$ with $\mathbb{H}$ by the Riesz theorem on isomorphism.

The case of a cylindrical Wiener process (with $Q = I$) requires special attention. In this case the term $Tr\,[F_{xx}(s, X(s))\Phi(s)Q\Phi^*(s)]$ is well defined since $\Phi(s) \in \mathcal{L}_{\mathsf{HS}}(\mathbb{H}, H)$ and hence $\Phi^*(s) \in \mathcal{L}_{\mathsf{HS}}(H, \mathbb{H})$.

The second and the fourth terms in (4.1.26) can be written in the form of scalar products, again due to the Riesz theorem. To make clear why the last term in (4.1.26) is expressed in the form of trace, first note that, due to the possibility of the cyclic permutation under the trace sign ([103]), it may be also written as

$$Tr\,[\Phi^*(s)F_{xx}(s, X(s))\Phi(s)Q] = Tr\left[(Q^*)^{1/2}\Phi^*(s)F_{xx}(s, X(s))\Phi(s)Q^{1/2}\right].$$

To explain this term in the formula, recall the finite-dimensional Itô formula and compare it with the infinite-dimensional one (4.1.26). Let

$$\{X(t) = X(0) + \int_0^t \varphi(s)ds + \int_0^t \Phi(s)\,dB(s), \quad t \geq 0\}$$

be an n-dimensional stochastic process, where $\varphi(t)$ is an n-dimensional stochastic process, $B(t)$ is a p-dimensional Brownian motion, and $\Phi(t)$ is an $(n \times p)$-matrix valued stochastic process. Let $F(t, x)$ be a twice continuously differentiable map from $[0, \infty) \times \mathbb{R}^n$ into $\mathbb{R}$; then the process $F = F(t, X(t))$ is again a stochastic process given by

$$dF(t) = \frac{\partial F}{\partial t}(t, X)\,dt \;+\; \sum_i \frac{\partial F}{\partial x_i}(t, X)\,dX_i(t)$$

$$+\; \frac{1}{2}\sum_{i,j} \frac{\partial^2 F}{\partial x_i \partial x_j}(t, X)\,dX_i(t)\,dX_j(t). \quad (4.1.27)$$

Substituting $dX_i(t)dX_j(t)$ into the last term of (4.1.27) and using the Brownian motion property $dB_i(t)dB_j(t) = \delta_{ij}dt$, we obtain that the sum of diagonal elements of the matrix

$$\frac{1}{2}\Phi^* \frac{\partial^2 F}{\partial x^2}\Phi\,dt \;\;\text{is equal to}\;\; \frac{1}{2}Tr\left[\Phi^* \frac{\partial^2 F}{\partial x^2}\Phi\right]dt.$$

4.2 Solutions to Cauchy problems for equations with additive noise and generators of regularized semigroups

Now, having at hand the necessary information from the theory of Hilbert space valued stochastic integrals and the details of the semi-group techniques set out in Part I, we proceed to study the stochastic Cauchy problem (4.0.1).

4.2. Cauchy problems with additive noise

The present section is devoted to the linear stochastic Cauchy problem

$$dX(t) = AX(t)\,dt + B\,dW(t), \quad t \in [0, T], \qquad X(0) = \zeta. \qquad (4.2.1)$$

First, we consider this basic problem with A generating a strongly continuous semi-group of solution operators $U = \{U(t), t \geq 0\}$, in particular a C_0-semi-group, and with additive noise, that is, with B independent of X. We define and construct weak solutions to (4.2.1). Then we consider solutions to the problem, where A is the generator of a more general regularized semi-group $S = \{S(t), t \in [0, \tau)\}$, with $\tau \leq \infty$. In this case we obtain weak regularized solutions to (4.2.1).

We show that all the solutions obtained have the form

$$X(t) = S(t)\zeta + W_A(t), \quad t \in [0, T], \quad T < \tau,$$

where the (regularized) stochastic convolution $W_A(t) := \int_0^t S(t-s)B dW(s)$ coincides with $\int_0^t U(t-s)B dW(s)$ if S is the semi-group of solution operators $\{U(t), t \geq 0\}$.

We pay special attention to the specificity of the stochastic convolution for the problems with differential operators $A = \mathbf{A}(i\partial/\partial x)$ generating R-semi-groups.

4.2.1 Definitions, existence, uniqueness, and properties of weak solutions

Let $(\Omega, \mathcal{F}, P)$ be a probability space with a given filtration $\{\mathcal{F}_t, t \geq 0\}$ and $H, \mathbb{H}$ be separable Hilbert spaces. We consider solutions to the stochastic Cauchy problem (4.2.1) where $\{W(t), t \geq 0\}$ is the $\mathbb{H}$-valued Q-Wiener or cylindrical Wiener process, $W(t)$ is measurable wrt $\{\mathcal{F}_t\}$ (often the given filtration is defined by W itself), $B \in \mathcal{L}(\mathbb{H}, H)$, ζ is an $\mathcal{F}_0$-measurable H-valued random value, and A is the generator of a strongly continuous at $t > 0$ semi-group of solution operators $\{U(t), t \geq 0\}$ in H, in particular, the generator of a C_0-semi-group.

By the Cauchy problem (4.2.1) we mean the integral equation

$$X(t) = \zeta + \int_0^t AX(s)ds + \int_0^t BdW(s), \quad t \in [0, T]. \qquad (4.2.2)$$

According to this, strong and weak solutions to (4.2.1) are defined.

Definition 4.2.1 *By a strong solution of (4.2.1) we call an H-valued predictable process $X = \{X(t), t \in [0, T]\}$, $X(t) = X(t, \omega)$, $\omega \in (\Omega, \mathcal{F}, P)$, such that*

(a) *for each $t \in [0, T]$, $X(t) \in \operatorname{dom} A$ and $\int_0^t \|AX(s)\|_H\, ds < \infty$ $P_{a.s.}$;*

140   *4. Itô integrated stochastic Cauchy problems in Hilbert spaces*

(b) X *a.s. pathwise satisfies Equation (4.2.2), that is,*

$$P\{\omega \in \Omega : X(t) = \zeta + \int_0^t AX(s)\, ds + BW(t), \quad t \geq 0\} = 1.$$

As follows from the definition, the existence of a strong solution necessitates the solution to be in *dom A*, otherwise certain strict conditions on A, like boundedness, should be imposed. Since unbounded operators A generally arise in applications, especially differential ones, we will pay more attention to weak solutions.

Definition 4.2.2 *By a weak solution of the problem (4.2.1) we call an H-valued predictable process $X = \{X(t),\, t \geq 0\}$ such that*

(a) *for each $t \in [0,T]$, $\int_0^t \|X(s)\|_H\, ds < \infty$ $P_{\text{a.s.}}$;*

(b) *X pathwise satisfies the equation*

$$\langle X(t), y \rangle = \langle \zeta, y \rangle + \int_0^t \langle X(s), A^*y \rangle\, ds + \langle BW(t), y \rangle \quad P_{\text{a.s.}},$$

$$t \in [0,T], \quad y \in dom\, A^*. \quad (4.2.3)$$

Before constructing weak solutions to (4.2.1), let us discuss how strong and weak solutions are related to solutions of the corresponding homogeneous problem. Since we consider here the stochastic Cauchy problem in the integral form and the supposed solutions of the stochastic problem depend on solution operators of the corresponding integral homogeneous Cauchy problem, it is natural to begin with relations between the "classical" homogeneous Cauchy problem

$$u'(t) = Au(t), \quad t \geq 0, \qquad u(0) = \zeta \in dom\, A, \qquad (4.2.4)$$

and the integral problems

$$u(t) = \zeta + \int_0^t Au(s)\, ds, \qquad t \geq 0, \quad \zeta \in dom\, A,$$

$$u(t) = \zeta + A \int_0^t u(s)\, ds, \qquad t \geq 0, \quad \zeta \in H. \qquad (4.2.5)$$

As shown in Section 1.1, for the case of a C_0-semi-group these problems are equivalent in the following sense: the unique solution $u(t) = U(t)\zeta,\, t \geq 0$, for each $\zeta \in dom\, A$ to the problem (4.2.4) exists if and only if $U(t)\zeta$ satisfies (4.2.5) for each $\zeta \in H$ and $U(t)A\zeta = AU(t)\zeta$ for $\zeta \in dom\, A$. Thus the following problems are equivalent for a semi-group $\{U(t),\, t \geq 0\}$ of class C_0:

$$U'(t)\zeta = AU(t)\zeta = U(t)A\zeta, \qquad \zeta \in dom\, A, \quad U(0) = I, \qquad (4.2.6)$$

4.2. Cauchy problems with additive noise

$$U(t)\zeta = \zeta + A \int_0^t U(t)\zeta, \qquad \zeta \in H. \tag{4.2.7}$$

It turns out that this equivalence holds for a wider class of strongly continuous semi-groups, namely, for Abel summable semi-groups. For A being the generator of a strongly continuous at $t > 0$ semi-group of solution operators $U(t)$, $t \geq 0$, the problem (4.2.5) is more general than (4.2.4).

Now, in view of the relations between the problems (4.2.6)–(4.2.7) and taking into account the presence of the stochastic term and H-valued random variable ζ in the stochastic problem, we can compare the following sequence of stochastic problems and notice their increasing generality, beginning from the problem corresponding to strong solutions and up to the problem corresponding to weak solutions:

$$X(t) = \zeta + \int_0^t AX(s)\,ds + BW(t) \quad P_{\text{a.s}},$$

$$X(t) = \zeta + A \int_0^t X(s)\,ds + BW(t) \quad P_{\text{a.s}},$$

$$\langle X(t), y \rangle = \langle \zeta, y \rangle + \langle \int_0^t X(s)\,ds, A^*y \rangle + \langle BW(t), y \rangle \quad P_{\text{a.s}}, \quad y \in \text{dom } A^*,$$

$$\langle X(t), y \rangle = \langle \zeta, y \rangle + \int_0^t \langle X(s), A^*y \rangle ds + \langle BW(t), y \rangle \quad P_{\text{a.s}}, \quad y \in \text{dom } A^*,$$

where $t \in [0, T]$ or $t \in [0, \tau), \tau \leq \infty$.

Comparing these problems, we see that the last one, that is, the Cauchy problem (4.2.3) in Definition 4.2.2, is weaker than the first one, i.e., the problem (4.2.2) in Definition 4.2.1, due to the following two reasons. The first one is the fact that we consider the problem where the operator A is not under the integral sign, but before it. Thus the solution X need not to be in $\text{dom } A$; only the integral of X needs to be. The second reason is the fact that, considering the problem in the weak sense, we change A by A^* applied to $y \in \text{dom } A^*$ and, due to this, the integral $\int_0^t X(s)\,ds$ does not need to be in $\text{dom } A$ too. Moreover, as shown in Section 4.1, the stochastic term BW in the first and second equations must be a BQB^*-Wiener process with a trace class operator BQB^*, while the term $\langle BW, y \rangle$ in the last two equations is well defined for a cylindrical process W and bounded B.

Thus we will study weak solutions to the stochastic problem (4.2.3) with different unbounded generators. In order to construct a weak solution to the linear additive stochastic Cauchy problem, where A generates a strongly continuous semi-group $\{U(t), t \geq 0\}$, we begin with the study of the Itô integral $W_A = \int_0^t U(t-s)B\,dW(s)$, $t \geq 0$, which is the main part of the process:

$$\{X(t) = U(t)\zeta + \int_0^t U(t-s)B\,dW(s), \quad t \geq 0\}. \tag{4.2.8}$$

142   4. Itô integrated stochastic Cauchy problems in Hilbert spaces

The latter is supposed to be a weak solution, due to analogy with the well-known formula for ordinary differential equations. Later in this section we will show that (4.2.8) is a unique weak solution to (4.2.1) in the sense of Definition 4.2.2, i.e., that it is a unique solution to (4.2.3).

Now we consider conditions for the existence of W_A based on the conditions given for the existence of stochastic integrals in Section 4.1.

Definition 4.2.3 *Let $\{W(t), t \geq 0\}$ be an $\mathbb{H}$-valued Wiener process and $\{U(t), t \geq 0\}$ be a strongly continuous semi-group in H satisfying the condition*

$$\int_0^t \|U(r)B\|_{HS}^2 \, dr < \infty, \quad t > 0, \tag{4.2.9}$$

where

$$\|U(r)B\|_{HS}^2 := \sum_{j=1}^{\infty} \|U(r)BQ^{\frac{1}{2}} e_j\|^2 = Tr(Q^{*\frac{1}{2}} B^* U^*(r) U(r) B Q^{\frac{1}{2}}).$$

Then $W_A(t) = \int_0^t U(t-s)B \, dW(s)$, $t \geq 0$, is called a stochastic convolution.

Let us show that the stochastic convolution W_A is well defined under the condition (4.2.9) wrt Q-Wiener and cylindrical Wiener processes W as a particular case of the stochastic integral (4.1.18) defined under the condition (4.1.19):

$$\mathbf{E}\left[\int_0^t \|\Phi(s)\|_{HS}^2 \, ds\right] < \infty, \tag{4.2.10}$$

where

$$\Phi(s) : \mathbb{H}_Q \to H, \qquad \|\Phi(s)\|_{HS}^2 := \mathrm{Tr}[\Phi(s)Q\Phi^*(s)].$$

As we explained in Section 4.1, the condition (4.2.10), being formally the same for Q-Wiener and cylindrical Wiener processes (where $TrQ = \infty$) turns out to be different in each case.

It is easy to see that, for a bounded operator $B : \mathbb{H} \to H$, a Q-Wiener process, and a semi-group satisfying the condition (4.2.9), the estimate

$$\mathbf{E}\int_0^t \|U(t-s)B\|_{HS}^2 \, ds = \int_0^t \|U(r)B\|_{HS}^2 \, dr < \infty \tag{4.2.11}$$

holds. The estimate is sufficient for $W_A(t)$ to be well defined in the case of a Q-Wiener process and a C_0-semi-group. The stochastic convolution is also well defined for more general strongly continuous as $t > 0$ semi-groups $\{U(t), t \geq 0\}$ with square-integrable singularities at $t = 0$ since

$$\int_0^t \|U(s)B\|_{HS}^2 \, ds = \int_0^t \sum_{j=1}^{\infty} \|U(s)BQ^{\frac{1}{2}} e_j\|^2 ds \leq \sum_{j=1}^{\infty} \|BQ^{\frac{1}{2}} e_j\|^2 \int_0^t \|U(s)\|^2 \, ds.$$

$$\tag{4.2.12}$$

4.2. Cauchy problems with additive noise

For the same semi-groups in the case of a cylindrical Wiener process, $W_A(t)$ is well defined under the additional condition $U(s)B \in \mathcal{L}_{\mathsf{HS}}(\mathbb{H}, H)$.

Now we present some properties of the stochastic convolution W_A. These properties will be used in studying solutions to stochastic problems in different settings. We begin with the case of a Q-Wiener process and a C_0-semi-group. After analyzing the stochastic convolution properties and properties of solutions to the stochastic Cauchy problem obtained in this case, we will use them for studying solutions to stochastic problems in the case of more general semi-groups and cylindrical Wiener processes.

Proposition 4.2.1 *Let $\{W(t),\, t \ge 0\}$ be an $\mathbb{H}$-valued Q-Wiener process, $B \in \mathcal{L}(\mathbb{H}, H)$, and $\{U(t),\, t \ge 0\}$ be a C_0-semi-group in H. Then the stochastic convolution W_A is a predictable process, which is continuous in the sense of mean square convergence:*

$$\lim_{t \to s} \mathbf{E}\, \|W_A(t) - W_A(s)\|^2 = 0.$$

Proof. Let $0 < s < t < \infty$. We have

$$W_A(t) - W_A(s) = \int_0^s [U(t - r) - U(s - r)] B\, dW(r) + \int_s^t U(t - r) B\, dW(r).$$

It follows that

$$\mathbf{E}\, \|W_A(t) - W_A(s)\|^2$$

$$= \mathbf{E}\, \Big\| \int_0^s [U(t - r) - U(s - r)] B\, dW(r) \Big\|^2 + \mathbf{E}\, \Big\| \int_0^{t-s} U(r) B\, dW(r) \Big\|^2$$

$$+ 2\mathbf{E}\, \Big\langle \int_0^s [U(t - r) - U(s - r)] B\, dW(r), \int_s^t U(r) B\, dW(r) \Big\rangle_H. \qquad (4.2.13)$$

The last term in (4.2.13) is equal to zero by the property of a stochastic integral wrt a Wiener process. We apply the Itô isometry for abstract stochastic integrals to the second term:

$$\mathbf{E}\, \Big\| \int_0^{t-s} U(r) B\, dW(r) \Big\|^2 = \mathbf{E}\, \int_0^{t-s} \|U(r)B\|_{\mathsf{HS}}^2\, dr = \int_0^{t-s} \|U(r)B\|_{\mathsf{HS}}^2\, dr.$$

Using the absolute continuity property of Bochner integrals for integrable functions, we obtain

$$\lim_{t-s \to 0} \int_0^{t-s} \|U(r)B\|_{\mathsf{HS}}^2\, dr = 0.$$

At last, by the semi-group continuity as $t \ge 0$, we have

$$\lim_{t-s \to 0} \mathbf{E}\, \Big\| \int_0^s [U(t - r) - U(s - r)] B\, dW(r) \Big\|^2$$

$$= \lim_{t-s \to 0} \int_0^s \|[U(r + t - s) - U(r)]B\|_{\mathsf{HS}}^2\, dr = 0.$$

144   4. Itô integrated stochastic Cauchy problems in Hilbert spaces

Thus the mean square continuity holds and consequently, as mentioned in the previous section, the stochastic convolution has a predictable version and can be used as part of the weak solution. $\qquad\square$

Note that for $B \in \mathcal{L}_{HS}(\mathbb{H}, H)$ Proposition 4.2.1 is also true for a cylindrical Wiener process since the estimate (4.2.11) holds for such operators.

Now we show the existence of weak solutions to the linear additive stochastic Cauchy problem (4.2.1) and consider probability characteristics of the solutions. As the first result we show that a weak solution to (4.2.1) with the generator of a C_0-semi-group exists and can be constructed in the form of a sum of two terms. The first one is the solution to the corresponding homogeneous Cauchy problem and the second one is the stochastic convolution. Then we will extend this result to more general strongly continuous semi-groups.

Theorem 4.2.1 *Let A be the generator of a C_0-semi-group $\{U(t),\ t \geq 0\}$ in a Hilbert space H and $\{W(t),\ t \geq 0\}$ be a Q-Wiener process in a Hilbert space $\mathbb{H}$, $B \in \mathcal{L}(\mathbb{H}, H)$. Then, for each $\mathcal{F}_0$-measurable H-valued random variable ζ, the process (4.2.8) is a weak solution to (4.2.1).*

Proof. We begin with the first term of the sum (4.2.8) and show that $\{U(t)\zeta,\ t \geq 0\}$ is $P_{\text{a.s.}}$ a weak solution for the corresponding homogeneous Cauchy problem. By the definition of a weak solution for the stochastic Cauchy problem, the process must be predictable.

Let us prove that $U\zeta$ is a predictable process for each H-valued $\mathcal{F}_0$-measurable ζ. The process $\{U(t)\zeta,\ t \geq 0\}$ is $\mathcal{F}_t$-measurable as the composition of the deterministic function $U(t, h) = U(t)h$, which is measurable as a function of the pair of variables $(t, h) \in [0, +\infty) \times H$ and $\mathcal{F}_0$-measurable function ζ. The paths of this process are $P_{\text{a.s.}}$ continuous on $[0, +\infty)$ due to the strong continuity of the semi-group. Hence the process $\{U(t)\zeta\}$ with ζ being an $\mathcal{F}_0$-measurable H-valued random variable is a measurable mapping from $([0, \infty) \times (\Omega, \mathcal{F}_\infty, P)$ into $(H, \mathcal{B}(H))$, i.e., predictable. Its paths are integrable due to the strong continuity as $t \geq 0$ of a C_0-semi-group. Hence, $\{U(t)\zeta\}$ can be used as part of a weak solution.

Now we prove that $X(t) = U(t)\zeta,\ t \geq 0$, for an H-valued $\mathcal{F}_0$-measurable ζ satisfies the homogeneous problem corresponding to (4.2.3). Due to the equivalence mentioned above of the homogeneous problems (4.2.6) and (4.2.7) which holds in the case of a C_0-semi-group $\{U(t),\ t \geq 0\}$, we obtain the homogeneous integral equation from (4.2.7) for each value of the H-valued random variable ζ. Given this equation and the continuity of the scalar product, we obtain for each $y \in dom\ A^*$

$$\langle \zeta, y \rangle + \int_0^t \langle U(r)\zeta, A^*y \rangle\, dr = \langle \zeta, y \rangle + \langle \int_0^t U(r)\zeta, A^*y \rangle\, dr$$

$$= \langle \zeta, y \rangle + \langle A \int_0^t U(r)\zeta\, dr, y \rangle = \langle U(t)\zeta, y \rangle, \quad t \geq 0\ \ P_{\text{a.s.}}.$$

4.2. Cauchy problems with additive noise

Hence, for $X(t) = U(t)\zeta$ with an H-valued random variable ζ, we have the equality

$$\langle X(t), y \rangle = \langle \zeta, y \rangle + \int_0^t \langle X(s)\, ds, A^* y \rangle, \quad t \geq 0 \ P_{a.s.}, \quad y \in dom\, A^*. \quad (4.2.14)$$

Thus $\{U(t)\zeta, t \geq 0\}$ satisfies (4.2.3) in the homogeneous case; hence $X = U\zeta$ is a weak solution to the corresponding homogeneous Cauchy problem with an H-valued $\mathcal{F}_0$-measurable initial data ζ.

Now we show that the stochastic convolution W_A is a weak solution to (4.2.1) with zero initial data. By Proposition 4.2.1, $\{W_A(t), t \geq 0\}$ is a predictable process. Next we show that the stochastic convolution has integrable paths, i.e., for each $0 < T < \infty$, $\int_0^T \|W_A(t)\|_H\, dt < \infty \ P_{a.s.}$. Then the estimate (a) from Definition 4.2.2 of weak solutions will be proved. Using the absolute continuity property of Bochner integrals for integrable functions, we obtain that the function $\int_0^t \|U(t-s)B\|_{HS}^2\, ds$ is continuous in $t \in [0, \infty)$, hence it is integrable on any $[0, T] \subset [0, \infty)$ and, by the abstract Itô identity (4.1.23) for H-valued stochastic integrals, we have

$$\int_0^T \int_0^t \|U(t-s)B\|_{HS}^2\, ds\, dt = \int_0^T \mathbf{E}\, \|W_A(t)\|_H^2\, dt = \mathbf{E} \int_0^T \|W_A(t)\|_H^2\, dt < \infty.$$

It remains to show that W_A satisfies Equation (4.2.3) with $\zeta = 0$:

$$\langle W_A(t), y \rangle = \int_0^t \langle W_A(s), A^* y \rangle\, ds + \int_0^t \langle B\, dW(s), y \rangle, \quad y \in dom\, A^*. \quad (4.2.15)$$

Due to the continuity of a scalar product and to the properties of stochastic integrals, we have

$$\int_0^t \langle W_A(s), A^* y \rangle\, ds = \int_0^t \langle \int_0^s U(s-r)B\, dW(r), A^* y \rangle\, ds$$

$$= \int_0^t \int_0^s \langle U(s-r)B\, dW(r), A^* y \rangle\, ds = \int_0^t \langle B\, dW(r), \int_r^t U^*(s-r)A^* y\, ds \rangle$$

$$= \int_0^t \langle B\, dW(r), \int_0^{t-r} U^*(s)A^* y\, ds \rangle, \quad y \in dom\, A^*. \quad (4.2.16)$$

As is known (see Section 1.1), if A is the generator of a C_0-semi-group, the adjoint operator A^* is the generator of the semi-group $\{U^*(t), t \geq 0\}$, which is a C_0-semi-group too. Hence, for each $y \in dom\, A^*$, we have

$$\int_0^{t-r} U^*(s)A^* y\, ds = \int_0^{t-r} A^* U^*(s)y\, ds = U^*(t-r)y - y,$$

146   *4. Itô integrated stochastic Cauchy problems in Hilbert spaces*

and as a result we obtain

$$
\begin{aligned}
\int_0^t \langle W_A(s), A^* y \rangle \, ds &= \int_0^t \langle B \, dW(r), U^*(t-r)y - y \rangle \\
&= \int_0^t \langle U(t-r)B \, dW(r), y \rangle - \int_0^t \langle B \, dW(r), y \rangle \\
&= \langle W_A(t), y \rangle - \int_0^t \langle B \, dW(r), y \rangle, \qquad y \in dom\, A^*.
\end{aligned}
$$

This means that the equality (4.2.15) is true for $X(t) = W_A(t)$, $t \geq 0$, and the process $\{W_A(t) = \int_0^t U(t-s)B \, dW(s), \ t \geq 0\}$ satisfies (4.2.3) with zero initial data. Consequently, the process (4.2.8) is a weak solution to (4.2.1) with general initial data $X(0) = \zeta$. $\qquad\square$

Remark 4.2.1 Under the condition $B \in \mathcal{L}_{HS}(\mathbb{H}, H)$, Theorem 4.2.1 remains valid for a cylindrical Wiener process since the estimate (4.2.11) holds for such operators.

Remark 4.2.2 It is easy to see from the proof of Theorem 4.2.1 that if in (4.2.2) we have an additional deterministic H-valued integrable inhomogeneity f

$$
X(t) = \zeta + \int_0^t AX(s) \, ds + \int_0^t f(s) \, ds + BW(t), \quad t \geq 0, \quad X(0) = \zeta,
$$

then the problem has the following weak solution:

$$
X(t) = U(t)\zeta + \int_0^t U(t-s)f(s) \, ds + \int_0^t U(t-s)B \, dW(s), \qquad t \geq 0.
$$

Now consider the question, what are the wider classes of strongly continuous in $t > 0$ solution operators, to which we can extend the result proved above. We will show that the existence result obtained for C_0-semi-groups can be extended to the class $(1, \mathcal{A})$ of Abel summable semi-groups (Definition 1.1.5). Recall that this is a strongly continuous family of solution operators $\{U(t), \ t \geq 0\}$ with the range of $U(t)$ dense in H, with the property $\int_0^t \|U(s)\| \, ds < \infty$, $t \geq 0$, and Abel summable:

$$
\lim_{\lambda \to \infty} \lambda \mathcal{R}_A(\lambda)f = f, \qquad f \in H.
$$

As noted in Section 1.1, for a strongly continuous exponentially bounded family of operators $\{U(t), \ t \geq 0\}$ satisfying these properties, there exists a closed densely defined operator A (the generator of the family) such that (4.2.7) holds:

$$
U(t)\zeta = \zeta + A \int_0^t U(s)\zeta \, ds, \qquad t \geq 0, \quad \zeta \in H.
$$

4.2. Cauchy problems with additive noise

147

Due to those properties of semi-groups of class $(1, \mathcal{A})$, which are similar to the properties of C_0-semi-groups, we can prove one more general existence result for the linear stochastic Cauchy problem.

Theorem 4.2.2 *Let $\{W(t), t \geq 0\}$ be a Q-Wiener process in a Hilbert space $\mathbb{H}$ and A be the generator of a semi-group $\{U(t), t \geq 0\}$ of class $(1, \mathcal{A})$ in a Hilbert space H satisfying (4.2.9). Then, for each $\mathcal{F}_0$-measurable H-valued random variable ζ, the process (4.2.8) is a weak solution to (4.2.1).*

Proof. We come to the proof by analyzing the usage of strong continuity as $t \geq 0$ of C_0-semi-groups in the proof of Theorem 4.2.1 and extending it to the case of strongly continuous as $t > 0$ semi-groups.

We begin with the first term in (4.2.8) and show that the process $\{U(t)\zeta, t \geq 0\}$ $P_{\text{a.s.}}$ is a solution to the corresponding homogeneous Cauchy problem. First we show that the process is predictable by proving that the strong continuity of the semi-group U is sufficient for predictability: let $\zeta \in dom\, A \,\, P_{\text{a.s.}}$; then the process $\{U(t)\zeta, t \geq 0\}$ is $\mathcal{F}_t$-measurable as the composition of deterministic function $U(t, h) = U(t)h$ measurable wrt the pair of variables $(t, h) \in (0, +\infty) \times dom\, A$ and the $\mathcal{F}_0$-measurable function ζ. The paths of this process are $P_{\text{a.s.}}$ continuous on $(0, +\infty)$ due to the strong continuity of the semi-group on $H = \overline{dom\, A}$ and are integrable on $[0, +\infty)$ due to the properties of Abel summable semi-groups. Hence the process $\{U(t)\zeta\}$ with an $\mathcal{F}_0$-measurable H-valued ζ is predictable. Its paths are integrable due to the integrability property of semi-groups of class $(1, \mathcal{A})$.

Now take an arbitrary $y \in dom\, A^*$ and prove (4.2.3) for $X(t) = U(t)\zeta$, where ζ is an $\mathcal{F}_0$-measurable H-valued random variable. As mentioned above, for the generator of a semi-group $\{U(t), t \geq 0\}$ of class $(1, \mathcal{A})$ the equality (4.2.7) as well as the equivalence of problems (4.2.6)–(4.2.7) hold true. Using these properties of semi-groups of class $(1, \mathcal{A})$) and the continuity of a scalar product, we obtain (4.2.14) for $X(t) = U(t)\zeta$:

$$\langle \zeta, y \rangle + \int_0^t \langle U(r)\zeta, A^*y \rangle \, dr = \langle U(t)\zeta, y \rangle \quad P_{\text{a.s.}}, \quad t \geq 0, \quad \zeta \in H.$$

Hence $\{U(t)\zeta, t \geq 0\}$ is a weak solution to the corresponding homogeneous Cauchy problem with an $\mathcal{F}_0$-measurable H-valued initial data ζ.

Now we show that the stochastic convolution W_A is a weak solution to (4.2.1) with zero initial data. First of all, by (4.2.12), the condition (4.2.9) is fulfilled and the stochastic convolution is well defined in the case of a Q-Wiener process and a strongly continuous semi-group with a square integrable singularity at zero; in the case of a cylindrical Wiener process the condition (4.2.9) is fulfilled and the stochastic convolution is defined under the additional condition for $U(t)B$ to be trace class.

Further, according to the definition of a weak solution, we need to prove that W_A is predictable. For semi-groups of class C_0 in Proposition 4.2.1 it is proved due to mean square continuity: $\lim_{t \to s} \mathbf{E}\|W_A(t) - W_A(s)\|^2 = 0$, which

148   4. Itô integrated stochastic Cauchy problems in Hilbert spaces

follows from the equality (4.2.13). It is easy to see that the arguments for the
terms in the second line of (4.2.13) are true for a semi-group of class $(1, \mathcal{A})$
too. As for the term in the first line of (4.2.13), let $\delta > 0$; then

$$\lim_{t-s \to 0} \mathbf{E} \| \int_0^s [U(t-r) - U(s-r)] B \, dW(r) \|^2$$

$$= \lim_{t-s \to 0} \int_0^s \| [U(r+t-s) - U(r)] B \|_{\text{HS}}^2 \, dr$$

$$= \lim_{t-s \to 0} (\int_0^\delta \| [U(r+t-s) - U(r)] B \|_{\text{HS}}^2 \, dr + \int_\delta^s \| [U(r+t-s) - U(r)] B \|_{\text{HS}}^2 \, dr) = 0$$

and the equalities hold true for strongly continuous square integrable at $t = 0$
semi-groups since the second term can be made small due to the semi-group
continuity as $t > 0$ for any $\delta > 0$ and the first term due to the choice of small
$\delta > 0$. Hence the square continuity is proved. By the mean square continuity, $\{W_A(t), t \geq 0\}$ is a predictable process. Thus the stochastic convolution
$\{W_A(t), t \geq 0\}$ can be used as part of a weak solution to (4.2.3) with generators of strongly continuous semi-groups.

It remains to show that for strongly continuous semi-groups and Wiener
processes under the condition (4.2.9) W_A satisfies Equation (4.2.3) with $\zeta =$
0. Similar to the proof of Theorem 4.2.1, due to the continuity of a scalar
product and properties of stochastic integrals, we have the equalities (4.2.16)
for the semi-groups and Wiener processes considered. Further, in the proof of
Theorem 4.2.1, we used that the adjoint operator A^* for A, the generator of a
C_0-semi-group, generates a C_0-semi-group $\{U^*(t), t \geq 0\}$. Here, as shown in
Section 1.1, for A, the generator of a semi-group of class $(1, \mathcal{A})$, the operator
A^* is the generator of the adjoint semi-group of bounded for each $t > 0$
operators $U^*(t)$ (and $\|U^*(t)\| = \|U(t)\|$), which is of class $(1, \mathcal{A})$ too. The
equalities (4.2.6) and (4.2.7), which hold for C_0-semi-groups, hold also for the
dual $(1, \mathcal{A})$ class semi-group $U^* = \{U^*(t), t \geq 0\}$:

$$A^* U^*(t) y = U^*(t) A^* y, \qquad y \in dom \, A^*;$$

$$U^*(t) y = A^* \int_0^t U^*(r) y \, dr + y, \qquad y \in \overline{dom \, A^*}, \quad t \geq 0. \tag{4.2.17}$$

(As a matter of fact, Abel summable semi-groups form the widest class of
semi-groups with this property [43].)

Using the equalities (4.2.17), similar to the case of C_0-semi-groups, we obtain the equality (4.2.15) true for $X(t) = W_A(t), t \geq 0$. Hence, the process
$\{W_A(t), t \geq 0\}$ is a weak solution to (4.2.1) with zero initial data $\zeta = 0$. Consequently we have proved that the process (4.2.8) is a weak solution to (4.2.1)
with general initial $\mathcal{F}_0$-measurable data $X(0) = \zeta$ and with the generator of
a square-integrable at zero semi-group of class $(1, \mathcal{A})$. $\qquad \square$

4.2. Cauchy problems with additive noise

Now let us present some more properties of the solution obtained:

$$X(t) = U(t)\zeta + W_A(t) = U(t)\zeta + \int_0^t U(t-s)B\,dW(s), \quad t \geq 0,$$

related to its probability characteristics. The probability characteristics of the process $\{U(t)\zeta, t \geq 0\}$ are determined by given characteristics of the random variable ζ: $\mathbf{E}\,U(t)\zeta = U(t)\mathbf{E}\,\zeta$. By the properties of stochastic integrals, the process $\{W_A(t), t \geq 0\}$ is Gaussian and $\mathbf{E}\,W_A(t) = 0$. Taking into account the definition of the covariance operator $\mathbf{Cov}$,

$$
\begin{aligned}
\mathbf{Cov}W_A(t)\,x &= \mathbf{E}\left(\langle W_A(t), x\rangle_{\mathbb{H}}\,W_A(t)\right) \\
&= \mathbf{E}\left\langle \int_0^t U(t-s)B\,dW(s), x\right\rangle \int_0^t U(t-s)B\,dW(s), \quad x \in \mathbb{H},
\end{aligned}
$$

and the definition of a convolution, in particular the stochastic convolution $W_A(t)$, defined as the limit of finite sums in $L^2(\Omega; H)$, and using the property of a BQB^*-Wiener process BW with the self-adjoint trace class operator BQB^*: $\mathbf{Cov}BW(t) = tBQB^*$, we obtain

$$\mathbf{Cov}[W_A(t)]x = \int_0^t U(\tau)BQB^*U^*(\tau)x\,d\tau. \tag{4.2.18}$$

Let us prove the equality in more detail. To evaluate the covariance operator $\mathbf{Cov}[X(t)] = \mathbf{Cov}[W_A(t)]$ we write the stochastic convolution W_A as the limit (in $L^2(\Omega; \mathbb{H})$) of integral sums:

$$W_A(t) = \int_0^t U(t-s)B\,dW(s) = \lim_{N\to\infty}\sum_{i=1}^N U(t-s_i)B[W(s_{i+1}) - W(s_i)].$$

Then

$$
\begin{aligned}
\mathbf{Cov}[X(t)]x &= \mathbf{Cov}[W_A(t)]x = \mathbf{E}\left[W_A(t)\langle W_A(t), x\rangle\right] \\
&= \lim_{N\to\infty}\sum_{i=1}^N U(t-s_i)B\mathbf{E}\Big[[W(s_{i+1}) - W(s_i)] \\
&\qquad\qquad\qquad\qquad \cdot\langle W(s_{i+1}) - W(s_i), B^*U^*(t-s_i)x\rangle\Big] \\
&= -\lim_{N\to\infty}\sum_{i=1}^N U(\tau_i)B\mathbf{E}\Big[[W(t-\tau_{i+1}) - W(t-\tau_i)] \\
&\qquad\qquad\qquad\qquad \cdot\langle W(t-\tau_{i+1}) - W(t-\tau_i), B^*U^*(\tau_i)x\rangle\Big] \\
&= -\lim_{N\to\infty}\sum_{i=1}^N U(\tau_i)B\mathbf{Cov}[(W(t-\tau_{i+1}) - W(t-\tau_i))B^*U^*(\tau_i)x] \\
&= \lim_{N\to\infty}\sum_{i=1}^N U(\tau_i)BQB^*U^*(\tau_i)(\tau_{i+1} - \tau_i)x = \int_0^t U(\tau)BQB^*U^*(\tau)x\,d\tau.
\end{aligned}
$$

150   *4. Itô integrated stochastic Cauchy problems in Hilbert spaces*

In the case of B commuting with $U(t)$,

$$\mathbf{Cov}W_A(t) = \int_0^t BU(\tau)QU^*(r)B^* \, d\tau, \qquad t \geq 0.$$

As mentioned above, the operator B can be taken from the space $\mathcal{L}_{HS}(\mathbb{H}_Q, H)$, in particular from $\mathcal{L}(\mathbb{H}, H)$ in the case of a Q-Wiener process and from $\mathcal{L}_{HS}(\mathbb{H}, H)$ in the case of a cylindrical Wiener process.

Thus, for the weak solution $X = U\zeta + W_A$ to the Cauchy problem (4.2.1), we have

$$\mathbf{E}X(t) = U(t)\mathbf{E}\zeta, \quad \mathbf{Cov}X(t) = \int_0^t U(\tau)BQB^*U^*(\tau)d\tau, \qquad t \geq 0.$$

Now we study the uniqueness of the weak solution obtained. In this connection it is important to note that in the case of linear stochastic problems we cannot proceed as is commonly done in the case of deterministic problems when one wants to prove the uniqueness of a solution, i.e., by considering the difference of two solutions $X_1 - X_2$ and proving that it is equal to zero. This is due to the fact that a solution of the stochastic problem is determined only up to a version.

Theorem 4.2.3 *Let A be the generator of a strongly continuous semi-group $\{U(t), t \geq 0\}$ of class $(1, \mathcal{A})$ in a Hilbert space H satisfying the condition (4.2.9) and let $\{W(t), t \geq 0\}$ be a Wiener process in a Hilbert space $\mathbb{H}$. Let $B \in \mathcal{L}(\mathbb{H}, H)$ in the case of a Q-Wiener process and $B \in \mathcal{L}_{HS}(\mathbb{H}, H)$ in the case of a cylindrical Wiener process. Then the process (4.2.8) is a unique weak solution to (4.2.1).*

Proof. As usual for a linear Cauchy problem, it is enough to prove the uniqueness of a weak solution with zero initial data. Thus we will show that any weak solution $X = \{X(t), t \geq 0\}$ to the problem (4.2.1) with $\zeta = 0 \, P_{a.s}$ is equal to W_A. To prove this we need the following *statement.*

Let X be a weak solution (4.2.1) with $\zeta = 0 \, P_{a.s}$. Then for each function $y(\cdot) \in C^1([0, \tau]; dom \, A^*)$, $\tau \in [0, \infty)$, the following equality holds:

$$\langle X(t), y(t) \rangle = \int_0^t \langle X(s), y'(s) + A^*y(s) \rangle \, ds + \int_0^t \langle B \, dW(s), y(s) \rangle. \quad (4.2.19)$$

To prove (4.2.19) we first consider a function of the form $y(s) = y_0 \varphi(s)$, $s \in [0, t]$, where $\varphi \in C^1([0, t])$, $y_0 \in dom \, A^*$, and similarly to the proof of the theorem in the case of a C_0-semi-group [20, 81], define the process

$$F_{y_0}(t) = \int_0^t \langle X(s), A^*y_0 \rangle \, ds + \langle BW(t), y_0 \rangle. \quad (4.2.20)$$

The integral in the defined process exists due to the properties of a weak

4.2. Cauchy problems with additive noise

solution. Taking into account that X is a weak solution to Equation (4.2.3) with $\zeta = 0$,

$$\langle X(t), y \rangle = \int_0^t \langle X(s), A^*y \rangle \, ds + \langle BW(t), y \rangle \ P_{\text{a.s}}, \ t \geq 0, \ y \in dom \, A^*,$$

and comparing with the definition of F_{y_0}, we obtain the equality

$$\varphi(t) F_{y_0}(t) = \langle X(t), \varphi(t)y_0 \rangle = \langle X(t), y(t) \rangle \ P_{a.s.} \quad \text{for } y(\cdot) = \varphi(\cdot)y_0 \in dom A^*.$$

Now we apply the Itô formula to the process $\{F_{y_0}(s), \ s \in (0, t)\}$ to obtain $\varphi(s) \, dF_{y_0}(s)$ in the equality

$$d[F_{y_0}\varphi(s)] = \varphi(s) \, dF_{y_0}(s) + \varphi'(s) F_{y_0}(s) \, ds. \tag{4.2.21}$$

Then from (4.2.20), which defines F_{y_0} via the process $\{X(t), t \geq 0\}$, we obtain

$$\varphi(s) \, dF_{y_0}(s) = \varphi(s)\big(\langle X(s), A^*y_0 \rangle \, ds + \langle B \, dW(s), y_0 \rangle\big). \tag{4.2.22}$$

Now integrate (4.2.21) and use the equality (4.2.22):

$$F_{y_0}(t)\varphi(t) = \int_0^t \varphi(s) dF_{y_0}(s) ds + \int_0^t \varphi'(s) F_{y_0}(s) ds$$

$$= \int_0^t \langle B \, dW(s), y(s) \rangle + \int_0^t \big(\langle X(s), A^*y(s) \rangle + \langle X(s), y'(s) \rangle \big) \, ds. \tag{4.2.23}$$

Here $y(s) = \varphi(s)y_0$. It follows from the equality (4.2.23) that the *statement* is proved for functions of the form $y(t) = y_0\varphi(t)$, $t \geq 0$. Since the linear combinations of these functions are dense in the space $C^1([0, \tau], dom \, A^*)$, we obtain the equality (4.2.19) in the general case and the *statement* is proved.

Now introduce the function $y(s) = U^*(t - s)y_0$, where $y_0 \in dom \, A^*$. Since the semi-group $\{U^*(t), t \geq 0\}$ satisfies the equalities (4.2.17) for each $t \geq 0$, the function $y(\cdot)$ is continuously differentiable on $[0, t]$ and takes values in $dom \, A^*$. Hence, $y(\cdot)$ satisfies the condition of the *statement* above and we have the equality (4.2.19) for $y(\cdot) = U^*(t - \cdot)y_0$. Since the adjoint semi-group U^* satisfies (4.2.17), we have for $y_0 \in dom A^*$

$$U^*(t - s)y_0 = \int_0^{t-s} A^*U^*(r)y_0 \, dr + y_0,$$

$$\frac{dU^*(t - s)y_0}{ds} = -A^*U^*(t - s)y_0, \qquad 0 \leq s \leq t.$$

Hence for $y(s) = U^*(t - s)y_0$ we have

$$y'(s) = \frac{dU^*(t - s)y_0}{ds} = -A^*U^*(t - s)y_0 = -Ay(s), \qquad 0 \leq s \leq t.$$

152   4. Itô integrated stochastic Cauchy problems in Hilbert spaces

Then for a solution $X(t)$, $t \geq 0$, and for $y(t) = U^*(t-t)y_0 = y_0$, where $y_0 \in dom A^*$, the proven equality (4.2.19) takes the form

$$\langle X(t), y_0 \rangle = \int_0^t \langle X(s), -A^*U^*(t-s)y_0 + A^*U^*(t-s)y_0 \rangle \, ds$$
$$+ \int_0^t \langle B \, dW(s), U^*(t-s)y_0 \rangle.$$

It follows that

$$\langle X(t), y_0 \rangle = \langle \int_0^t U(t-s)B \, dW(s), y_0 \rangle = \langle W_A(t), y_0 \rangle \; P_{\text{a.s}},$$

$$t \geq 0, \quad y_0 \in dom \, A^*.$$

Thus a weak solution of the linear problem (4.2.8) with zero initial data is unique and equal to $W_A(t)$, $t \geq 0$. Hence, the solution $X(t) = U(t)\zeta + W_A(t)$, $t \geq 0$, with general initial data $X(0) = \zeta$ is unique. $\qquad \square$

In Sections 3.1 and 3.2 we introduced many examples of generators of different semi-groups in Hilbert spaces. Now we can take them as A in the stochastic problems that we consider. These semi-groups, including ones generated by differential operators in the Gelfand–Shilov systems (2.3.1), give us numeral examples of regularized semi-groups, as well as examples of strongly continuous in $t \geq 0$ C_0-semi-groups and semi-groups strongly continuous only in $t > 0$.

An important example of a C_0-semi-group is the semi-group of solution operators to the Cauchy problem for the heat equation

$$\frac{\partial u(x;t)}{\partial t} = \frac{\partial^2 u(x;t)}{\partial x^2}, \quad t \geq 0, \quad x \in \mathbb{R},$$

with the generator $A = d^2/dx^2$ in the space $L^2(\mathbb{R})$ and

$$dom \, A = \{u \in L^2(\mathbb{R}) : d^2/dx^2 u \in L^2(\mathbb{R})\}.$$

However, the solution operators to the same problem with the generator in the space $C(\mathbb{R})$ do not generate a C_0-semi-group and provide a simple example of a semi-group that is strongly continuous only as $t > 0$.

In order to compare the strongly continuous in $t > 0$ semi-groups with those strongly continuous in $t \geq 0$, we give one more example.

Example 4.2.1 *Consider a semi-group of class $(1, \mathcal{A})$ that is not a semi-group of class C_0. It is constructed in the space of pairs of sequences:*

$$x = \{(\chi_n)_{n \in \mathbb{N}}, (\eta_n)_{n \in \mathbb{N}}\} : \quad \sup_n |\chi_n| < \infty, \quad \sum_{n=1}^{\infty} n^{1/2} |\eta_n| < \infty$$

with a specially chosen norm $\|x\| := \sup_n |\chi_n| + \sum_{n=1}^{\infty} n^{1/2} |\eta_n|$.

It is shown in [43] that the family of bounded operators $\{U(t), \, t > 0\}$ defined by $U(t)x = \{(\hat{\chi}_n)_{n \in \mathbb{N}}, (\hat{\eta}_n)_{n \in \mathbb{N}}\}$, where

$$\hat{\chi}_n = e^{-(n^{3/2}+in^2)t}(\chi_n \cos n^{1/2}t - \eta_n \sin n^{1/2}t),$$

$$\hat{\eta}_n = e^{-(n^{3/2}+in^2)t}(\chi_n \sin n^{1/2}t + \eta_n \cos n^{1/2}t),$$

generates a semi-group of class $(1, \mathcal{A})$, which is not a C_0-semi-group.

4.2.2 Weak regularized solutions. Some properties and examples of R-solutions

We continue to study linear stochastic Cauchy problems with additive noise in the setting that extends the Itô approach to Hilbert spaces. Now we suppose that A generates a regularized semi-group $\{S(t), \, t \in [0, \tau)\}$ in a Hilbert space H, $B \in \mathcal{L}(\mathbb{H}, H)$, and $\{W(t), \, t \geq 0\}$ is an $\mathbb{H}$-valued Wiener process.

Definition 4.2.4 *An H-valued predictable process X is called a weak regularized solution of (4.2.1) if it satisfies the following conditions:*

(a) *for each $t \in [0, T]$* $\displaystyle\int_0^t \|X(s)\|_H ds < \infty \; P_{a.s.}$*;*

(b) *for each $y \in dom \, A^*$ and $t \in [0, T]$*

$$\langle X(t), y \rangle = \langle R(t)\zeta, \, y \rangle + \int_0^t \langle X(s), A^*y \rangle \, ds + \langle \int_0^t R(t-s)B \, dW(s), y \rangle \quad P_{a.s.}. \tag{4.2.24}$$

In the particular case of a K-convoluted semi-group, X is called *a weak convoluted solution* of (4.2.1) if

$$\langle X(t), y \rangle = \langle \int_0^t K(s) \, ds \, \zeta, \, y \rangle + \int_0^t \langle X(s), A^*y \rangle \, ds$$

$$+ \langle \int_0^t \int_0^{t-s} K(r) \, dr B \, dW(s), \, y \rangle, \tag{4.2.25}$$

a weak n-times integrated solution if $K(r) = r^{n-1}/(n-1)!$:

$$\langle X(t), y \rangle = \langle \frac{t^n}{n!} \, \zeta, \, y \rangle + \int_0^t \langle X(s), A^*y \rangle \, ds + \langle \int_0^t \frac{(t-s)^n}{n!} B \, dW(s), \, y \rangle,$$

and *a weak R-solution* if

$$\langle X(t), y \rangle = \langle R\zeta, y \rangle + \int_0^t \langle X(s), A^*y \rangle \, ds + \langle RBW(t), \, y \rangle.$$

154   4. Itô integrated stochastic Cauchy problems in Hilbert spaces

Transforming the last term in the equality (4.2.25), we can write it in another form:

$$\langle X(t), y \rangle = \langle \int_0^t K(s)\, ds\, \zeta,\, y \rangle + \int_0^t \langle X(s),\, A^* y \rangle\, ds + \langle \int_0^t K(t-s)BW(s)\, ds,\, y \rangle.$$

Now we show the existence and uniqueness of solutions to the linear additive stochastic Cauchy problem with the generator of a regularized semi-group in H and an $\mathbb{H}$-valued Q-Wiener or cylindrical process W.

In the beginning of this section we showed that if A is the generator of a strongly continuous semi-group U of solution operators in H, then the stochastic convolution $W_A(t) := \int_0^t U(t-s)B\, dW(s), t \geq 0$, is well defined in $L^2(\Omega; H)$ for a Q-Wiener or cylindrical ($Q = I$) Wiener process W and the process $X(t) = U(t)\zeta + W_A(t), t \geq 0$, is the unique weak solution of (4.2.1).

If A is the generator of a regularized semi-group $\{S(t), t \in [0, \tau)\}$ in H, we can obtain only a weak regularized solution of (4.2.1). For such operators A we will show that the H-valued process

$$X(t) = S(t)\zeta + W_A(t), \quad t \in [0, T], \quad W_A(t) := \int_0^t S(t-s)B\, dW(s), \quad (4.2.26)$$

is a weak regularized solution to (4.2.1) for any $T < \tau \leq \infty$. The process $W_A(t)$ is called the *regularized stochastic convolution*.

Before proving theorems on the existence and uniqueness of the solution to the stochastic Cauchy problem (4.2.1) with the generator of a regularized semi-group, we give two propositions necessary for studying the problem. The first one concerns properties of adjoint regularized semi-groups and the second one is the extension of the *statement* in Theorem 4.2.3 to the case of regularized semi-groups. A regularized with a family $\{R(t)\}$ semi-group we will call an R-regularized semi-group.

Proposition 4.2.2 *Let A be the generator of an R-regularized semi-group $\{S(t), t \in [0, \tau)\}$ on a Hilbert space H. Suppose that $\{R(t)\}$ is a strongly differentiable family and $\overline{dom\, A} = H$. Then $\{S^*(t), t \in [0, \tau)\}$ is an R^*-regularized semi-group on H with the generator A^*. If operators $R(t)$ are invertible and have dense ranges, then the adjoint operators $R^*(t)$ have the same properties.*

Proof. First, we note that since the operators $R(t)$ and $S(t)$ are bounded for each t, then their adjoint operators $R^*(t)$ and $S^*(t)$ are bounded too. Second, since A is a closed densely defined operator, then the operator A^* is closed as well and $\overline{dom\, A^*} = H$.

Next we show that the family $\{S^*(t), t \in [0, \tau)\}$ forms an R^*-regularized semi-group with the generator A^*. The commutativity of operators $S^*(t)$ with A^* on dom A^* follows from the commutativity of $S(t)$ with A. Further, we need to prove that the family $\{S^*(t), t \in [0, \tau)\}$ is strongly continuous in t and the

4.2. Cauchy problems with additive noise

following equality holds:

$$S(t)^*y - R^*(t)y = \int_0^t S^*(s)A^*y\,ds, \qquad y \in dom\,A^*. \qquad (4.2.27)$$

Due to the continuity of the scalar product, Equation (4.2.27) implies

$$\langle S(t)f - R(t)f, y\rangle\langle f, S(t)^*y - R^*(t)y\rangle = \langle \int_0^t S(s)Af\,ds, y\rangle$$

$$= \int_0^t \langle AS(s)f, y\rangle\,ds = \int_0^t \langle S(s)f, A^*y\rangle\,ds \qquad (4.2.28)$$

for each $y \in dom\,A^*$ and $f \in H = \overline{dom\,A}$. Then we have

$$\frac{d}{dt}\langle f, S^*(t)y\rangle = \lim_{\Delta t \to 0}\left\langle f, \frac{S^*(t+\Delta t) - S^*(t)}{\Delta t}y\right\rangle$$

$$= \langle f, S^*(t)A^*y\rangle + \langle R'(t)f, y\rangle$$

$$= \langle S(t)f, A^*y\rangle + \langle R'(t)f, y\rangle, \quad f \in H, \quad y \in dom\,A^*,$$

which implies the pointwise convergence of the functionals

$$\frac{S^*(t+\Delta t) - S^*(t)}{\Delta t}y \quad \text{as} \quad \Delta t \to 0.$$

Hence, for any $t \in [0, \tau)$ and $y \in dom\,A^*$, the norms $\left\|\frac{S^*(t+\Delta t)-S^*(t)}{\Delta t}y\right\|$ are uniformly bounded for all Δt such that $t + \Delta t \in [0, \tau_1]$, $\tau_1 < \tau$. Therefore,

$$\|S^*(t+\Delta t)y - S^*(t)y\| \to 0 \quad \text{for} \quad y \in dom\,A^* \quad \text{as} \quad \Delta t \to 0.$$

Since the norms $\|S^*(t+\Delta t)\| = \|S(t+\Delta t)\|$ are also uniformly bounded for $t + \Delta t \in [0, \tau_1]$, $\tau_1 < \tau$, and the convergence takes place on $dom\,A^*$, then, by the Banach–Steinhaus theorem, $S^*(t)y$ is continuous in t for any $y \in H = \overline{dom\,A^*}$.

Strong continuity of $\{S^*(s), s \in [0, \tau)\}$ and equality (4.2.28) imply

$$\langle f, S(t)^*y - R^*(t)y\rangle = \int_0^t \langle f, A^*S^*(s)y\rangle\,ds$$

$$= \langle f, \int_0^t S^*(s)A^*y\,ds\rangle, \qquad f \in H, \quad y \in dom\,A^*,$$

which proves (4.2.27).

Finally, we show that the operators $R^*(t)$, $t \in [0, \tau)$, are invertible and have dense ranges if such are the operators $R(t)$. For any bounded operator R we have the equality

$$(ker\,R)^\perp = \overline{ran\,R^*}.$$

156   4. Itô integrated stochastic Cauchy problems in Hilbert spaces

If $R(t)$ is invertible, then $ker\, R(t) = \{0\}$ and therefore $\overline{ran\, R^*(t)} = H$. Thus,

$$(ker\, R^*(t))^{\perp} = \overline{ran\, R^{**}(t)} = H,$$

and $R^*(t)$ is invertible. $\hspace{4cm}\square$

Proposition 4.2.3 *Let A be the generator of an R-regularized semi-group $\{S(t), t \in [0, \tau)\}$ in a Hilbert space H and X be a weak regularized solution of (4.2.1) with $\zeta = 0$. Then for any $y \in C^1([0, T], dom\, A^*)$ the following equality holds:*

$$\langle X(t), y(t)\rangle = \int_0^t \langle X(s), y'(s) + A^* y(s)\rangle\, ds + \int_0^t \langle \int_0^s R'(s-r)B\, dW(r), y(s)\rangle\, ds$$

$$+ \langle BW(t), R^*(0)y(s)\rangle, \qquad t \in [0, T]. \quad (4.2.29)$$

Proof. As in the case of a strongly continuous semi-group, to prove (4.2.29) we define the following function of $t \in [0, T]$:

$$F_{y_0}(t) := \langle X(t), y_0\rangle = \int_0^t \langle X(s), A^* y_0\rangle\, ds + \langle \int_0^t R(t-s)B\, dW(s), y_0\rangle$$

for arbitrary $y_0 \in dom\, A^*$. Let $\varphi \in C^1(\mathbb{R})$. By the Itô formula we obtain

$$d(\varphi(s)F_{y_0}(s)) = \varphi'(s)F_{y_0}(s)\, ds + \varphi(s)\, dF_{y_0}(s).$$

Differentiating the Itô integral $\int_0^t R(t-s)B\, dW(s)$ wrt t, we have

$$\begin{aligned}
d(\varphi(s)F_{y_0}(s)) =\;& \varphi(s)\, dF_{y_0}(s) + \varphi'(s)F_{y_0}(s)\, ds \\[4pt]
=\;& \varphi(s)\left[\langle X(s), A^* y_0\rangle + \langle \int_0^s R'(s-r)B\, dW(r), y_0\rangle\right] ds \\[4pt]
& + \langle R(0)B\, dW(s), \varphi(s)y_0\rangle + \varphi'(s)\langle X(s), y_0\rangle \\[4pt]
=\;& \langle X(s), \varphi'(s)y_0 + A^*\varphi(s)y_0\rangle\, ds \\[4pt]
& + \langle \int_0^s R'(s-r)B\, dW(r), \varphi(s)y_0\rangle\, ds + \langle R(0)BdW(s), y(s)\rangle \\[4pt]
=\;& \langle X(s), y'(s) + A^* y(s)\rangle\, ds \\[4pt]
& + \langle \int_0^s R'(s-r)B\, dW(r), y(s)\rangle\, ds + \langle R(0)B\, dW(s), y(s)\rangle.
\end{aligned}$$

By the definition of F_{y_0} we have $d(\langle X(t), \varphi(s)y_0 \rangle) = d(\varphi(s)F_{y_0}(s))$. It follows that

$$d(\langle X(t), \varphi(s)y_0 \rangle) = d(\varphi(s)F_{y_0}(s)) = \langle X(s), y'(s) + A^*y(s) \rangle \, ds$$

$$+ \langle \int_0^s R'(s-r)B \, dW(r), y(s) \rangle \, ds + \langle R(0)B \, dW(s), y(s) \rangle.$$

Integrating this equality from 0 to t we obtain (4.2.29) for $y(s) = \varphi(s)y_0$, $s \in [0, T]$. Since the set of such functions is dense in $C^1([0, T], dom\, A^*)$, the equality (4.2.29) holds for each $y \in C^1([0, T], dom\, A^*)$. $\square$

Theorem 4.2.4 *Let a densely defined operator A generate a regularized semi-group $\{S(t), t \in [0, \tau)\}$ and $\{W(t), t \geq 0\}$ be a Wiener process. Suppose the condition*

$$\int_0^t \|S(r)B\|_{HS}^2 \, dr < \infty \tag{4.2.30}$$

is fulfilled. Then for each $\mathcal{F}_0$-measurable H-valued ζ the stochastic process (4.2.26) is a weak regularized solution to (4.2.1).

Proof. We first show that the process $S\zeta = \{S(t)\zeta, t \in [0, \tau)\}$ is a weak R-regularized solution for the corresponding homogeneous equation. The process $S\zeta$ is $\mathcal{F}_t$-measurable as the composition of $S(t)h$, a deterministic function of two variables $(t, h) \in [0, \tau) \times H$ and an $\mathcal{F}_0$-measurable random variable ζ. The trajectories of this process are continuous in $t \in [0, \tau)$ almost surely. Let $y \in dom\, A^*$. Then

$$\int_0^t \langle S(s)\zeta, A^*y \rangle \, ds = \langle \int_0^t S(s)\zeta \, ds, A^*y \rangle$$

$$= \langle A \int_0^t S(s)\zeta \, ds, y \rangle = \langle S(t)\zeta - R(t)\zeta, y \rangle.$$

Now consider W_A, the second term of the solution. Using the definition of the stochastic integral as the limit of integrals of step functions, we can readily verify that W_A is a predictable process. The function $\int_0^t \|S(t-s)B\|_{HS}^2 \, ds$ is continuous wrt $t \in [0, \tau)$, and hence integrable. Furthermore,

$$\int_0^r \int_0^t \|S(t-s)B\|_{HS}^2 \, ds \, dt = \int_0^r \int_0^t \|S(s)B\|_{HS}^2 \, ds \, dt.$$

Since the inner product is continuous, using the stochastic version of the Fubini theorem (Theorem 4.1.3) and taking into account the properties of the adjoint

158   4. Itô integrated stochastic Cauchy problems in Hilbert spaces

semi-group $\{S^*(t),\, t \in [0,\tau)\}$ given by Proposition 4.2.2, we obtain

$$\int_0^t \langle \int_0^s S(s-r)B\,dW(r), A^*y \rangle \, ds = \langle \int_0^t \int_0^s B\,dW(r), S^*(s-r)A^*y\,ds \rangle$$

$$= \langle \int_0^t B\,dW(r), \int_r^t S^*(s-r)A^*y\,ds \rangle = \langle \int_0^t B\,dW(r), \int_0^{t-r} S^*(\sigma)A^*y\,d\sigma \rangle$$

$$= \langle \int_0^t B\,dW(r), S^*(t-r)y - R^*(t-r)y \rangle$$

$$= \langle \int_0^t S(t-r)B\,dW(r), y \rangle - \langle \int_0^t R(t-r)B\,dW(r), y \rangle, \qquad t \in [0,\tau).$$

This means that W_A is a weak R-regularized solution of problem (4.2.1) with the initial condition $\zeta = 0$. Hence, $X(t) = S(t)\zeta + W_A(t)$, $t \in [0,\tau)$, is a weak R-regularized solution of (4.2.1). $\qquad\square$

We noted earlier that the condition of the existence of the stochastic convolution being formally written in the same form (4.2.30) for both Q-Wiener and cylindrical Wiener processes in fact gives different conditions for these processes. Namely, for a Q-Wiener process (4.2.30) holds for any regularized semi-group S and $B \in \mathcal{L}(\mathbb{H}, H)$, while for a cylindrical process (4.2.30) becomes more restrictive and holds for $S(r)B \in \mathcal{L}_{\mathrm{HS}}(\mathbb{H}, H)$.

Let us demonstrate on the basic example of the backward Cauchy problem that the condition on the operators $S(r)B$ to be Hilbert–Schmidt cannot be fulfilled.

Example 4.2.2 *Consider the local R-semi-groups related to the backward Cauchy problem in $H = L^2(\mathfrak{G})$, $\mathfrak{G} = \{x \in \mathbb{R}^n : 0 < x_k < a_k,\ k = 1,\ldots,n\}$:*

$$S(t)\zeta := \sum_{k=1}^{\infty} \langle \zeta, e_k \rangle e^{\mu_k(t-T)} e_k, \quad t \in [0,T), \quad \text{where} \quad R\zeta = \sum_{k=1}^{\infty} \langle \zeta, e_k \rangle e^{-\mu_k T} e_k.$$

The semi-group is bounded and invertible in H and the generator A is defined as

$$Au := -\triangle u, \qquad dom\, A := \mathcal{H}^2(\mathfrak{G}) \cap \mathcal{H}_0^1(\mathfrak{G}) \subset L^2(\mathfrak{G}).$$

For this operator we have

$$Sp(A) \;=\; \Big\{ \sum_{i=1}^{n} \frac{k_i^2 \pi^2}{a_i^2};\ k_i \in \mathbb{N} \Big\},$$

$$\{\text{set of eigenfunctions}\} \;=\; \Big\{ \prod_{i=1}^{n} \Big(\frac{2}{a_i}\Big)^{1/2} \cdot \sin \frac{k_i \pi x_i}{a_i};\ k_i \in \mathbb{N} \Big\}.$$

Denote by $\{\mu_k\}_{k=1}^\infty$, $\{e_k\}_{k=1}^\infty$ an ordering of the eigenvalues and eigenbasis of A; then we have the equalities

$$\int_0^T Tr\left(S(s)S^*(s)\right) ds = \int_0^T \sum_{k=1}^\infty \langle S(t)e_k, S^*(t)e_k \rangle\, dt = \int_0^T \sum_{k=1}^\infty \|S(t)e_k\|^2\, ds$$

$$= \sum_{k=1}^\infty \int_0^T e^{2\mu_k(t-T)}\, dt = \sum_{k=1}^\infty \left[\frac{e^{2\mu_k(t-T)}}{2\mu_k}\right]_{t=0}^T = \sum_{k=1}^\infty \frac{1}{2\mu_k}\left(1 - e^{-2\mu_k T}\right).$$

The series obtained is convergent if $n = 1$; hence the condition (4.2.30) with $B \in \mathcal{L}(\mathbb{H}, H)$ is fulfilled in this case.

Now we prove the uniqueness result for the case of regularized semi-groups.

Theorem 4.2.5 *Let a densely defined operator A generate a regularized semi-group $\{S(t), t \in [0, \tau)\}$ in H and $\{W(t), t \geq 0\}$ be a Q-Wiener process. Suppose that the condition (4.2.30) is fulfilled. Then if $R(t)$ is a strongly continuously differentiable operator-function, the solution of (4.2.1) is unique up to such an $\eta \in H$ that $(R(0) + R'*)\eta = 0$. If S is a K-convoluted or R-semi-group, the solution is unique.*

Proof. The uniqueness is based on the equality obtained in Proposition 4.2.3. Suppose that X is a weak regularized solution of (4.2.1) with $\zeta = 0$ and let $y \in C^1([0, T], dom\, A^*)$ be equal to $y(s) = S^*(t - s)y_0$, $y_0 \in dom\, A^*$. Then, by the equality (4.2.3) and by the properties of adjoint regularized semi-groups proved in Proposition 4.2.2, we have

$$\begin{aligned}
\langle X(t), R^*(0)y_0 \rangle &= \int_0^t \langle X(s), -R^{*'}(t - s)y_0 \rangle\, ds \\
&+ \int_0^t \langle S(t - s)\int_0^s R'(s - r)B\, dW(r)\, ds, y_0 \rangle \\
&+ \langle \int_0^t S(t - s)B\, dW(s), R^*(0)y_0 \rangle. \qquad (4.2.31)
\end{aligned}$$

The equality is true for any solution of (4.2.24) with $\zeta = 0$. In particular, for $X = W_A$ we have

$$\langle R(0)\int_0^t S(t - s)B\, dW(s), y_0 \rangle + \langle \int_0^t R'(t - s)\int_0^s S(s - r)B\, dW(r)\, ds, y_0 \rangle$$

$$= \langle R(0)\int_0^t S(t - s)B\, dW(s), y_0 \rangle + \langle \int_0^t S(t - s)\int_0^s R'(s - r)B\, dW(r)\, ds, y_0 \rangle.$$

From this equality, taking into consideration $\overline{dom\, A^*} = H^*$, we obtain

$$\int_0^t R'(t - s)\int_0^s S(s - r)B\, dW(r)\, ds = \int_0^t S(t - s)\int_0^s R'(s - r)B\, dW(r)\, ds.$$

160   4. Itô integrated stochastic Cauchy problems in Hilbert spaces

By the same reasoning, we can write (4.2.31) on the whole space H, that is, the following equality holds:

$$R(0)X(t) + \int_0^t R'(t-s)X(s)\,ds$$

$$= R(0)\int_0^t S(t-s)B\,dW(s) + \int_0^t R'(t-s)\int_0^s S(s-r)B\,dW(r)\,ds.$$

This equality can be written as $(R(0) + R'*)X = (R(0) + R'*)W_A$; therefore,

$$X(t) = \int_0^t S(t-r)B\,dW(r) + \eta(t), \quad t \in [0,T],$$

where η is a solution of the equation $(R(0) + R'*)\eta = 0$.

In particular, if the semi-group S in (4.2.31) is an R-semi-group, that is, $R(t) \equiv R$ and R is invertible, then we have the equality $R\eta = 0$. Hence $\eta = 0$ and the solution (4.2.26) is unique in the case of R-semi-groups.

If S is a K-convoluted semi-group, that is, $R'(t) = K(t)$ and $R(0) = 0$, then the solution is unique up to such an η that $KI * \eta = (K * \eta)I = 0$. Hence it is unique up to such an η that $\mathcal{L}[K * \eta]I = 0$. Since $K(0) = 0$, the Laplace transform $\mathcal{L}[K * \eta]$ is equal to the product of the Laplace transforms $\mathcal{L}[K]\mathcal{L}[\eta]$, where $\mathcal{L}[K] \neq 0$ for any convoluted semi-group. It follows that $\mathcal{L}[\eta] = 0$; hence $\eta = 0$, and the solution is unique in the case of a convoluted (in particular integrated) semi-group too.   □

Note that, if $\zeta \in L^2(\Omega; H)$, then for the unique solution obtained

$$X(t) = S(t)\zeta + \int_0^t S(t-r)B\,dW(r) \in L^2(\Omega; H), \quad t \in [0,T],$$

we have $\mathbf{E}[X(t)] = S(t)\mathbf{E}[\zeta]$ and similarly to the case of a strongly continuous semi-group

$$\mathbf{Cov}[X(t)] = S(t)\mathbf{Cov}[\zeta]S^*(t) + \int_0^t S(t-s)BQ[S(t-s)B]^*\,ds.$$

Now consider the Cauchy problem for Gelfand–Shilov systems (2.3.1):

$$\frac{\partial u(x;t)}{\partial t} = \mathbf{A}\left(i\frac{\partial}{\partial x}\right)u(x;t), \quad t \in [0,T],\ x \in \mathbb{R}^n, \quad u(x;0) = \zeta(x), \quad (4.2.32)$$

as an example important in applications of problems with generators of R-semi-groups. We show that all types of the systems generate R-semi-groups S in

$$H = L_m^2(\mathbb{R}^n) := L^2(\mathbb{R}^n) \times \cdots \times L^2(\mathbb{R}^n).$$

Then, by Theorems 4.2.4 and 4.2.5, X defined by (4.2.26) is the unique R-solution to (4.2.1) with $A = \mathbf{A}\,(i\partial/\partial x)$.

4.2. Cauchy problems with additive noise

Applying the generalized Fourier transform to (4.2.32) as in Section 2.3, we obtain

$$\frac{\partial \widetilde{u}(t,s)}{\partial t} = \mathbf{A}(s)\widetilde{u}(s,t), \quad t \in [0,T], \quad s \in \mathbb{C}^n, \quad \widetilde{u}(s,t) = e^{t\mathbf{A}(s)}\widetilde{\zeta}(s),$$

and

$$
\begin{aligned}
u(x;t) = \mathcal{F}^{-1}\left[e^{t\mathbf{A}(\cdot)}\widetilde{\zeta}(\cdot)\right](x) &= \left(\mathcal{F}^{-1}[e^{t\mathbf{A}(\cdot)}] * \mathcal{F}^{-1}[\widetilde{\zeta}(\cdot)]\right)(x) \\
&\equiv (G_t * \zeta)(x) =: U(t)\zeta(x), \qquad x \in \mathbb{R}^n,
\end{aligned}
$$

where G_t is the Green function of the problem (4.2.32) and the equalities are understood in the sense of distributions. We show that the operators $\mathbf{A}\left(i\frac{\partial}{\partial x}\right)$ generate different R-semi-groups in dependence on the type of the system (4.2.32) in the Gelfand–Shilov classification.

Recall that $e^{t\mathbf{A}(\cdot)}$ satisfies the estimates

$$e^{t\Lambda(s)} \leq \left\|e^{t\mathbf{A}(s)}\right\|_m \leq C(1+|s|)^{p(m-1)}e^{t\Lambda(s)}, \qquad t \geq 0, \quad s \in \mathbb{C}^n, \quad (4.2.33)$$

and according to Definition 2.3.1, the system is called

- *Petrovsky correct* if there exists a $C > 0$ such that $\Lambda(\sigma) \leq C$, $\sigma = Res \in \mathbb{R}^n$;

- *conditionally-correct* if there exist constants $C, C_1 > 0$, $0 < h < 1$ such that
$$\Lambda(\sigma) \leq C|\sigma|^h + C_1, \; \sigma \in \mathbb{R}^n;$$

- *incorrect* if the function $\Lambda(\cdot)$ grows for real $s = \sigma$ in the same way as for complex ones: $\Lambda(\sigma) \leq C|\sigma|^{p_0} + C_1$, $\sigma \in \mathbb{R}^n$, where p_0 is the exact order of the system.

In the following theorem we construct R-semi-groups $\{S(t), t \in [0,\tau)\}$, choosing special regularizing functions $K : \mathbb{R}^n \to \mathbb{R}^n$ in dependence on the class of the system.

Theorem 4.2.6 *Let $\tau > 0$ and $K(\sigma)$, $\sigma \in \mathbb{R}^n$, satisfy the conditions*

1) *all functions in the matrix-function $K(\cdot)e^{T\mathbf{A}(\cdot)}$ belong to $L^2(\mathbb{R}^n)$;*

2) *all the functions are bounded as functions of $(n+1)$ variables on $[0,T] \times \mathbb{R}^n$, $T < \tau$.*

Then the family of convolution operators $S(t)$:

$$[S(t)f](x) := G_R(t,x) * f(x), \qquad t \in [0,\tau), \; x \in \mathbb{R}^n, \qquad (4.2.34)$$

where

$$G_R(t,x) := \frac{1}{(2\pi)^n}\int_{\mathbb{R}^n} e^{i\sigma x}K(\sigma)e^{t\mathbf{A}(\sigma)}\,d\sigma, \qquad t \in [0,\tau), \; x \in \mathbb{R}^n, \quad (4.2.35)$$

forms an R-semi-group in $L_m^2(\mathbb{R}^n)$ with the generator $\mathbf{A}\left(i\frac{\partial}{\partial x}\right)$ and

$$Rf(x) = \frac{1}{(2\pi)^n}\int_{\mathbb{R}^n} e^{i\sigma x}K(\sigma)\widetilde{f}(\sigma)\,d\sigma, \qquad x \in \mathbb{R}^n. \qquad (4.2.36)$$

162   4. Itô integrated stochastic Cauchy problems in Hilbert spaces

Proof. For the sake of simplicity we give the proof for $n = 1$. First note that to prove the operators $S(t)$, $t \in [0, \tau)$, form an R-semi-group, it is sufficient to prove they are strongly continuous and the equalities

$$S(s)Af = AS(t)f, \qquad t \in [0, \tau), \quad f \in dom\, A, \qquad (4.2.37)$$

$$A \int_0^t S(s)f\, ds = S(t)f - Rf, \qquad t \in [0, \tau), \quad f \in H, \qquad (4.2.38)$$

hold true on arbitrary $[0, T] \subset [0, \tau)$ (see Definition 1.2.12).

Thus, take $T \in (0, \tau)$. Due to condition 1, the integral in (4.2.35),

$$\frac{1}{2\pi} \int_{-\infty}^{\infty} e^{i\sigma x} K(\sigma) e^{t\mathbf{A}(\sigma)}\, d\sigma, \qquad t \in [0, T],$$

is convergent. Moreover, this convergence is uniform with respect to $t \in [0, T]$ and the matrix-function $G_R(t, x)$ obtained, called the regularized Green function, is well defined. Since, in addition, functions of the matrix-function $K(\sigma) e^{t\mathbf{A}(\sigma)}$ are bounded, the integral

$$\int_{-\infty}^{\infty} e^{i\sigma x} K(\sigma) e^{t\mathbf{A}(\sigma)} \widetilde{f}(\sigma)\, d\sigma, \qquad t \in [0, T], \qquad (4.2.39)$$

is an element of $L_m^2(\mathbb{R})$ for each $\widetilde{f} \in L_m^2(\mathbb{R})$.

Now we are ready to check that the family (4.2.34) forms a local R-semi-group in $L_m^2(\mathbb{R})$. First, we verify the strong continuity property of the family $\{S(t), t \in [0, T]\}$, i.e., for each $f \in L_m^2(\mathbb{R})$ and $T < \tau$ we show that $\|S(t)f - S(t_0)f\| \to 0$ as $t \to t_0$, $t_0 \in [0, T]$.[3]

$$\|S(t)f - S(t_0)f\|^2$$

$$= \int_{\mathbb{R}} \left(\frac{1}{2\pi} \int_{-\infty}^{\infty} e^{i\sigma x} K(\sigma) \left[e^{t\mathbf{A}(\sigma)} \widetilde{f}(\sigma) - e^{t_0 \mathbf{A}(\sigma)} \widetilde{f}(\sigma) \right] d\sigma \right)^2 dx.$$

Let us split the inner integral into the following three:

$$\int_{|\sigma| \geq N} e^{i\sigma x} K(\sigma) e^{t\mathbf{A}(\sigma)} \widetilde{f}(\sigma)\, d\sigma - \int_{|\sigma| \geq N} e^{i\sigma x} K(\sigma) e^{t_0 \mathbf{A}(\sigma)} \widetilde{f}(\sigma)\, d\sigma$$

$$+ \int_{|\sigma| \leq N} e^{i\sigma x} K(\sigma) \left[e^{t\mathbf{A}(\sigma)} - e^{t_0 \mathbf{A}(\sigma)} \right] \widetilde{f}(\sigma)\, d\sigma. \qquad (4.2.40)$$

Here the functions $h_N(x, t) := \int_{|\sigma| \geq N} e^{i\sigma x} K(\sigma) e^{t\mathbf{A}(\sigma)} \widetilde{f}(\sigma)\, d\sigma$ and

$$g_N(x, t) := \int_{|\sigma| \leq N} e^{i\sigma x} K(\sigma) \left[e^{t\mathbf{A}(\sigma)} - e^{t_0 \mathbf{A}(\sigma)} \right] \widetilde{f}(\sigma)\, d\sigma$$

---

[3]Throughout this proof the norm $\| \cdot \|$ denotes the norm in $L_m^2(\mathbb{R})$.

$$4.2. \text{ Cauchy problems with additive noise} \qquad 163$$

are elements of $L_2^m(\mathbb{R})$ for each $t \in [0,T]$ as the inverse Fourier transforms of the functions from $L_2^m(\mathbb{R})$

$$\widetilde{h}_N(\sigma, t) = \begin{cases} 0, & |\sigma| \le N, \\ K(\sigma)e^{t\mathbf{A}(\sigma)}\widetilde{f}(\sigma), & |\sigma| > N, \end{cases}$$

and

$$\widetilde{g}_N(\sigma, t) = K(\sigma)e^{t\mathbf{A}(\sigma)}\widetilde{f}(\sigma) - \widetilde{h}_N(\sigma, t) - K(\sigma)e^{t_0\mathbf{A}(\sigma)}\widetilde{f}(\sigma) + \widetilde{h}_N(\sigma, t_0),$$

respectively. The integral (4.2.39) is convergent uniformly with respect to both $x \in \mathbb{R}$ and $t \in [0,T]$. This is due to condition 1 and $\widetilde{f}(\cdot) \in L_m^2(\mathbb{R})$. Hence, for any $\varepsilon > 0$,

$$|h_N(x,t)| < \varepsilon/4, \quad x \in \mathbb{R}, \quad t \in [0,T],$$

due to the choice of N, and the sum of norms of the first two integrals in (4.2.40) is less than $\varepsilon/2$. Now fix N. Since $\left(e^{(t-t_0)\mathbf{A}(\sigma)} - 1\right) \to 0$ as $t \to t_0$ uniformly with respect to $\sigma \in [-N, N]$, it follows that $|g_N(x,t)| < \varepsilon/2$ for $x \in \mathbb{R}$ and $t \to t_0$. To obtain the estimate for

$$\|S(t)f - S(t_0)f\|^2 = \frac{1}{4\pi^2} \int_{\mathbb{R}} \left(h_N(x,t) - h_N(x,t_0) + g_N(x,t)\right)^2 dx$$

consider the difference $h_N(x,t) - h_N(x,t_0) =: \Delta_N(x,t,t_0)$, $t, t_0 \in [0,T]$, as a single function. Then $\Delta_N(\cdot, t, t_0) \in L_m^2(\mathbb{R})$ and for a fixed N we obtain $|\Delta_N(x,t,t_0)| < \varepsilon/2$ for $x \in \mathbb{R}$. In these notations we have

$$4\pi^2 \|S(t)f - S(t_0)f\|^2 = \int_{\mathbb{R}} \Delta_N^2(x,t,t_0)\, dx$$

$$+ 2 \int_{\mathbb{R}} \Delta_N(x,t)g_N(x,t,t_0)\, dx + \int_{\mathbb{R}} g_N^2(x,t)\, dx.$$

In the way described above one can show that each of these three integrals is an infinitesimal value. This is indeed the case since the integrals over the infinite intervals $|x| > M$ are small due to the choice of M. This follows from their uniform convergence with respect to $t \in [0,T]$. The integrals over compacts $[-M, M]$ are small since the integrands are small. This can be shown by the sequential choice of M and $t \to t_0$. This completes the proof of the strong continuity of the operators of the family (4.2.34).

Next we show that the operators obtained commute with $\mathbf{A}\left(i\frac{\partial}{\partial x}\right)$ on the domain of $\mathbf{A}\left(i\frac{\partial}{\partial x}\right)$. By the properties of convolution, a differential operator may be applied to any component of the convolution. We apply $\mathbf{A}\left(i\frac{\partial}{\partial x}\right)$ to $f \in dom\, \mathbf{A}\left(i\frac{\partial}{\partial x}\right)$; then

$$\mathbf{A}\left(i\frac{\partial}{\partial x}\right)[S(t)f](x) = G_R(t,x) * \mathbf{A}\left(i\frac{\partial}{\partial x}\right)f(x) = S(t)\mathbf{A}\left(i\frac{\partial}{\partial x}\right)f(x).$$

164   4. Itô integrated stochastic Cauchy problems in Hilbert spaces

Thus the first equality in (4.2.37) holds. Now we verify the second one. For an arbitrary $f \in dom\, \mathbf{A}\left(i\frac{\partial}{\partial x}\right)$ consider the equality

$$\frac{\partial}{\partial t}[S(t)f](x) = \frac{\partial}{\partial t}[G_R(t,x) * f(x)] = \frac{1}{2\pi}\frac{\partial}{\partial t}\int_{-\infty}^{\infty} e^{i\sigma x}K(\sigma)e^{t\mathbf{A}(\sigma)}\widetilde{f}(\sigma)\,d\sigma.$$

To differentiate under the integral sign we apply the dominated convergence theorem. Due to conditions on $K(\cdot)$, the difference quotient is uniformly bounded wrt $t \in [0, T]$:

$$\left|e^{i\sigma x}K(\sigma)\left(\frac{e^{t\mathbf{A}(\sigma)} - e^{t_0\mathbf{A}(\sigma)}}{t - t_0}\right)\widetilde{f}(\sigma)\right|$$
$$= \left|e^{i\sigma x}K(\sigma)e^{(t_0+\theta(t-t_0))\mathbf{A}(\sigma)}\mathbf{A}(\sigma)\widetilde{f}(\sigma)\right| \le C\left|\mathbf{A}(\sigma)\widetilde{f}(\sigma)\right|.$$

The condition $f \in dom\, \mathbf{A}\left(i\frac{\partial}{\partial x}\right)$ provides $\mathbf{A}(\cdot)\widetilde{f}(\cdot) \in L_m^2(\mathbb{R})$. Hence, the conditions of the dominated convergence theorem hold and

$$\frac{\partial}{\partial t}[S(t)f](x) = \frac{1}{2\pi}\int_{-\infty}^{\infty} e^{i\sigma x}K(\sigma)e^{t\mathbf{A}(\sigma)}\mathbf{A}(\sigma)\widetilde{f}(\sigma)\,d\sigma.$$

Taking into account that the inverse Fourier transform of $\mathbf{A}(\sigma)\widetilde{f}(\sigma)$ is $\mathbf{A}\left(i\frac{\partial}{\partial x}\right)f(x)$, we obtain

$$\begin{aligned}
\frac{\partial}{\partial t}[S(t)f](x) &= G_R(t,x) * \mathbf{A}\left(i\frac{\partial}{\partial x}\right)f(x) \\
&= \mathbf{A}\left(i\frac{\partial}{\partial x}\right)[G_R(t,x) * f(x)] = \mathbf{A}\left(i\frac{\partial}{\partial x}\right)[S(t)f](x).
\end{aligned}$$

Integration with respect to t gives the equality

$$[S(t)f](x) - [S(0)f](x) = \int_0^t \mathbf{A}\left(i\frac{\partial}{\partial x}\right)[S(\tau)f](x)\,d\tau.$$

Since $\mathbf{A}\left(i\frac{\partial}{\partial x}\right)$ is closed in $L_m^2(\mathbb{R})$ and differentiable functions form a dense subset of $L_m^2(\mathbb{R})$, the equality holds for any $f \in L_m^2(\mathbb{R})$:

$$[S(t)f](x) - [S(0)f](x) = \mathbf{A}\left(i\frac{\partial}{\partial x}\right)\int_0^t [S(\tau)f](x)\,d\tau, \qquad t \in [0, T].$$

Let $R : L_m^2(\mathbb{R}) \to L_m^2(\mathbb{R})$ be equal to $S(0)$; then, by the strong continuity property of S,

$$Rf(x) = \frac{1}{2\pi}\int_{-\infty}^{\infty} e^{i\sigma x}K(\sigma)\widetilde{f}(\sigma)\,d\sigma, \qquad x \in \mathbb{R}.$$

Thus we have proved that operators (4.2.34) form an R-semi-group generated by $\mathbf{A}\left(i\frac{\partial}{\partial x}\right)$ in $L_m^2(\mathbb{R})$ with R defined by (4.2.36).

4.2. Cauchy problems with additive noise

Note that the inverse to R (unbounded) operator is obtained by solving the equation $g(x) = Rf(x) = \mathcal{F}^{-1}[K(\sigma)\tilde{f}(\sigma)](x)$:

$$R^{-1}g = \mathcal{F}^{-1}\left[\frac{\tilde{g}(\sigma)}{K(\sigma)}\right]. \qquad \square$$

Corollary 4.2.1 *If the system (4.2.32) is Petrovsky correct, then the function $K(\sigma) = 1/(1+\sigma^2)^{d/2+1}$ satisfies the conditions of the theorem with $d = p(m-1)$. If the system is conditionally correct, we can choose $K(\sigma) = e^{-a|\sigma|^h}$, and if the system is incorrect, we can choose $K(\sigma) = e^{-a|\sigma|^{p_0}}$ with $a > a_0 T$, where parameters h and $a_0 = a_0(p, m, C, C_1)$ are defined by the corresponding parameters of (4.2.33) and the estimates in the classification.*

In conclusion of the section we give one more example of a regularized semi-group, which in dependence on the generator spectrum can be an integrated, convoluted, or R-semi-group. The example is interesting for its geometric clarity.

Example 4.2.3 *Let $a(x)$, $x \in \mathbb{R}$, be a continuous $\mathbb{C}$-valued function. Consider the operator $A\colon H = L^2(\mathbb{R}) \to H$ defined by*

$$[Au](x) = a(x)u(x), \quad x \in \mathbb{R}, \quad dom\, A = \{u \in L^2(\mathbb{R}) : au \in L^2(\mathbb{R})\}.$$

We have $Sp(A) = \{$range of $a(\cdot)\}$ and the following semi-groups generated by A.

1. If the set $\{$range of $a(\cdot)\}$ is not a subset of $\Lambda_{n,\gamma,\varpi}^{\ln}$ for some parameters n, γ, ϖ and

$$dist\left(\{\text{range of } a(\cdot)\}, \Lambda_{n,\gamma,\varpi}^{\ln}\right)$$

does not tend to zero as $|\lambda| \to \infty$, then the estimate (R2) (see Section 1.2) holds for the resolvent of A and hence A is the generator of an n-times integrated semi-group defined by

$$[S(t)u](x) = u(x)\int_0^t \frac{(t-s)^{n-1}}{(n-1)!}\, e^{a(x)s}\, ds, \quad x \in \mathbb{R}, \quad t \in [0, \tau).$$

2. If $\{$range of $a(\cdot)\}$ is not a subset of $\Lambda_{\alpha,\gamma,\varpi}^{M}$ for some parameters α, γ, ϖ and the distance between $\{$range of $a(\cdot)\}$ and $\Lambda_{\alpha,\gamma,\varpi}^{M}$ does not tend to zero, then the estimate (R3) holds for the resolvent of A and hence A is the generator of a K-convoluted semi-group. It is defined by

$$[S(t)u](x) = u(x)\int_0^t K(t-s)\, e^{a(x)s}\, ds, \quad x \in \mathbb{R}, \quad t \in [0, \tau),$$

where K is a continuous function with the Laplace transform satisfying the estimate $|\mathcal{L}[K](\lambda)| = \mathcal{O}_{|\lambda|\to\infty}\left(e^{-\beta M(\gamma|\lambda|)}\right)$, $\beta > \varpi$, which is consistent with the estimate (R3) for the resolvent.

3. If there exists a sequence x_n such that $a(x_n) \in \mathbb{R}$ and $a(x_n) \to +\infty$, then A can generate only an R-semi-group.

Now we pass to semi-linear problems.

166   *4. Itô integrated stochastic Cauchy problems in Hilbert spaces*

4.3   Solutions to Cauchy problems for semi-linear equations with multiplicative noise

In the present section we continue to investigate stochastic Cauchy problems in infinite-dimensional spaces, which is the main focus of the book. Under consideration is the Cauchy problem for semi-linear stochastic equations with multiplicative noise. More precisely, we consider the integral equation

$$X(t) = \zeta + \int_0^t AX(s)\,ds + \int_0^t F(s, X(s))\,ds + \int_0^t B(s, X(s))\,dW(s), \quad t \in [0, T],$$

where A is the generator of a semi-group in a Hilbert space H, $F : H \to H$ is a non-linear mapping, $B(t, X) : \mathbb{H} \to H$ is a linear mapping, and the stochastic integral is taken wrt an $\mathbb{H}$-valued Wiener process W. This setting extends the Itô approach to the case of Hilbert spaces. Similarly to the linear case, the problem is written in the short form (4.0.1):

$$dX(t) = AX(t)\,dt + F(t, X(t))\,dt + B(t, X(t))\,dW(t),$$
$$t \in [0, T], \quad X(0) = \zeta. \quad (4.3.1)$$

4.3.1   Statement of the problem. Non-linear Cauchy problems from the point of view of the theory of non-linear semi-groups

We consider (4.3.1) where A is the generator of a strongly continuous semi-group of solution operators $\{U(t), t \in [0, \infty)\}$ in a Hilbert space H or of a regularized semi-group $\{S(t), t \in [0, \tau)\}$. Since, generally, such an operator A is unbounded, we consider a weak solution to (4.3.1):

$$\langle X(t), y \rangle = \langle \zeta, y \rangle + \int_0^t \langle X(s)\,ds, A^* y \rangle + \int_0^t \langle F(s, X(s))\,ds, y \rangle$$
$$+ \int_0^t \langle B(s, X(s))\,dW(s), y \rangle, \quad t \in [0, T], \quad X(0) = \zeta, \quad y \in dom\,A^*. \quad (4.3.2)$$

In addition to weak solutions, for semi-linear equations we introduce the concept of a *mild solution*. This is a stochastic process X satisfying the following equation:

$$X(t) = U(t)\zeta + \int_0^t U(t - s)F(s, X(s))\,ds + \int_0^t U(t - s)B(s, X(s))\,dW(s)$$
$$=: \mathcal{K}(X)(t). \quad (4.3.3)$$

4.3. Cauchy problems with multiplicative Wiener processes 167

Taking into account numerous applications of the problem with differential operators A generating R-semi-groups (see Sections 2.3 and 4.2), we also present weak R-solutions to (4.3.1) with A generating an R-semi-group $\{S(t), t \in [0, \tau)\}$, $\tau \leq \infty$:

$$\langle X(t), y \rangle = \langle R\zeta, y \rangle + \int_0^t \langle X(s)\, ds, A^* y \rangle + \int_0^t \langle F(s, X(s))\, ds, y \rangle$$
$$+ \int_0^t \langle B(s, X(s))\, dW(s), y \rangle$$

and a mild R-solution. This is a stochastic process X satisfying the equation

$$X(t) = S(t)\zeta + \int_0^t S(t-s)F(s, X(s))\, ds + \int_0^t S(t-s)B(s, X(s))\, dW(s)$$
$$=: \mathcal{K}_R(X)(t). \quad (4.3.4)$$

In the next subsection we prove the existence and uniqueness of mild solutions for the classes of semi-groups identified above and show the relationships between weak and mild solutions. Namely, we prove that if under certain conditions on F and B a mild solution exists, then the unique weak solution exists and coincides with the mild solution. Under the Lipschitz conditions on $F = F(t, X)$ and $B = B(t, X)$ we construct a mild solution by the successive approximations method based on the contraction operators technique, which is conventional for non-linear problems. In this case, for a small $T > 0$, the operator $\mathcal{K}$ defined by (4.3.3) and $\mathcal{K}_R$ defined by (4.3.4) are proved to be contractions:

$$\|\mathcal{K}(Y_1) - \mathcal{K}(Y_2)\|_{\mathcal{H}_2} \leq const\, \|Y_1 - Y_2\|_{\mathcal{H}_2}, \quad where \quad const < 1,$$

in a space $\mathcal{H}_2$ with the norm $\|Y\|_{\mathcal{H}_2} = \left(\mathbf{E}\int_0^T \|Y(t)\|_H^2\, dt\right)^{\frac{1}{2}}$. The condition is sufficient for application of the successive approximations method. This method gives the way of construction of approximated solutions as well. In the case of dissipative non-linear mapping F approximations can be constructed via Lipschitz non-linearities.

Before considering solutions to semi-linear stochastic problems in detail, let us discuss the issues and ideas stipulating solution methods, types of solutions, and conditions on the non-linear term $F(t, X(t))$ to be Lipschitz or dissipative. Some elements of the theory of semi-groups of non-linear operators will be helpful here.

Recall that a mapping $\mathcal{A} : dom\, \mathcal{A} \subset H \to ran\, \mathcal{A} \subset H$, generally non-linear, satisfies the *Lipschitz condition* if there exists $C > 0$ such that

$$\|\mathcal{A}(x) - \mathcal{A}(y)\| \leq C\|x - y\|, \quad x, y \in H.$$

Operator $\mathcal{A}$ is called m-dissipative if

$$\|x - y - h(\mathcal{A}(x) - \mathcal{A}(y))\| \geq \|x - y\|, \quad h > 0, \quad x, y \in H. \quad (4.3.5)$$

168   4. Itô integrated stochastic Cauchy problems in Hilbert spaces

An m-dissipative operator $\mathcal{A}$ is *dissipative* if its range $ran\,\mathcal{A}$ coincides with the whole H. Lipschitz non-linear operators, in a sense, can be considered as a generalization of linear bounded operators and dissipative non-linear operators as a generalization of negative linear operators.

Now, at the conceptual level, we comment on the results obtained further for the semi-linear Cauchy problem with F satisfying the Lipschitz condition and on the results that can be extended to the case of dissipative F. We do it from the standpoint of generating non-linear semi-groups by non-linear generators. We begin with comparison of the known results for the homogeneous Cauchy problem with a linear operator A:

$$du(t)/dt = Au(t), \qquad u(0) = \zeta \in dom\,A,$$

and the one with a non-linear operator $\mathcal{A}$:

$$dv(t)/dt = \mathcal{A}(v(t)), \qquad v(0) = \zeta \in dom\,\mathcal{A}. \tag{4.3.6}$$

Roughly speaking, the Crandall–Ligette theorem on generation of non-linear semi-groups states the following. With any operator $\mathcal{A}$ from a certain class of non-linear operators with $dom\,\mathcal{A}$ in a Banach space H, one can associate a family of non-linear operators

$$\{\mathcal{U}(t),\, t \geq 0\}, \qquad \mathcal{U}(t) : \overline{dom\,\mathcal{A}} \;\to\; \overline{dom\,\mathcal{A}},$$

which are solution operators to the non-linear homogeneous Cauchy problem (4.3.6) and which are extended from $dom\,\mathcal{A}$ to $\overline{dom\,\mathcal{A}}$. It turns out that for Lipschitz and dissipative operators $\mathcal{A}$ the solution operators $\mathcal{U}(t)$ form a non-linear semi-group on $\overline{dom\,\mathcal{A}}$. In the case of a densely defined $\mathcal{A}$ the semi-group is defined on the whole space (see, e.g., [17, 87]).

The reason for this, as noted above, is that Lipschitz non-linear operators are in certain sense generalizations of linear bounded operators and dissipative non-linear operators are generalizations of linear negative operators, and as is known (see Section 1.1), both types of linear operators, bounded and negative, generate C_0-semi-groups. Moreover, negative operators generate contraction semi-groups.

The generator A of a contraction semi-group in the linear case has the resolvent $\mathcal{R}$ that satisfies the MFPHY conditions (1.1.8):

$$\exists\, M > 0,\, \omega \in \mathbb{R} \;:\; \|\mathcal{R}^n(\lambda)\| \leq \frac{M}{(\lambda - \omega)^n}, \qquad \lambda > \omega,\; n = 0, 1, 2, \ldots,$$

with $M = 1$ and $\omega = 0$. This particular case of MFPHY conditions can be written in the following equivalent forms:

$$\|\mathcal{R}(\lambda)\| \leq \frac{1}{\lambda}, \qquad \lambda > 0;$$

$$\|I - hA\| \geq 1, \qquad h > 0 \;\; \left(h = \frac{1}{\lambda}\right); \tag{4.3.7}$$

$$\|x - hAx\| \geq \|x\|, \qquad h > 0,\; x \in H.$$

4.3. Cauchy problems with multiplicative Wiener processes

They can be naturally generalized and used for construction of non-linear semi-groups $\{\mathcal{U}(t),\ t \geq 0\}$ with dissipative generators. In fact, we can see that the inequality (4.3.5) in the definition of dissipative operators written for a linear operator A takes the form of the last inequality in (4.3.7).

One way constructing approximations to the non-linear Cauchy problem (4.3.6) under these conditions is using the implicit Euler scheme for solving the problem. Namely, we replace $dv(t)/dt$ in (4.3.6) by $h^{-1}(v(t+h) - v(t))$ and, starting with $t = 0$, obtain an approximate solution v_a from the equality

$$\frac{1}{h}(v_a(h) - \zeta) = \mathcal{A}(v_a(h)) \quad \text{or} \quad \zeta = (I - h\mathcal{A})(v_a(h)), \quad h \geq 0.$$

If we suppose that the equality has a unique solution $v_a(h) = (I - h\mathcal{A})^{-1}\zeta$, then under the condition that operators $(I - \frac{t}{n}\mathcal{A})^{-n}$ are contractions (which is satisfied for a dissipative $\mathcal{A}$) there exists

$$\lim_{n \to \infty} \left(I - \frac{t}{n}\mathcal{A}\right)^{-n} \zeta =: \mathcal{U}(t)\zeta. \tag{4.3.8}$$

It follows that for each $t \geq 0$ operators $\mathcal{U}(t)$ are contractions and satisfy the semi-group relation

$$\mathcal{U}(t+s) = \mathcal{U}(t)\mathcal{U}(s), \quad t, s \geq 0, \quad \mathcal{U}(0) = I \quad \text{on} \quad \overline{dom\,\mathcal{A}}.$$

The equality (4.3.8) gives one way of constructing approximations to the non-linear Cauchy problem (4.3.6) and hence to the stochastic Cauchy problem under consideration with $\mathcal{A} := A + F$.

Another way of constructing approximations to (4.3.6) is generalization to the case of non-linear $\mathcal{A}$ of the Yosida approximations A_n defined for A, the generator of a C_0-semi-group, as follows:

$$A_n := -\lambda_n I + \lambda_n^2 R_A(\lambda_n) = \lambda_n R_A(\lambda_n)A, \quad A_n x \underset{\lambda_n \to \infty}{\longrightarrow} Ax, \quad x \in H, \tag{4.3.9}$$

or

$$A_h := (I - hA)^{-1}A = \frac{1}{h}\left[(I - hA)^{-1} - I\right], \quad A_h x \underset{h \to 0}{\longrightarrow} Ax, \quad x \in H.$$

For a linear operator A generating a C_0-semi-group, the operators A_h are bounded and bounded operators e^{tA_h} give one of the known approximations to the semi-group (see Section 1.1). These ideas can be used in constructing solutions to semi-linear stochastic Cauchy problems with dissipative non-linearities. The Yosida approximations can be defined for a non-linear term F in (4.3.1) by

$$F_h := \frac{1}{h}\left[(I - hF)^{-1} - I\right], \quad F_h(x) \underset{h \to 0}{\longrightarrow} F(x).$$

It is easy to see here that the operators F_h satisfy Lipschitz conditions.

170   4. Itô integrated stochastic Cauchy problems in Hilbert spaces

Now we begin the detailed study of weak and mild solutions to semi-linear stochastic Cauchy problems, where the non-linearity F satisfies the Lipschitz condition. First we construct a mild solution and later we show that a mild solution is a weak solution. Thus the construction of a weak solution will be given as well.

4.3.2 Existence and uniqueness of mild solutions to the semi-linear problem. Relations between mild and weak solutions

Let $(\Omega, \mathcal{F}, P)$ be a probability space with a given filtration $\{\mathcal{F}_t, t \geq 0\}$. Let H and $\mathbb{H}$ be separable Hilbert spaces and $\{W(t), t \geq 0\}$ be an $\mathbb{H}$-valued Q-Wiener or cylindrical (with $Q = I$) Wiener process.

Consider a mild solution to the Cauchy problem (4.3.1), i.e., a process X satisfying (4.3.3). More precisely.

Definition 4.3.1 *Let H-valued predictable process $X = \{X(t), t \in [0, T]\}$ satisfy the conditions for the existence of the convolutions in (4.3.3):*

$$\boldsymbol{E}\left(\int_0^T \|U(t - s)F(s, X(s))\| \, ds\right) < \infty,$$

$$\boldsymbol{E}\left(\int_0^T \|U(t - s)B(s, X(s))\|_{HS}^2 \, ds\right) < \infty,$$

where $\| \cdot \|_{HS}$ is the norm in the space of the Hilbert–Schmidt operators from $\mathbb{H}_Q = Q^{1/2}\mathbb{H}$ to H (see Section 4.1). Then X satisfying (4.3.3) is called a mild solution to (4.3.1).

The concept of a mild solution appears when we are trying to find a solution of the semi-linear stochastic problem (4.3.1) using, as was done in the previous section, the method of variation of the parameters, or the Cauchy formula from the theory of ordinary differential equations. Contrary to the linear case, here (4.3.3) is not a formula for a solution like (4.2.8), but the equation itself. Its solution is called a mild solution to (4.3.1). Further we formulate conditions on A, F, W, and B and prove that a mild solution exists and coincides with a weak solution under these conditions.

Similar to the case of a linear Cauchy problem studied in the previous section, for the semi-linear stochastic problem we construct solutions for the case of semi-groups of class $(1, \mathcal{A})$, here under the additional condition of fourth degree integrability. We denote this class of semi-groups by $(1, \mathcal{A})_4$. It is defined as the class of the families $\{U(t), t \geq 0\}$ satisfying the semi-group conditions (U1)–(U2) from Definition 1.1.2, the condition (U3$'$) from Definition 1.1.4, and above that the following conditions:

(U4) $\displaystyle\int_0^1 \|U(t)\|^4 dt < \infty;$

4.3. Cauchy problems with multiplicative Wiener processes 171

(U5) convergence in the sense of Abel:

$$\lim_{\lambda \to \infty} \lambda \int_0^{+\infty} e^{-\lambda t} U(t) f \, dt = f, \qquad f \in H.$$

As noted in Section 4.2, semi-groups of class $(1, \mathcal{A})$ follow C_0-semi-groups in the sense of more general behavior at $(t + 0)$, keeping at the same time "good" properties of C_0-semi-groups at $t > 0$. An example of a semi-group of class $(1, \mathcal{A})$ is given in Section 4.2. The example is slightly artificial, but it shows a class of semi-groups which is wider than that of C_0-semi-groups. These semi-groups allowed us to define the stochastic convolution in the case of additive noise in Section 4.2 and we show that they can be used in the case of multiplicative noise in this section. Unlike the previous section, we will not start here with constructing solutions for the case of C_0-semi-groups and continue with the case of more general strongly continuous at $t > 0$ semi-groups. Instead we construct solutions in the case of semi-groups of class $(1, \mathcal{A})_4$ first and obtain solutions in the case of C_0-semi-groups as a consequence. In addition, we show what must be changed in the proof of these constructions in order to get an R-solution for A generating an R-semi-group.

Thus we consider the semi-linear Cauchy problem (4.3.1):

$$dX(t) = AX(t)dt + F(t, X(t))dt + B(t, X(t))dW(t), \quad t \in [0, T], \quad X(0) = \zeta,$$

where $\{X(t), t \in [0, T]\}$ is a sought-for solution to the problem with

$$F(t, X(t)) : \; [0, T] \times \Omega \times H \to H \text{ and } B(t, X(t)) : \; [0, T] \times \Omega \times H \to \mathcal{L}(\mathbb{H}, H).$$

Theorem 4.3.1 *Let A be the generator of a semi-group of the class $(1, \mathcal{A})_4$ in H, W be an $\mathbb{H}$-valued Q-Wiener or weak Wiener process, and the following conditions for F and B are fulfilled:*

1) F is $(\mathcal{P}_T \times H)| \mathcal{B}(H)$-measurable and B is $(\mathcal{P}_T \times H)| \mathcal{B}(\mathcal{L}_{HS}(\mathbb{H}_Q, H))$-measurable;

2) there exists $C > 0$ such that

$$\|F(t, \omega; x) - F(t, \omega; y)\|^2 + \|B(t, \omega; x) - B(t, \omega; y)\|_{HS}^2 \le C^2 \|x - y\|^2,$$

$$\|F(t, \omega; x)\|^2 + \|B(t, \omega; x)\|_{HS}^2 \le C^2 (1 + \|x\|^2),$$

for arbitrary $x, y \in H$, $t \in [0, T]$ and $\omega \in \Omega$;

3) initial value ζ is $\mathcal{F}_0$-measurable.

Then there exists a mild solution to (4.3.1) and it is unique.

Proof. Recall that $\mathcal{P}_T$ is the σ-field consisting of the sets $\{(s, t] \times G, \; 0 \le s < t < \infty, \; G \in \mathcal{F}_s\}$, where $\mathcal{F}_s$ is a normal filtration on $(\Omega, \mathcal{F}, P)$.

172   *4. Itô integrated stochastic Cauchy problems in Hilbert spaces*

Let us consider the following mappings:

$$\mathcal{K}(Y)(t) := U(t)\xi + \int_0^t U(t-s)F(s,Y(s))\,ds + \int_0^t U(t-s)B(s,Y(s))\,dW(s)$$
$$=: U(t)\xi + \mathcal{K}_1(Y)(t) + \mathcal{K}_2(Y)(t), \qquad t \in [0,T], \quad Y \in \mathcal{H}_2,$$

and show that for T small enough, $\mathcal{K}$ is a contraction on $\mathcal{H}_2$, where $\mathcal{H}_2$ is a Banach space of all H-valued predictable processes $\{Y(t),\, t \in [0,T]\}$ such that $\|Y\|_{\mathcal{H}_2} = \left(\mathbf{E}\int_0^T \|Y(t)\|_H^2\, dt\right)^{\frac{1}{2}} < \infty$.

Due to condition 1 the mappings $\mathcal{K}_1$ and $\mathcal{K}_2$ are correctly defined as superpositions of measurable mappings and predictable processes.

To continue, first, consider the case when $\mathbf{E}\left(\|\zeta\|^2\right) < \infty$ and show that $\mathcal{K}_1$ and $\mathcal{K}_2$ are continuous mappings from $\mathcal{H}_2$ to $\mathcal{H}_2$, hence $\mathcal{K}$ continuously maps $\mathcal{H}_2$ to $\mathcal{H}_2$. The following estimates hold true:

$$\|\mathcal{K}_1(Y)\|_{\mathcal{H}_2}^2 = \mathbf{E}\int_0^T dt \left\|\int_0^t U(t-s)F(s,Y(s))\,ds\right\|_H^2$$
$$\leq \mathbf{E}\int_0^T dt \left(\int_0^T \|U(t-s)F(s,Y(s))\|\,ds\right)^2$$
$$\leq T\mathbf{E}\int_0^T dt \int_0^T \|U(t-s)F(s,Y(s))\|^2\,ds$$
$$\leq T\mathbf{E}\int_0^T \|F(s,Y(s))\|^2\,ds \int_0^T \|U(\tau)\|_{\mathcal{L}(H)}^2\,d\tau. \qquad (4.3.10)$$

Denote $\int_0^T \|U(t)\|^4\,dt =: M^2$, then $\int_0^T \|U(t)\|^2\,dt \leq M\sqrt{T}$. From the estimates (4.3.10) and condition 2 we have

$$\|\mathcal{K}_1(Y)\|_{\mathcal{H}_2}^2 \leq MT^{3/2}\mathbf{E}\int_0^T \|F(s,Y(s))\|^2 ds$$
$$\leq MT^{3/2}\mathbf{E}\int_0^T C^2(1+\|Y(s)\|_H^2)\,ds = MT^{3/2}C^2\left(T + \mathbf{E}\int_0^T \|Y(s)\|^2\,ds\right)$$
$$= MT^{3/2}C^2(T + \|Y\|_{\mathcal{H}_2}^2). \qquad (4.3.11)$$

That is, $\mathcal{K}_1$ acts from $\mathcal{H}_2$ to $\mathcal{H}_2$. Now, using the theorem conditions on B and

4.3. Cauchy problems with multiplicative Wiener processes 173

the Itô equality, we show that $\mathcal{K}_2$ acts from $\mathcal{H}_2$ to $\mathcal{H}_2$ too:

$$
\begin{aligned}
\|\mathcal{K}_2(Y)\|_{\mathcal{H}_2}^2 &= \mathbf{E} \int_0^T dt \left\| \int_0^t U(t-s)B(s,Y(s))\,dW(s) \right\|_H^2 \\
&= \mathbf{E} \int_0^T dt \int_0^t \|U(t-s)B(s,Y(s))\|_{\mathrm{HS}}^2\,ds \\
&\le \mathbf{E} \int_0^T \|B(s,Y(s))\|_{\mathrm{HS}}^2\,ds \int_s^T \|U(t-s)\|^2\,dt \\
&\le MT^{1/2}\mathbf{E} \int_0^T \|B(s,Y(s))\|_{\mathrm{HS}}^2\,ds = MT^{1/2}C^2\mathbf{E} \int_0^T (1+\|Y(s)\|_H^2)\,ds \\
&= MT^{1/2}C^2(T+\|Y\|_{\mathcal{H}_2}^2).
\end{aligned} \tag{4.3.12}
$$

Finally, using the estimates obtained for $\mathcal{K}_1$ and $\mathcal{K}_2$, we show that $\mathcal{K}$ is continuous and, moreover, for T small enough it is a contraction in $\mathcal{H}_2$. Let Y_1 and Y_2 be $\mathcal{H}_2$-valued stochastic processes. Then

$$
\|\mathcal{K}(Y_1) - \mathcal{K}(Y_2)\|_{\mathcal{H}_2} \le \|\mathcal{K}_1(Y_1) - \mathcal{K}_1(Y_2)\|_{\mathcal{H}_2} + \|\mathcal{K}_2(Y_1) - \mathcal{K}_2(Y_2)\|_{\mathcal{H}_2},
$$

and taking into account the estimates (4.3.11) we have

$$
\begin{aligned}
\|\mathcal{K}_1(Y_1) - \mathcal{K}_1(Y_2)\|_{\mathcal{H}_2}^2 &= \mathbf{E} \int_0^t dt \left\| \int_0^t U(t-s)(F(s,Y_1(s)) - F(s,Y_2(s)))\,ds \right\|^2 \\
&\le T\mathbf{E} \int_0^T dt \int_0^T \|U(t-s)(F(s,Y_1(s)) - F(s,Y_2(s)))\|^2\,ds \\
&\le T\mathbf{E} \int_0^T \|F(s,Y_1(s)) - F(s,Y_2(s))\|^2\,ds \int_s^T \|S(t-s)\|^2\,dt \\
&\le MT^{3/2}C^2\mathbf{E} \int_0^T \|Y_1(s) - Y_2(s)\|^2\,ds = MT^{3/2}C^2\|Y_1 - Y_2\|_{\mathcal{H}_2}^2.
\end{aligned} \tag{4.3.13}
$$

Thus the continuity of $\mathcal{K}_1$ is proved. Let us prove the continuity of $\mathcal{K}_2$. Taking into account (4.3.12), we have

$$
\begin{aligned}
\|\mathcal{K}_2(Y_1) &- \mathcal{K}_2(Y_2)\|_{\mathcal{H}_2}^2 \\
&= \mathbf{E} \int_0^t dt \left\| \int_0^t U(t-s)(B(s,Y_1(s)) - B(s,Y_2(s)))\,dW(s) \right\|^2 \\
&\le \mathbf{E} \int_0^T \|B(s,Y_1(s)) - B(s,Y_2(s))\|_{\mathrm{HS}}^2\,ds \int_s^T \|U(t-s)\|^2\,dt \\
&\le MT^{1/2}C^2\mathbf{E} \int_0^T \|Y_1(s) - Y_2(s)\|^2\,ds = MT^{1/2}C^2\|Y_1 - Y_2\|_{\mathcal{H}_2}^2.
\end{aligned} \tag{4.3.14}
$$

From (4.3.2) and (4.3.14) it follows that

$$
\|\mathcal{K}(Y) - \mathcal{K}(Y)\|_{\mathcal{H}_2} \le C(MT^{1/2}(T+1))^{1/2}\|Y_1 - Y_2\|_{\mathcal{H}_2}
$$

174   4. Itô integrated stochastic Cauchy problems in Hilbert spaces

and in addition to continuity we obtain that under a suitable choice of parameters, namely, if

$$C(MT^{1/2}(T+1))^{1/2} < 1 \quad \text{or} \quad C^2 MT^{1/2}(T+1) < 1, \tag{4.3.15}$$

the operator $\mathcal{K}$ is a contraction in $\mathcal{H}_2$. Hence, for T satisfying (4.3.15) the operator $\mathcal{K}$ has a unique fixed point $X \in \mathcal{H}_2$, which is a solution to the semi-linear Cauchy problem (4.3.1). We can satisfy the condition (4.3.15) considering the problem successively on intervals $[0, \widetilde{T}]$, $[\widetilde{T}, 2\widetilde{T}]$, ..., such that $\widetilde{T}$ satisfies (4.3.15).

For constructing a solution in the general case $\mathbf{E}\left(\|\zeta\|^2\right) \leq \infty$ we use the scheme from [23] and prove that if ζ and η are initial data such that $\mathbf{E}\left(\|\zeta\|^2\right)$, $\mathbf{E}\left(\|\eta\|^2\right) < \infty$, then for the corresponding solutions X, Y to (4.3.1) we have $I_\Gamma X(t) = I_\Gamma Y(t)$, where $\Gamma = \{\omega \in \Omega : \zeta(\omega) = \eta(\omega)\}$ and $I_\Gamma = 1$ if $\omega \in \Gamma$ and $I_\Gamma = 0$ otherwise.

Consider the sequence $\{X^k\}$:

$$X^0(t) = U(t)\zeta, \quad \ldots, \quad X^{k+1}(t) = \mathcal{K}(X^k)(t), \quad t \in [0, T], \quad k = 0, 1, 2, \ldots,$$

that is,

$$X^{k+1}(t) = U(t)\zeta + \int_0^t U(t-s)F(s, X^k(s))\,ds + \int_0^t U(t-s)B(s, X^k(s))\,dW(s).$$

Since the initial data ζ and η are $\mathcal{F}_0$-measurable, the random variable I_Γ is $\mathcal{F}_0$-measurable too. Hence, I_Γ is $\mathcal{F}_t$-measurable for any $t \in [0, T]$. Since $B(t, X^k(t))$ is an $\mathcal{L}_{\mathrm{HS}}$-measurable process and a superposition of measurable processes is a measurable process, then $I_\Gamma B(t, X^k(t))$ is an $\mathcal{L}_{\mathrm{HS}}$-predictable process. It follows that the integral $\int_0^t U(t-s)I_\Gamma B(s, X^k(s))\,dW(s)$ exists and

$$\int_0^t U(t-s)I_\Gamma B(s, X^k(s))\,dW(s) = I_\Gamma \int_0^t U(t-s)B(s, X^k(s))\,dW(s),$$
$$t \in [0, T].$$

Then for $t \in [0, T]$ we have

$$I_\Gamma X^{k+1}(t) = U(t)I_\Gamma \zeta + \int_0^t U(t-s)I_\Gamma F(s, X^k(s))\,ds$$

$$+ \int_0^t U(t-s)I_\Gamma B(s, X^k(s))\,dW(s).$$

Define in a similar way the sequence $\{Y^k\}$:

$$Y^0(t) = U(t)\eta, \quad Y^{k+1}(t) = \mathcal{K}(Y^k)(t), \quad t \in [0, T].$$

Then we have the equalities

$$I_\Gamma F(t, X^k(t)) = I_\Gamma F(t, Y^k(t)), \qquad I_\Gamma B(t, X^k(t)) = I_\Gamma B(t, Y^k(t))$$

4.3. Cauchy problems with multiplicative Wiener processes 175

for k satisfying $I_\Gamma X^k(t) = I_\Gamma Y^k(t)$. Hence the equality $I_\Gamma X^{k+1}(t) = I_\Gamma Y^{k+1}(t)$ holds for each k. Since X and Y are limits of X^k and Y^k, then passing to the limit we obtain $I_\Gamma X(t) = I_\Gamma Y(t)$.

Now we are ready to construct a mild solution in the general case. Let

$$\zeta_n = \begin{cases} \zeta, & \text{if } \|\zeta\| \leq n, \\ 0, & \text{if } \|\zeta\| > n, \end{cases}$$

and let $X_n(t)$ be the solution corresponding to the initial data ζ_n. Then $X_n(t) = X_{n+1}(t)$ on $\{\omega \in \Omega : \|\zeta\| \leq n\}$. Passing to the limit we obtain a solution to (4.3.1). Hence, the existence of a mild solution is proved for the general case $\mathbf{E}\left(|\zeta|^2\right) \leq \infty$. The mild solution obtained is unique as a solution of an equation with a contractive operator. $\qquad\square$

Thus, using the contraction operators technique and successive approximations method conventional for non-linear problems, we have constructed a mild solution to the stochastic semi-linear Cauchy problem (4.3.1) with A being the generator of a semi-group of class $(1, \mathcal{A})_4$ in H and Lipschitz non-linearities, in particular with A being the generator of a C_0-semi-group.

Notice once again (see the discussion in Section 4.1) the difference between the case of a Q-Wiener and that of a cylindrical Wiener process. In Section 4.1 we commented on the difference between the conditions for the existence of stochastic integrals with respect to a Q-Wiener and a cylindrical Wiener process: being formally the same, the conditions are in fact different. In the statement of Theorem 4.3.1 the norm $\|\cdot\|_{\mathrm{HS}}$ in condition 2 in the case of a Q-Wiener process is the norm of the space of Hilbert–Schmidt operators acting from $\mathbb{H}_Q$ to H, while in the case of a cylindrical Wiener process (with $Q = I$) it is the norm of the space of Hilbert–Schmidt operators from $\mathbb{H}$ to H. The stochastic integral with respect to a cylindrical Wiener process W is defined since the Hilbert–Schmidt operator B from $\mathbb{H}$ to H is bounded as an operator from $\mathbb{H}$ to a space $H_1 \supset H$, where W is well defined.

Remark 4.3.1 (*On the particular case of a C_0-semi-group.*) If, instead of being the generator of a semi-group of class $(1, \mathcal{A})_4$ in H, the operator A satisfies a stronger condition, namely, it is the generator of a C_0-semi-group, then the estimates in Theorem 4.3.1 become better. Instead of $\int_0^T \|U(t)\|^4 \, dt =: M^2$ and $\int_0^T \|U(t)\|^2 \, dt \leq M\sqrt{T}$, we have $\sup_{0 \leq t \leq T} \|U(t)\| \leq M_1$. In this case the estimate (4.3.15) has the following form: $C^2 M_1^2 T(T + 1) < 1$.

Remark 4.3.2 (*On the case of an R-semi-group and R-solution.*) If, instead of generating a C_0-semi-group $\{U(t), t \in [0, \infty)\}$ in H, the operator A generates an R-semi-group $\{S(t), t \in [0, \tau)\}$, then for $T < \tau$ we can obtain estimates similar to the ones obtained in Theorem 4.3.1 and get a mild R-solution (4.3.4).

Remark 4.3.3 (*On a dissipative non-linearity F.*) As noted above, if the non-linearity F is dissipative, we can construct a mild solution via approximating

176   4. Itô integrated stochastic Cauchy problems in Hilbert spaces

F by Lipschitz non-linearities F_n, in particular by Yosida approximations (4.3.9).

Now we consider examples of the non-linear terms F and operators B satisfying the conditions of Theorem 4.3.1 on existence and uniqueness.

Let Q be a trace class operator in $\mathbb{H}$ and, as usual, $\{e_i\}$ be a basis of its eigenvectors, $Qe_i = \sigma_i^2 e_i$, $\sum_1^\infty \sigma_i^2 < \infty$. Let $\{g_i = \sigma_i e_i\}$, $\{f_j\}$ be orthonormal base in spaces $\mathbb{H}_Q$, H, respectively.

Define the operator $B(X) : \mathbb{H}_Q \to \mathbb{H}$ as follows:

$$B(X)g_i = \sum_j b_{ij}(X)f_j, \quad \text{where} \quad \sum_i b_{ij}^2(X) < \infty.$$

For example, let

$$b_{ij} = \sigma_i \sin X_j, \quad \text{where} \quad X = \sum_j X_j f_j, \quad \sum_j X_j^2 < \infty.$$

Then the condition $\sum_i \sigma_i^2 < \infty$ is sufficient for such $B(X)$ to be a Hilbert–Schmidt operator from $\mathbb{H}_Q$ to H:

$$\|B(X)\|_{\mathrm{HS}}^2 = \sum_i \|B(X)g_i\|_{\mathrm{H}}^2 = \sum_i \sum_j b_{ij}^2(X) = \sum_{ij} \sigma_i^2 \sin^2 X_j \le C \sum_i \sigma_i^2.$$

Let us verify the Lipschitz condition:

$$\|B(X) - B(Y)\|_{\mathrm{HS}}^2 = \sum_i \|(B(X) - B(Y))g_i\|_{\mathrm{H}}^2$$

$$= \sum_i \sum_j (b_{ij}(X) - b_{ij}(Y))^2 = \sum_{ij} \sigma_i^2 (\sin X_j - \sin Y_j)^2$$

$$= \sum_{ij} \sigma_i^2 \, 4 \sin^2\left(\frac{X_j - Y_j}{2}\right) \cos^2\left(\frac{X_j + Y_j}{2}\right) \le \sum_{ij} \sigma_i^2 (X_j - Y_j)^2 \le C\|X - Y\|_{\mathrm{H}}^2$$

and the condition of linear growth:

$$\|B(X)\|_{\mathrm{HS}}^2 = \sum_i \|B(X)g_i\|_{\mathrm{H}}^2$$

$$= \sum_i \sum_j b_{ij}(X)^2 = \sum_i \sum_j \sigma_i^2 \sin^2 X_j \le C(1 + \|X\|_{\mathrm{H}}^2).$$

By analogy, we can take $F : H \to H$ as $F(X) = \sum_j a_j(X)f_j$ with suitable functions $a_j : H \to \mathbb{R}$.

To clarify the importance of the concept of a mild solution we will show the connection between mild and weak solutions. Namely, we will show that if a mild solution exists, then it is a weak solution. Thus the construction given for mild solutions can be used for obtaining weak solutions, both exact and approximate.

4.3. Cauchy problems with multiplicative Wiener processes

Theorem 4.3.2 *Let the conditions of Theorem 4.3.1 be fulfilled. Then a mild solution to the semi-linear stochastic Cauchy problem (4.3.1) exists and is a weak solution.*

Proof. Let $X(t)$, $t \in [0, T]$, be a mild solution to the problem (4.3.1). Let us show that X is a weak solution, i.e., X satisfies (4.3.2):

$$\langle X(t), y \rangle = \langle \zeta, y \rangle + \int_0^t \langle X(s), A^* y \rangle \, ds$$
$$+ \int_0^t \langle F(s, X(s)) \, ds, y \rangle + \int_0^t \langle B(s, X(s)) \, dW(s), y \rangle$$

for each $y \in dom \, A^*$.

We begin with a solution to the corresponding homogeneous problem. In Theorem 4.2.1 we proved that the process $\{U(t)\zeta, \, t \in [0, T]\}$ is a solution to the homogenous Cauchy problem

$$\langle \zeta, y \rangle + \int_0^t \langle U(r)\zeta, A^* y \rangle \, dr = \langle U(t)\zeta, y \rangle \qquad (4.3.16)$$

for each $\mathcal{F}_0$-measurable H-valued ζ. Further, let us show that the sum of convolutions

$$\int_0^t U(t-s)F(s, X(s)) \, ds + \int_0^t U(t-s)B(s, X(s)) \, dW(s), \quad t \in [0, T], \quad (4.3.17)$$

is a weak solution to (4.3.2) with initial data $X(0) = 0$, i.e., the sum satisfies the equation

$$\langle X(t), y \rangle = \int_0^t \langle X(s), A^* y \rangle ds + \int_0^t \langle F(s, X(s)) \, ds, y \rangle + \int_0^t \langle B(s, X(s)) dW(s), y \rangle.$$
$$(4.3.18)$$

Since U is a semi-group of class $(1, \mathcal{A})_4$, using the condition 2 of Theorem 4.3.1, we obtain the following estimates:

$$\int_0^T \|U(t)F(t, X(t))\|^2 \, dt < \infty, \qquad \int_0^T \|U(t)B(t, X(t))\|_{\text{HS}}^2 \, dt < \infty.$$

We have shown in Theorem 4.3.1 that these conditions imply that the paths of the processes defined by the convolutions are integrable:

$$\int_0^T dt \left\| \int_0^t U(t - s)F(s, Y(s)) \, ds \right\| < \infty,$$
$$\int_0^T dt \left\| \int_0^t U(t - s)B(s, X(s)) \, dW(s) \right\| < \infty.$$

178   4. Itô integrated stochastic Cauchy problems in Hilbert spaces

Now we show that the process X defined by the sum (4.3.17) satisfies (4.3.18) for each $y \in dom\,(A^*)$ ($P_{a.s.}$, of course). By the equalities valid for U, we have

$$\langle \int_0^t U(t-s)F(s, X(s))\, ds + \int_0^t U(t-s)B(s, X(s))\, dW(s), y \rangle$$

$$= \int_0^t \langle \int_0^s U(s-\tau)F(\tau, X(\tau))\, d\tau + \int_0^s U(s-\tau)B(\tau, X(\tau))\, dW(\tau), A^*y \rangle\, ds$$

$$+ \int_0^t \langle F(s, X(s))\, ds, y \rangle + \int_0^t \langle B(s, X(s))\, dW(s), y \rangle, \qquad (4.3.19)$$

and for the first term in the right-hand side the following equalities are true:

$$\int_0^t \langle \int_0^s U(s-\tau)F(\tau, X(\tau))\, d\tau, A^*y \rangle\, ds$$

$$= \int_0^t \int_0^s \langle U(s-\tau)F(\tau, X(\tau))\, d\tau, A^*y \rangle\, ds$$

$$= \int_0^t \langle F(\tau, X(\tau))\, d\tau, \int_\tau^t U^*(s-\tau)A^*y\, ds \rangle$$

$$= \int_0^t \langle F(\tau, X(\tau))\, d\tau, \int_0^{t-\tau} U^*(s)A^*y\, ds \rangle.$$

As we have already mentioned, if A is the generator of a semi-group $\{U(t)\}$ of class $(1, \mathcal{A})$ in H, its conjugate operator A^* is the generator of the semi-group $\{U^*(t)\}$ which belongs to the same class. Since the class $(1, \mathcal{A})_4$ belongs to $(1, \mathcal{A})$, for $y \in dom\,A^*$ the equality

$$\int_0^{t-\tau} U^*(s)A^*y\, ds = U^*(t-\tau)y - y$$

is true and hence

$$\int_0^t \langle \int_0^s U(s-\tau)F(\tau, X(\tau))\, d\tau, A^*y \rangle\, ds = \int_0^t \langle F(\tau, X(\tau))\, d\tau, U^*(t-\tau)y - y \rangle$$

$$= \int_0^t \langle U(t-\tau)F(\tau, X(\tau))\, d\tau, y \rangle - \int_0^t \langle F(\tau, X(\tau))\, d\tau, y \rangle.$$

4.4. Extension of the Feynman–Kac theorem to Hilbert spaces   179

For the second term of the sum of convolutions we similarly obtain

$$\int_0^t \langle \int_0^s U(s-\tau)B(\tau, X(\tau))\, dW(\tau), A^* y \rangle\, ds$$

$$= \int_0^t \int_0^s \langle U(s-\tau)B(\tau, X(\tau))\, dW(\tau), A^* y \rangle\, ds$$

$$= \int_0^t \langle B(\tau, X(\tau))\, dW(\tau), \int_\tau^t U^*(s-\tau)A^* y\, ds \rangle$$

$$= \int_0^t \langle B(\tau, X(\tau))\, dW(\tau), \int_0^{t-\tau} U^*(s)A^* y\, ds \rangle$$

$$= \int_0^t \langle B(\tau, X(\tau))\, dW(\tau), U^*(t-\tau)y - y \rangle \int_0^t \langle U(t-\tau)B(\tau, X(\tau))\, dW(\tau), y \rangle$$

$$- \int_0^t \langle B(\tau, X(\tau))\, dW(\tau), y \rangle.$$

Thus the equality (4.3.19) for the sum of convolutions is true, hence the process

$$\{ \int_0^t U(t-s)F(s, X(s))\, ds + \int_0^t U(t-s)B(s, X(s))\, dW(s), \quad t \in [0, T] \}$$

is a weak solution to (4.3.1) with zero initial data. Since $\{U(t)\zeta,\, t \geq 0\}$ is a weak solution to (4.3.16), we obtain that the mild solution

$$X(t) = U(t)\zeta + \int_0^t U(t-s)F(s, X(s))\, ds + \int_0^t U(t-s)B(s, X(s))\, dW(s)$$

is a weak solution to (4.3.1). $\qquad\qquad\qquad\qquad\qquad\qquad\qquad\qquad\qquad\square$

Thus we have constructed a mild solution to (4.3.1) and have proved that if a mild solution X for the semi-linear stochastic Cauchy problem (4.3.1) exists, then it is a weak solution to (4.3.1). Hence under the conditions of Theorem 4.3.1 there exists a unique weak solution equal to the mild solution obtained. As for the converse result, we can see that if X is a weak solution for (4.3.1), then it is represented in the form (4.3.3) and hence it is a mild solution.

4.4 Extension of the Feynman–Kac theorem to the case of relations between stochastic equations and PDEs in Hilbert spaces

In the previous sections we studied solutions to stochastic Cauchy problems in Hilbert spaces. Along with the study of the solutions themselves, the important problem is to obtain probabilistic characteristics of these solutions.

180   *4. Itô integrated stochastic Cauchy problems in Hilbert spaces*

It turns out that there exists a direct connection between solutions of stochastic problems and solutions of deterministic problems for their probabilistic characteristics of a certain type. In finite-dimensional spaces such a deterministic problem is, for example, the well-known backward Cauchy problem for the Kolmogorov equation. If we refer to its solution as to temperature of a medium, it can be treated as a deterministic characteristic of the solution of the corresponding stochastic problem describing the motion of particles in the medium.

4.4.1   Statement of the problem. Applications

The finite-dimensional theorem concerning the connection of stochastic equations with partial differential equations for certain probabilistic characteristics is known as the Feynman–Kac theorem. Its particular one-dimensional case establishes the relationship between solutions of the Cauchy problem for stochastic equation in $\mathbb{R}$ of the form

$$dX(\tau) = a(t, X(\tau))\, d\tau + b(\tau, X(\tau))\, d\beta(\tau), \qquad \tau \in [t, T],$$
$$0 \le t \le T, \quad X(t) = x, \quad x \in \mathbb{R}, \qquad (4.4.1)$$

where $\{\beta(t),\, t \ge 0\}$ is a Brownian motion, and solutions of the backward Cauchy problem for the deterministic partial differential equation

$$g_t(t, x) + a(t, x)g_x(t, x) + \frac{1}{2}b^2(t, x)g_{xx}(t, x) = 0, \quad t \in [0, T],$$
$$g(T, x) = h(x), \quad (4.4.2)$$

where h is a Borel function from H to $\mathbb{R}$, $g(t, x) := \mathbf{E}^{t,x}h(X(T))$ and $\mathbf{E}^{t,x}$ denotes the expectation of $h(X(T))$, where $X(T)$ is the solution to the Cauchy problem (4.4.1) at $\tau = T$.

The study of the relationship between the problems (4.4.1) and (4.4.2) was initiated by the needs of physics. For example, if the process $\{X(\tau)\}$ describes random motion of a particle in a liquid or gas, the temperature $g(t, x)$ of the medium is described by the Kolmogorov equation. The importance of the relationship between stochastic and deterministic problems became even more relevant recently due to the development of numerical methods which use this relationship in both directions: from stochastic problems to deterministic ones and from deterministic to stochastic (see, e.g., [86]). It also has found extensive applications in financial mathematics. For example, if X, a solution to (4.4.1), describes a stock price at time τ then $g(t, x)$ is the value of the stock option determined by the famous Black–Scholes equation (see, e.g., [108]).

Moreover, there exist recent applications of infinite-dimensional stochastic equations in financial mathematics. As an example we can mention the relationship between a process $\{X(\tau),\, \tau \ge t > 0\}$ describing evolution of a coupon bond price and $\mathbf{E}^{t,x}h(X(T))$ describing prices of bond options with the maturity date T.

4.4. Extension of the Feynman–Kac theorem to Hilbert spaces   181

Let us consider these applications in more detail. Let $X(\tau, T)$, $0 \le \tau \le T$, be the price at time τ of a coupon bond with maturity date T parametrized by $X(\tau, \tau) = 1$ and let $f(\tau, T)$, $\tau \le T$, be the forward curve, i.e.,

$$X(\tau, T) = \exp\left(-\int_\tau^T f(\tau, s)\, ds\right).$$

Then the Musiela reparameterization $r(\tau, z) := f(\tau, z + \tau)$, $z \ge 0$, in the special case of zero Heath–Jarrow–Morton shift, satisfies the following Cauchy problem in a Hilbert space H of functions acting from $\mathbb{R}_+$ to $\mathbb{R}$ (see, e.g., [32, 33]):

$$dr(\tau) = Ar(\tau)d\tau + \sigma(\tau, r(\tau))dW(\tau), \quad \tau \ge 0, \quad r(0) = \zeta,$$

where A is the generator of the right-shift semi-group in H, W is an $\mathbb{H}$-valued Q-Wiener process, and σ is a random mapping from $\mathbb{H}$ to H. Here the value of bond options may be calculated, at least numerically, via $g(t, x)$ defined as $g(t, x) := \mathbf{E}^{t,x}h(r(T))$ and the relationship between bonds and bond options is an important particular case of the relationship which the present section is devoted to.

In the present section we introduce a generalization of the Feynman–Kac theorem to the case of Hilbert spaces. Namely, we establish a relationship between solutions of Cauchy problems for the linear stochastic equation $dX(t) = AX(t)dt + B\, dW(t)$ studied in detail in the previous sections and solutions to the backward and direct Cauchy problems for deterministic partial differential equations for the probabilistic characteristics $\mathbf{E}^{t,x}h(X(T))$ and $\mathbf{E}^{0,x}h(X(t))$ with derivatives in Hilbert spaces.

We prove the relationship on the basis of two different approaches. One is based on the usage of the Itô formula in Hilbert spaces and is a generalization of the finite-dimensional approach from [11, 108]. The other is based on the usage of properties of the semi-group $\mathbf{U} = \{\mathbf{U}_t, t \ge 0\}$ defined via the solution to the stochastic problem. In addition, in this subsection we pay enough attention to the statement of the problem and interpretation of objects in the infinite-dimensional equations introduced.

Generalization of the Feynman–Kac theorem to the infinite-dimensional case, besides the proof of the theorem itself, raises many questions related to the formulation of the problem in infinite-dimensional spaces, definitions, and rigorous justifications of relevant objects. We are going to start with these matters.

We consider the linear stochastic Cauchy problem in a Hilbert space with a Wiener process W, which is the infinite-dimensional generalization of (4.4.1) in the linear case:

$$dX(\tau) = AX(\tau)\, d\tau + B\, dW(\tau), \quad \tau \in [t, T], \quad X(t) = x. \tag{4.4.3}$$

We prove the infinite-dimensional extension of the Feynman–Kac theorem

182   4. Itô integrated stochastic Cauchy problems in Hilbert spaces

under the following standard conditions. Let A be the generator of a C_0-semi-group in a Hilbert space H, B be a bounded linear operator from a Hilbert space $\mathbb{H}$ to H, and $W = \{W(t),\, t \geq 0\}$ be an $\mathbb{H}$-valued Q-Wiener process or a cylindrical Wiener process, which at the same time is an $\mathbb{H}_{Q_1}$-valued Q_1-Wiener process for a trace class operator Q_1 (see Section 4.1).

We associate with the Cauchy problem (4.4.3) a problem for the following deterministic partial differential equation:

$$\frac{\partial g}{\partial t}(t, x) + \frac{\partial g}{\partial x}(t, x)Ax + \frac{1}{2}Tr\left[B^*\frac{\partial^2 g}{\partial x^2}(t, x)BQ\right] = 0, \quad t \in [0, T], \quad (4.4.4)$$

which is the generalization of the equation in (4.4.2) for the case of Hilbert spaces. We will show that $g(t, x) := \mathbf{E}^{t,x}h(X(T))$ satisfies the following infinite-dimensional deterministic backward Cauchy problem:

$$\frac{\partial g}{\partial t}(t, x) + \frac{\partial g}{\partial x}(t, x)Ax + \frac{1}{2}Tr\left[B^*\frac{\partial^2 g}{\partial x^2}(t, x)BQ\right] = 0, \quad t \in [0, T],$$

$$g(T, x) = h(x), \quad x \in dom\, A, \quad (4.4.5)$$

and that $g(t, x) := \mathbf{E}^{0,x}h(X(t))$ satisfies the corresponding forward Cauchy problem with initial data $g(0, x) = h(x)$.

First we give a rigorous interpretation of the objects included in the stochastic and deterministic equations. Then we prove the connection between their solutions using the two approaches mentioned above – the Itô approach and the semi-group approach.

The Itô approach consists in successive application of the following steps. The first is proving the Markov property for the solution X to the stochastic Cauchy problem, the second is proving the martingale property for the process $g(t, X(t)) = g(t, x)|_{x=X(t)},\, t \geq 0$, and the last is applying the infinite-dimensional Itô formula to $g(t, X(t))$. Particular attention is paid to the subtle transition from zero expectation for functions of g to the equality (4.4.5) for g itself.

The semi-group approach consists in applying semi-group techniques to the operator family $\mathbf{U} = \{\mathbf{U}_t,\, t \geq 0\}$ defined as $[\mathbf{U}_t h](x) := \mathbf{E}^{0,x}h(X(t))$ on a subspace of the space of bounded functions $h : H \to \mathbb{R}$. We calculate the infinitesimal generator for the semi-group and write the Cauchy problem for the generator of $\mathbf{U}$, which coincides with the closure of the infinitesimal generator.

We start with interpretation of the objects included in the stochastic problem (4.4.3). For simplicity we let the operator A be the generator of a C_0-semi-group $\{U(t),\, t \geq 0\}$ in a Hilbert space H. This ensures uniform well-posedness of the Cauchy problem for the corresponding homogeneous equation and the existence of strongly continuous solution operators $U(t),\, t \geq 0$, to the homogeneous problem, as well as the existence and uniqueness of the weak solution to the stochastic problem

$$dX(t) = AX(t)\,dt + B\,dW(t), \quad t \geq 0, \quad X(0) = \zeta. \quad (4.4.6)$$

4.4. Extension of the Feynman–Kac theorem to Hilbert spaces

The solution is given by the formula

$$X(t) = U(t)\zeta + W_A(t) = U(t)\zeta + \int_0^t U(t-s) B dW(s), \quad t \geq 0, \quad (4.4.7)$$

(see Section 4.1, where we discussed the conditions that ensure the existence of the stochastic integral $\int_0^t \Phi(s) \, dW(s)$, $t \geq 0$ for Q-Wiener and cylindrical Wiener processes W, in particular, the existence of stochastic convolution). The stochastic convolution $W_A(t)$ with respect to both types of Wiener processes formally is defined under the same condition (4.2.11). In the case of a Q-Wiener process with a trace class operator Q to satisfy (4.2.11) it is sufficient for the operators $U(s)B$ to be bounded from $\mathbb{H}$ to H, while in the case of a cylindrical Wiener process with a bounded Q, in particular for $Q = I$ ($Tr\, Q = \infty$), the operators $U(s)B$ have to be Hilbert–Schmidt operators from $\mathbb{H}$ to H.

Now we give the interpretation for the objects of the deterministic partial differential equation (4.4.4) in Hilbert space H. Here the derivatives $\frac{\partial g}{\partial x}$ and $\frac{\partial^2 g}{\partial x^2}$ are understood in the Frechet sense: for any fixed $t \in [0,T]$ and $x \in H$

$$\frac{\partial g}{\partial x}(t,x)(\cdot): \; H \; \to \; \mathbb{R}, \qquad \frac{\partial^2 g}{\partial x^2}(t,x)(\cdot): \; H \; \to \; H^*,$$

and we have $\frac{\partial g}{\partial x}(t,x) \in H^*$, $\frac{\partial^2 g}{\partial x^2}(t,x) \in \mathcal{L}(H, H^*)$ if h satisfies certain smoothness conditions on h (see, e.g., [53, 106]).

The term $Tr\left[B^* \frac{\partial^2 g}{\partial x^2} BQ\right]$ in (4.4.4), where $Q : \mathbb{H} \to \mathbb{H}$ is a trace class operator if W is a Q-Wiener process and $Q = I$ if W is a cylindrical Wiener process, requires special attention since the expression Tr is usually defined as the trace of an operator acting in the same Hilbert space but here the operator $B^* \frac{\partial^2 g}{\partial x^2} BQ$ under the trace sign maps Hilbert space $\mathbb{H}$ to its adjoint $\mathbb{H}^*$.

Using the traditional definition of the trace (see Section 4.1), we can give meaning to it, using the Riesz theorem on the isomorphism of $\mathbb{H}$ and $\mathbb{H}^*$ and identifying $\mathbb{H}^*$ with $\mathbb{H}$. The isomorphism allows us to consider operators BQ, $\frac{\partial^2 g}{\partial x^2}$, and B^* as mappings from $\mathbb{H}$ to H, from H to H, and from H to $\mathbb{H}$, respectively. Then the operator $B^* \frac{\partial^2 g}{\partial x^2} BQ$ maps the Hilbert space $\mathbb{H}$ into $\mathbb{H}$ and its trace can be understood in the usual sense. Namely, in the case of a Q-Wiener process with a trace class operator Q and bounded operators $B: \mathbb{H} \to H$ and $\frac{\partial^2 g}{\partial x^2}: H \to H$ we have

$$\left| Tr\left[B^* \frac{\partial^2 g}{\partial x^2} BQ\right]\right| \leq \sum_{j=1}^{\infty} \left| \langle \frac{\partial^2 g}{\partial x^2} BQ e_j, B e_j \rangle \right| \leq \sum_{j=1}^{\infty} \sigma_j^2 \|B\|^2 \left\| \frac{\partial^2 g}{\partial x^2} \right\|^2 < \infty.$$

In the case of a cylindrical Wiener process with $Q = I$ the estimate $Tr\left[B^* \frac{\partial^2 g}{\partial x^2} B\right] < \infty$ holds if the operator $\frac{\partial^2 g}{\partial x^2}$ is bounded and B is Hilbert–Schimidt.

184   4. Itô integrated stochastic Cauchy problems in Hilbert spaces

One can give meaning to the concept of trace without identifying $\mathbb{H}$ with $\mathbb{H}^*$ by considering a wider class of operators, namely, linear nuclear operators acting from a separable Hilbert space $\mathbb{H}$ to $\mathbb{H}^*$. As is well known, any operator G from this class can be written in the form $Gz = \sum_{j=1}^{\infty} a_j(z)\, b_j$, $z \in \mathbb{H}$, where $a_j, b_j \in \mathbb{H}^*$ and $\sum_{j=1}^{\infty} \|a_j\|\|b_j\| < \infty$ (see Section 4.1). In our case for $G = B^* \frac{\partial^2 g}{\partial x^2} BQ : \mathbb{H} \to H^*$ we can take $a_j(e_k) = \delta_{jk}$ and $b_j = Ge_j$, where $\{e_j\}$ is an orthonormal basis in $\mathbb{H}$ consisting of eigenvectors of the trace class operator Q $(Qe_j = \sigma^2 e_j)$. Then $\|b_j\| \le c\sigma_j^2$ and $Tr[\mathfrak{Q}]$ is well defined and can be understood as

$$Tr[G] = \sum_{j=1}^{\infty} \mathfrak{Q} e_j(e_j). \qquad (4.4.8)$$

Now we prove that $g(t, x) := \mathbf{E}^{t,x} h(X(T)) : [0, T] \times H \to \mathbb{R}$, where h is a measurable function from H to $\mathbb{R}$ satisfying the conditions ensuring the existence and boundedness of the derivatives in (4.4.4), satisfies the infinite-dimensional deterministic problem (4.4.5).

4.4.2   Proof of the relations on the basis of the Itô approach

To determine the relationship between solutions of problems (4.4.3) and (4.4.5) we need some properties of $g(t, X(t)) = g(t, x)\big|_{x=X(t)}$, where the process X is a solution of (4.4.6). Let us obtain the required properties for the class of diffusion processes, which contains solutions of (4.4.3).

An H-valued Itô process $X = \{X(t), t \ge 0\}$ is called a *diffusion* if it is a solution to the Cauchy problem for the equation which can be written in the form

$$dX(t) = A(X(t))\, dt + B(X(t))\, dW(t), \qquad t \ge 0. \qquad (4.4.9)$$

We consider the diffusion processes being solutions of the Cauchy problem for Equations (4.4.9) under the condition that the solution of the problem exists and is unique.[4] In the particular case of the problem (4.4.3) with the generator of a C_0-semi-group $\{U(t), t \ge 0\}$ its unique solution can be written as the sum of the term depending on initial data and the stochastic convolution, i.e., in the form (4.4.7).

To prove the relationship under study, we first prove that if X is the unique solution to the stochastic Cauchy problem, then $g(t, x)$ satisfies (4.4.5). To prove this it is important to establish the Markov property for a solution of the Cauchy problem. The following statement is the generalization of the finite-dimensional result on the Markov property (see, e.g., [95]) to the case of Hilbert spaces.

Proposition 4.4.1 *Let* $X = \{X(t), t \ge 0\}$ *be the unique solution to the*

---

[4]It may be guaranteed, for example, by the following estimate of the mappings A and B: $\|A(y_1) - A(y_2)\| + \|B(z_1) - B(z_2)\| \le c(\|y_1 - y_2\| + \|z_1 - z_2\|)$, $y_1, y_2 \in H$, $z_1, z_2 \in \mathbb{H}$, $c \in \mathbb{R}$ (see Section 4.3).

4.4. Extension of the Feynman–Kac theorem to Hilbert spaces   185

Cauchy problem for (4.4.9), in particular, a solution to (4.4.6) and $h : H \to \mathbb{R}$ be a Borel-measurable square integrable function. Then the process X satisfies the Markov property with respect to a filter $\mathcal{F}_t$ defined by the Wiener process W:

$$\mathbf{E}\left[h(X(t+s))|\mathcal{F}_t\right] = \mathbf{E}^{0,X(t)}[h(X(s))], \qquad t,s \geq 0. \tag{4.4.10}$$

Proof. Let $X^{t,x}(r)$ ($r \geq t$, $t \in [0,T]$, $x \in H$) be the unique solution of (4.4.9) satisfying the condition $X(t) = x$. By the uniqueness of a solution to the Cauchy problem for Equation (4.4.9) we have $X(r) = X^{t,X(t)}(r)$, $r \geq t$ almost surely, i.e., $X(r)(\omega) = X^{t,X(t)(\omega)}(r)$ for $\omega \; P_{a.s.}$.

Fix some $m \in \mathbb{N}$ and consider the partition of the segment $[t,T]$ by the points $t_k = t + \frac{k(T-t)}{m}$, $k = 0, \dots, m$. Consider

$$v_m^{t,X(t)}(\tau,\omega) := \sum_{k=0}^{m} h(X^{t,X(t)(\omega)}(t_{k+1}))\chi_{\tau\in[t_k,t_{k+1})}$$

$$= \sum_{k=0}^{m} h(X(t_{k+1})(\omega))\chi_{\tau\in[t_k,t_{k+1})}, \tag{4.4.11}$$

where $\chi_{\tau\in[t_k,t_{k+1})}$ is a characteristic function of the semi-open interval $[t_k,t_{k+1})$. Then we obtain the equalities for the functions defined on the intervals:

$$\mathbf{E}\left[v_m^{t,X(t)}(\tau,\omega)\,\middle|\,\mathcal{F}_t\right] = \mathbf{E}\left[\sum_{k=0}^{m} h(X(t_{k+1}))\chi_{\tau\in[t_k,t_{k+1})}\,\middle|\,\mathcal{F}_t\right]$$

$$= \sum_{k=0}^{m}\chi_{\tau\in[t_k,t_{k+1})}\mathbf{E}\left[h(X(t_{k+1}))|\,\mathcal{F}_t\right] = \sum_{k=0}^{m}\chi_{\tau\in[t_k,t_{k+1})}\mathbf{E}\left[h(X(t_{k+1}))\right]$$

$$= \mathbf{E}\left[\sum_{k=0}^{m}\chi_{\tau\in[t_k,t_{k+1})}h(X(t_{k+1}))\right] = \mathbf{E}\left[v_m^{t,X(t)}(\tau,\omega)\right]. \tag{4.4.12}$$

Here the first and the last equalities follow from the equalities (4.4.11), which define $v_m^{t,X(t)}$. The second one holds since the characteristic functions of intervals do not depend on the variable $\omega \in \Omega$. The third equality holds since $X(t_{k+1})$ is independent of the σ-algebra $\mathcal{F}_t$ for all $t_{k+1} > t$ and by the uniqueness equality: $X(r) = X^{t,X(t)}(r)$, $r \geq t$.

Further, for a Borel-measurable h we have $v_m^{t,X(t)}(\tau,\cdot) \xrightarrow[m\to\infty]{} h(X(\tau))$ in $L^2(\Omega;\mathbb{H})$ and, letting m tend to infinity in the equalities (4.4.12), we obtain

$$\mathbf{E}\left[h(X(\tau))|\mathcal{F}_t\right] = \mathbf{E}\left[h(X(\tau))\right].$$

Since $\tau = t + s$, we conclude that

$$\mathbf{E}\left[h(X(t+s))|\mathcal{F}_t\right] = \mathbf{E}\left[h(X(t+s))\right] = \mathbf{E}\left[h(X^{t,X(t)}(t+s))\right]$$

$$= \mathbf{E}\left[h(X^{t,z}(t+s))\right]_{z=X(t)}. \tag{4.4.13}$$

186   4. Itô integrated stochastic Cauchy problems in Hilbert spaces

Using the diffusion property of the Itô process $\{X(t)\}$, we obtain

$$\mathbf{E}\left[h(X^{t,z}(t+s))\right]_{z=X(t)} = \mathbf{E}\left[h(X^{0,z}(s))\right]_{z=X(t)}. \qquad (4.4.14)$$

The equalities (4.4.13) and (4.4.14) imply (4.4.10).   □

Note that, if a process X is a solution to (4.4.3), it is a diffusion and, as proved above, has the Markov property. Hence, by the homogeneity of diffusions with respect to time, the following relation is fulfilled for a solution of (4.4.3):

$$\mathbf{E}^{0,X(t)}[h(X(s))] = \mathbf{E}^{t,X(t)}[h(X(t+s))]. \qquad (4.4.15)$$

As a consequence of (4.4.10) and (4.4.15) we obtain the following.

Remark 4.4.1 The Markov property (4.4.10) can be written as follows

$$\mathbf{E}\left[h(X(t+s))|\mathcal{F}_t\right] = \mathbf{E}^{t,X(t)}[h(X(t+s))].$$

Now we prove that the process $\{g(t, X(t))\}$ is a martingale, which is an important property.

Proposition 4.4.2 *Let $\{X(t),\, t \geq 0\}$ and h satisfy the conditions of Proposition 4.4.1. Then the process $\{g(t, X(t)) := \mathbf{E}^{t,x}h(X(T))|_{x=X(t)},\, 0 \leq t \leq T\}$ is a martingale, i.e.,*

$$\mathbf{E}\left[g(t, X(t))|\mathcal{F}(s)\right] = g(s, X(s)), \qquad 0 \leq s \leq t \leq T.$$

Proof. According to Proposition 4.4.1, $\{X(t)\}$ has the Markov property. Therefore,

$$\mathbf{E}\left[h(X(T))|\mathcal{F}(t)\right] = \mathbf{E}^{t,X(t)}h(X(T)) = g(t, X(t))$$

and we obtain the following equalities:

$$\begin{aligned}
\mathbf{E}\left[g(t, X(t))|\mathcal{F}(s)\right] &= \mathbf{E}\left[\mathbf{E}[h(X(T))|\mathcal{F}(t)]|\mathcal{F}(s)\right] = \mathbf{E}\left[h(X(T))|\mathcal{F}(s)\right] \\
&= \mathbf{E}^{s,X(s)}h(X(T)) = g(s, X(s)).
\end{aligned}$$

The first equality follows from the representation obtained for the process $\{g(t, X(t))\}$ via the conditional expectation. The second equality follows from the properties of conditional expectations. The third one is the direct consequence of the Markov property for $g(t, X(t))$. The last equality follows from the definition of the process $g(t, X(t))$ and completes the proof.   □

Now we can proceed to the proof of the connection between Cauchy problems for stochastic and deterministic equations. We begin with the relationship between the Cauchy problem (4.4.3) and the problem (4.4.5).

4.4. Extension of the Feynman–Kac theorem to Hilbert spaces 187

Theorem 4.4.1 *Let $\mathbb{H}$ and H be separable Hilbert spaces, X be the unique solution to the stochastic Cauchy problem (4.4.3), where A is the generator of a C_0-semi-group and $B \in \mathcal{L}(\mathbb{H}, H)$ if W is a Q-Wiener process and $B \in \mathcal{L}_{HS}(\mathbb{H}, H)$ if W is a cylindrical Wiener process. Let $g(t, x) := \mathbf{E}^{t,x} h(X(T))$, and suppose that $\frac{\partial g}{\partial x}(t, x)$, $\frac{\partial^2 g}{\partial x^2}(t, x)$, $t \geq 0$, $x \in \mathbb{R}$ exist and are bounded. Then g is a solution of the infinite-dimensional backward Cauchy problem (4.4.5).*

Proof. Note that under the conditions of the theorem the unique solution X to (4.4.3) exists and all terms in (4.4.5) are well defined. Applying the Itô formula in Hilbert spaces (Theorem 4.1.4) to $g(\tau, X(\tau))$, where $\{X(\tau), \tau \geq t\}$ is the solution to (4.4.3), we obtain

$$dg(\tau, X(\tau)) = \frac{\partial g}{\partial x}(\tau, X(\tau)) B \, dW(\tau) + \left(\frac{\partial g}{\partial \tau}(\tau, X(\tau)) \right.$$
$$\left. + \frac{\partial g}{\partial x}(\tau, X(\tau)) AX(\tau) + \frac{1}{2} Tr \left[B^* \frac{\partial^2 g}{\partial x^2}(\tau, X(\tau)) BQ \right] \right) d\tau.$$

This equality is written in the form of differentials. In the integral form it can be written as

$$g(T, X(T)) = g(t, X(t)) + \int_t^T \frac{\partial g}{\partial x}(s, X(s)) B \, dW(s) + \int_t^T \left(\frac{\partial g}{\partial s}(s, X(s)) \right.$$
$$\left. + \frac{\partial g}{\partial x}(s, X(s)) AX(s) + \frac{1}{2} Tr \left[B^* \frac{\partial^2 g}{\partial x^2}(s, X(s)) BQ \right] \right) ds.$$

Taking the expectation of both sides of the equation, we obtain

$$\mathbf{E} \left[\int_0^t \frac{\partial g}{\partial x}(s, X(s)) B \, dW(s) \right] = 0.$$

This can be proved by approximating the integrand by step processes (see Section 4.1) and using the equality $\mathbf{E}(\Delta W) = 0$. Further, since the process $g(t, X(t))$ is a martingale and $g(T, X(T))$ is independent of $\mathcal{F}_t$, we have

$$\mathbf{E}[g(t, X(t))] = \mathbf{E}[g(T, X(T))|\mathcal{F}_t] = \mathbf{E}[g(T, X(T))].$$

Using the Tonelli–Fubini theorem in Hilbert spaces (Theorem 4.1.3), we can change the order of integration and obtain

$$0 = \mathbf{E} \left[\int_t^T \left(\frac{\partial g}{\partial s}(s, X(s)) + \frac{\partial g}{\partial x}(s, X(s)) AX(s) \right. \right.$$
$$\left. \left. + \frac{1}{2} Tr \left[B^* \frac{\partial^2 g}{\partial x^2}(s, X(s)) BQ \right] \right) ds \right]$$
$$= \int_t^T \mathbf{E} \left[\frac{\partial g}{\partial s}(s, X(s)) + \frac{\partial g}{\partial x}(s, X(s)) AX(s) \right.$$
$$\left. + \frac{1}{2} Tr \left[B^* \frac{\partial^2 g}{\partial x^2}(s, X(s)) BQ \right] \right] ds.$$

188   4. Itô integrated stochastic Cauchy problems in Hilbert spaces

The last equality is true for all $t \in [0, T]$. Therefore,

$$\mathbf{E}\left[\frac{\partial g}{\partial s}(s, X(s)) + \frac{\partial g}{\partial x}(s, X(s))AX(s) + \frac{1}{2}Tr\left[B^*\frac{\partial^2 g}{\partial x^2}(s, X(s))BQ\right]\right] = 0$$

for $s \in [0, T]$. Rewrite this equality at a point (t, x),

$$\mathbf{E}\left[\frac{\partial g}{\partial t}(t, x) + \frac{\partial g}{\partial x}(t, x)Ax + \frac{1}{2}Tr\left[B^*\frac{\partial^2 g}{\partial x^2}(t, x)BQ\right]\right] = 0,$$

that is,

$$\mathbf{E}\left[\frac{\partial g}{\partial t}(t, x)\right] + \mathbf{E}\left[\frac{\partial g}{\partial x}(t, x)Ax\right] + \frac{1}{2}\mathbf{E}\left[Tr\left[B^*\frac{\partial^2 g}{\partial x^2}(t, x)BQ\right]\right] = 0.$$

Since Ax does not depend on $\omega \in \Omega$, we have

$$\mathbf{E}\left[\frac{\partial g}{\partial x}(t, x)Ax\right] = \mathbf{E}\left[\frac{\partial g}{\partial x}(t, x)\right]Ax.$$

Using the Lebesgue dominated convergence theorem, the fact that the mappings $\frac{\partial}{\partial t}$, $\frac{\partial}{\partial x}$, and $\frac{\partial^2}{\partial x^2}$ are independent of the variable $\omega \in \Omega$, and that the expectation $\mathbf{E}$ is an integral with respect to ω, we conclude that all these operations commute with $\mathbf{E}$. Furthermore, using the interpretation given above of the trace (4.4.8), we show that it also commutes with $\mathbf{E}$ due to the following equalities:

$$\mathbf{E}\left[Tr\left[B^*\frac{\partial^2 g}{\partial x^2}(t, x)BQ\right]\right] = \mathbf{E}\left[\sum_{j=1}^{\infty}B^*\frac{\partial^2 g}{\partial x^2}(t, x)BQe_j(e_j)\right]$$

$$= \sum_{j=1}^{\infty}\mathbf{E}\left[B^*\frac{\partial^2 g}{\partial x^2}(t, x)BQe_j(e_j)\right] = Tr\,\mathbf{E}\left[B^*\frac{\partial^2 g}{\partial x^2}(t, x)BQ\right].$$

Note that if we identify $\mathbb{H}^*$ with $\mathbb{H}$ and H^* with H, then

$$B^*\frac{\partial^2 g}{\partial x^2}(t, x)BQe_j(e_j) = \langle B^*\frac{\partial^2 g}{\partial x^2}(t, x)BQe_j, e_j\rangle.$$

Further, the definition of $g(t, x)$ implies

$$\mathbf{E}g(t, x) = \mathbf{E}\left[\mathbf{E}^{t,x}[h(X(T))]\right] = \mathbf{E}^{t,x}[h(X(T))] = g(t, x).$$

It follows from the equalities

$$\frac{\partial g}{\partial t}(t, x) + \frac{\partial g}{\partial x}(t, x)Ax + \frac{1}{2}Tr\left[B^*\frac{\partial^2 g}{\partial x^2}(t, x)BQ\right] = 0, \quad t \in [0, T], \quad x \in H.$$

4.4. Extension of the Feynman–Kac theorem to Hilbert spaces 189

Thus Equation (4.4.5) is proved. It remains to verify the Cauchy condition at $t = T$. We have

$$g(T, x) = \mathbf{E}^{T,x} h(X(T)) = h(X(T))|_{X(T)=x} = h(x),$$

hence $g(t, x)$ satisfies (4.4.5), which completes the proof. $\qquad\square$

Remark 4.4.2 The requirement on the function $g(t, x) = \mathbf{E}^{t,x} h(X(T))$ to have bounded derivatives which is necessary for the proof of Theorem 4.4.1 is not usually fulfilled in applications. If the function h is not smooth enough one has to consider some type of generalized solutions to the problem (4.4.5) with h being a limit of smooth functions h_n.

Remark 4.4.3 In Theorem 4.4.1 the backward Kolmogorov equation (4.4.4) for the probabilistic characteristics $g(t, x) = \mathbf{E}^{t,x}[h(X(T))]$ is obtained. The fact that the resulting deterministic problem is the backward problem is not the specifics of the relationship studied. The proof, similar to the proof of Theorem 4.4.1, shows that the probabilistic characteristics $\hat{g}(t, x) :=$ $\mathbf{E}^{0,x}[h(X(t))] : [0, T] \times H \to \mathbb{R}$ lead to the Cauchy problem for the forward Kolmogorov equation

$$\frac{\partial \hat{g}}{\partial t}(t, x) = \frac{\partial \hat{g}}{\partial x}(t, x)Ax + \frac{1}{2}Tr\left[B^* \frac{\partial^2 \hat{g}}{\partial x^2}(t, x)BQ\right], \quad t \in [0, T], \quad \hat{g}(0, x) = h(x).$$

We will prove this relation in the next subsection on the basis of the semigroup approach.

Now, in addition to the result obtained in Theorem 4.4.1, we establish the connection between solutions to the Cauchy problem (4.4.3) and the problem (4.4.5) in the opposite direction, namely, from (4.4.5) to (4.4.3).

Theorem 4.4.2 *Let* $g = g(t, x)$ *be a solution of the infinite-dimensional backward Cauchy problem (4.4.5), where* A *is the generator of a* C_0-*semigroup,* $B \in \mathcal{L}(\mathbb{H}, H)$ *in the case of* Q-*Wiener process* W, *and* $B \in \mathcal{L}_{HS}(\mathbb{H}, H)$, $Q = I$ *in the case of cylindrical Wiener process* W. *Then* $g(t, x) = \mathbf{E}^{t,x} h(X(T))$, *where* $X(\tau)$, $\tau \in [t, T]$, *is the unique solution to the stochastic Cauchy problem (4.4.3).*

Proof. Let $t \geq 0$ and $\{X(\tau), \tau \in [t, T]\}$ be a weak solution of the Cauchy problem (4.4.3). Applying the infinite-dimensional Itô formula to $g(\tau, X(\tau))$, $\tau \in [t, T]$, where $g = g(t, x)$ is a solution to the Cauchy problem (4.4.3), we obtain

$$g(T, X(T)) = g(t, X(t)) + \int_t^T \frac{\partial g}{\partial x}(s, X(s))B\, dW(s) + \int_t^T \left(\frac{\partial g}{\partial s}(s, X(s))\right.$$

$$\left. + \frac{\partial g}{\partial x}(s, X(s))AX(s) + \frac{1}{2}Tr\left[B^* \frac{\partial^2 g}{\partial x^2}(s, X(s))BQ\right]\right) ds.$$

190   *4. Itô integrated stochastic Cauchy problems in Hilbert spaces*

Since g is a solution of the Cauchy problem (4.4.5), we have

$$g(T, X(T)) = g(t, X(t)) + \int_t^T \frac{\partial g}{\partial x}(s, X(s))B\, dW(s).$$

Now apply the mathematical expectation to the terms in this equality. Since the mathematical expectation of an Itô's integral is equal to zero, we have

$$\mathbf{E}^{t,x}(g(T, X(T)) = \mathbf{E}^{t,x}(g(t, X(t))). \tag{4.4.16}$$

Rewrite both sides of the equality (4.4.16). Since T is the end point of the Cauchy problem (4.4.5), we can write it in another way:

$$\mathbf{E}^{t,x}(g(T, X(T))) = \mathbf{E}^{t,x}(h(X(T)).$$

On the other hand, we have the following equality for the right-hand side:

$$\mathbf{E}^{t,x}(g(t, X(t))) = \mathbf{E}^{t,x}g(t, x) = g(t, x)$$

and as the result we obtain $\mathbf{E}^{t,x}(h(X(T))) = g(t, x)$.

Thus we have obtained the connection between solutions to (4.4.3) and (4.4.5) in the opposite direction: if g is a solution of the Cauchy problem (4.4.5) and X is a weak solution of the Cauchy problem (4.4.3) with operator coefficients defined by (4.4.5), then $g(t, x) = \mathbf{E}^{t,x}(h(X(T)))$. That means that the postulated connection between the solutions of the problems (4.4.5) and (4.4.3) is proved. $\qquad\square$

Now show that the term $Tr[\cdot]$ in the deterministic partial differential equation (4.4.5) obtained in Theorems 4.4.1–4.4.2 can also be written in another form.

Let Q be a trace class operator in $\mathbb{H}$ such that $\{e_i\}_{i=1}^{\infty}$ is an orthonormal basis in $\mathbb{H}$ and $Qe_i = \sigma_i^2 e_i$. For operators $L, M \in \mathcal{L}_{\mathsf{HS}}(\mathbb{H}, H)$ their scalar product is defined as

$$\begin{aligned}
\langle L, M \rangle_{\mathcal{L}_{\mathsf{HS}}(\mathbb{H}, H)} &:= \sum_i \langle Le_i, Me_i \rangle \\
&= \sum_i \langle M^*Le_i, e_i \rangle = Tr[M^*L] = Tr[LM^*]. \tag{4.4.17}
\end{aligned}$$

The last equality is true since the trace of an operator is equal to the trace of its conjugate. Let $\mathbb{H}_Q = Q^{1/2}\mathbb{H}$ with the norm $\|y\|_{\mathbb{H}_Q} = \|Q^{-1/2}y\|_{\mathbb{H}}$; then $\{g_i = \sigma_i e_i\}$ is the orthonormal basis in $\mathbb{H}_Q$. Using the equalities (4.4.17), we have

$$\begin{aligned}
\|L\|^2_{\mathcal{L}_{\mathsf{HS}}(\mathbb{H}_Q, H)} &= \sum_i \|Lg_i\|^2_H = \sum_{i,j}\langle Lg_i, e_j \rangle^2_H = \sum_{i,j}\langle L(\sigma_i e_i), e_j \rangle^2_H \\
&= \sum_{i,j}\langle LQ^{1/2}e_i, e_j \rangle^2_H = \|LQ^{1/2}\|^2_{\mathcal{L}_{\mathsf{HS}}(\mathbb{H}, H)} = Tr[(LQ^{1/2})(LQ^{1/2})^*].
\end{aligned}$$

4.4. Extension of the Feynman–Kac theorem to Hilbert spaces   191

Similarly to (4.4.17) for the scalar product of operators from $\mathcal{L}_{\mathsf{HS}}(\mathbb{H}_Q, H)$, we have the equality

$$\langle L, M \rangle_{\mathcal{L}_{\mathsf{HS}}(\mathbb{H}_Q, H)} = Tr[(LQ^{1/2})(MQ^{1/2})^*]. \qquad (4.4.18)$$

Proposition 4.4.3 *Let $K \in \mathcal{L}(H)$ and $B \in \mathcal{L}(\mathbb{H}, H)$; then the following equality holds:*

$$Tr[KBB^*] = Tr[B^*KB].$$

Proof. We have the following chain of equalities:

$$
\begin{aligned}
Tr[KBQB^*] &= Tr[KBQ^{1/2}Q^{1/2}B^*] = Tr[(KBQ^{1/2})(BQ^{1/2})^*] \\
&= \langle KB, B \rangle_{\mathcal{L}_{\mathsf{HS}}(\mathbb{H}_Q, H)} = \langle B^*KB, I \rangle_{\mathcal{L}_{\mathsf{HS}}(\mathbb{H}_Q, H)} \\
&= Tr[B^*KBQ^{1/2}Q^{1/2}] = Tr[B^*KBQ].
\end{aligned}
$$

The first and the second equalities follow from the properties of the operator Q. The relation (4.4.18) for $L = KB$, $M = B$ is used in the third one. The fourth equality follows from the properties of a scalar product and in the fifth one the equality (4.4.18) is used again, but for operators $L = B^*KB$, $M = I$. This completes the proof. $\square$

Corollary 4.4.1 *The following equalities hold true:*

$$
Tr\left[\frac{\partial^2 g}{\partial x^2}(t, x)BQB^*\right] = Tr\left[\frac{\partial^2 g}{\partial x^2}(t, x)(BQ^{1/2})(BQ^{1/2})^*\right]
$$

$$
= Tr\left[B^*\frac{\partial^2 g}{\partial x^2}(t, x)BQ\right].
$$

The equalities obtained are consistent with the trace property to allow a cyclic permutation [103] and the term $Tr\left[B^*\frac{\partial^2 g}{\partial x^2}(t, x)BQ\right]$ in the deterministic PDE (4.4.5) in Theorems 4.4.1–4.4.2 can also be written in any of the forms proven to be equivalent.

4.4.3 Proof of the relations on the basis of the semi-group approach

We continue to study the relationship between Cauchy problems for stochastic equations and Cauchy problems for deterministic partial differential equations in Hilbert spaces. In contrast to (4.4.3), where "the final part" of the Cauchy problem on $[0, T]$ was considered, now we consider "the initial part":

$$dX(\tau) = AX(\tau)\,dt + B\,dW(\tau), \quad \tau \in [0, t], \quad t \geq 0, \quad X(0) = x, \qquad (4.4.19)$$

where A is, as above, the generator of a C_0-semi-group $\{U(t),\ t \geq 0\}$ in a Hilbert space H, the operator $B : \mathbb{H} \to H$ is linear bounded, and W is an

192   4. Itô integrated stochastic Cauchy problems in Hilbert spaces

$\mathbb{H}$-valued Q-Wiener process. (As we have already mentioned, a cylindrical Wiener process can be treated as a Q_1-Wiener process in a suitably chosen space $\mathbb{H}_{Q_1}$.)

Let the operator family $\{\mathbf{U}_t,\, t \geq 0\}$ be defined by

$$[\mathbf{U}_t h](x) := \mathbf{E}^{0,x}[h(X(t))] \tag{4.4.20}$$

in the space $\mathbb{B}(H)$ of bounded functions from H to $\mathbb{R}$ with the norm $\|h\|_{\mathbb{B}(H)} = \sup_{x \in H} |h(x)|$. We will show the relationship under consideration using the semi-group property of the family $\mathbf{U}$. The process X in (4.4.20) is the unique solution to (4.4.19). As proved in the previous subsections, it may be written in the form (4.4.7) and due to the properties of the stochastic convolution, $X(t)$ is a Gaussian random variable for each $t \geq 0$. For the expectation of X we have

$$\mathbf{E}[X(t)] = \mathbf{E}[U(t)x] + \mathbf{E}[W_A(t)] = U(t)x.$$

For the covariance operator $\mathbf{Cov}[X(t)]$, by the equality (4.2.18), we have

$$\mathbf{Cov}[X(t)]x = \mathbf{Cov}[W_A(t)]x = \int_0^t U(\tau)BQB^*U^*(\tau)x\,d\tau =: Q_t x.$$

For the Gaussian random value $X(t)$, it means that $X(t) \sim \mathcal{N}_{U(t)x,Q_t}$ for each $t \in [0, T]$. Then for $h \in \mathbb{B}(H)$ we have

$$[\mathbf{U}_t h](x) = \mathbf{E}^{0,x}[h(X(t))] = \int_H h(r)\mathcal{N}_{U(t)x,Q_t}(r)\,dr$$

$$= \int_H h(U(t)x + z)\,\mathcal{N}_{0,Q_t}(z)\,dz. \tag{4.4.21}$$

Recall that if H is n-dimensional, then for an $a \in \mathbb{R}$ and $(n \times n)$-matrix Q,

$$\mathcal{N}_{a,Q}(z) = (2\pi)^{-\frac{n}{2}} (\det Q)^{-\frac{1}{2}} e^{-\frac{1}{2}\langle Q^{-1}(z-a), z-a\rangle}, \qquad z \in H,$$

and

$$\int_H e^{i\langle z,y\rangle} \mathcal{N}_{a,Q}(z)\,dz = e^{i\langle a,y\rangle} e^{-\frac{1}{2}\langle Qy,y\rangle}, \qquad y \in H. \tag{4.4.22}$$

In the infinite-dimensional case the equality (4.4.22) is taken as the definition of Gaussian probability distribution for an H-valued random variable z with expectation a and covariation Q, which is denoted by $z \sim \mathcal{N}_{a,Q}$.

Now using (4.4.21), we show that the family $\{\mathbf{U}_t,\, t \geq 0\}$ defined by (4.4.20) has the semi-group property. We will obtain its generator in the space $C(H)$ of uniformly continuous and bounded real-valued functions on H.

Proposition 4.4.4 *The family $\{\mathbf{U}_t,\, t \geq 0\}$ possesses the semi-group property*

$$\mathbf{U}_{t+s}h = \mathbf{U}_t\,\mathbf{U}_s h, \quad t, s > 0, \quad h \in C(H).$$

4.4. Extension of the Feynman–Kac theorem to Hilbert spaces 193

Proof. Let Z be the linear span of functions $h_y(x) = e^{i\langle x,y\rangle}$, $x \in H$, $y \in H$. Taking $h = h_y(x)$ in the equalities (4.4.21) we have

$$[\mathbf{U}_t h_y](x) = \int_H h_y(U(t)x + z)\mathcal{N}_{0,Q_t}(z)\,dz = \int_H e^{i\langle U(t)x+z,y\rangle}\mathcal{N}_{0,Q_t}(z)\,dz$$

$$= e^{i\langle U(t)x,y\rangle}\int_H e^{i\langle z,y\rangle}\mathcal{N}_{0,Q_t}(z)\,dz = e^{i\langle U(t)x,y\rangle}e^{i\langle 0,y\rangle}e^{-\frac{1}{2}\langle Q_t y,y\rangle}$$

$$= e^{i\langle U(t)x,y\rangle}e^{-\frac{1}{2}\langle Q_t y,y\rangle} = e^{-\frac{1}{2}\langle Q_t y,y\rangle}h_{U^*(t)y}(x). \tag{4.4.23}$$

It follows that $[\mathbf{U}_{t+s} h_y](x) = e^{-\frac{1}{2}\langle Q_{t+s} y,y\rangle}h_{U^*(t+s)y}(x)$. On the other hand, we have

$$[\mathbf{U}_t \mathbf{U}_s h_y](x) = \mathbf{U}_t\left[e^{i\langle U(s)x,y\rangle}e^{-\frac{1}{2}\langle Q_s y,y\rangle}\right]$$

$$= \int_H e^{i\langle U(s)U(t)x,y\rangle}e^{i\langle U(s)z,y\rangle}e^{-\frac{1}{2}\langle Q_s y,y\rangle}\mathcal{N}_{0,Q_t}(z)\,dz$$

$$= e^{i\langle U(s)U(t)x,y\rangle}e^{-\frac{1}{2}\langle Q_s y,y\rangle}\int_H e^{i\langle U(s)z,y\rangle}\mathcal{N}_{0,Q_t}(z)\,dz$$

$$= e^{i\langle U(t+s)x,y\rangle}e^{-\frac{1}{2}\langle Q_s y,y\rangle}e^{i\langle 0,U^*(s)y\rangle}e^{-\frac{1}{2}\langle Q_t U^*(s)y,U^*(s)y\rangle}$$

$$= e^{i\langle U(t+s)x,y\rangle}e^{-\frac{1}{2}\langle(Q_s+U(s)Q_t U^*(s))y,y\rangle}$$

$$= e^{-\frac{1}{2}\langle(Q_s+U(s)Q_t U^*(s))y,y\rangle}h_{U^*(t+s)y}(x), \quad x,y \in H, \quad t,s \geq 0. \tag{4.4.24}$$

To prove that (4.4.23) coincides with (4.4.24) we have to show that $Q_s + U(s)Q_t U^*(s) = Q_{t+s}$. By the definition of Q_t in (4.2.18), we have

$$(Q_s + U(s)Q_t U^*(s))y = \int_0^s U(\tau)BQB^*U^*(\tau)y\,d\tau$$

$$+ \int_0^t U(s+\tau)BQB^*U^*(s+\tau)y\,d\tau$$

$$= \int_0^{t+s} U(r)BQB^*U^*(r)y\,dr = Q_{t+s}y, \quad y \in H.$$

Hence $[\mathbf{U}_{t+s} h_y](x) = [\mathbf{U}_t \mathbf{U}_s h_y](x)$. Thus the semi-group property holds for the family $\{\mathbf{U}_t, t \geq 0\}$ on elements h_y and, as a consequence, on their linear span, i.e., on Z. Since the space $C(H)$ of uniformly continuous and bounded functions h from H to $\mathbb{R}$ (with the same norm $\|h\| = \sup_{x \in H}|h(x)|$ as in $\mathbb{B}(H)$) can be approximated by elements of Z ([20], Proposition 1.2), it follows that $\{\mathbf{U}_t, t \geq 0\}$ possesses the semi-group property on $C(H)$. $\qquad\square$

Now we can proceed to the proof of the infinite-dimensional extension of the Feynman–Kac theorem for the "initial" Cauchy problems. We will limit ourselves to the case of a Q-Wiener process, as we have shown in the previous theorems what changes if we replace it with a cylindrical Wiener process.

194   4. Itô integrated stochastic Cauchy problems in Hilbert spaces

Theorem 4.4.3 *Let H and $\mathbb{H}$ be separable Hilbert spaces, X be a unique solution to the stochastic Cauchy problem*

$$dX(t) = AX(t)dt + B\,dW(t), \quad t \geq 0, \quad X(0) = x, \qquad (4.4.25)$$

where A is the generator of a C_0-semi-group in H, B is a bounded operator from $\mathbb{H}$ into H, and W is a Q-Wiener process in $\mathbb{H}$. Suppose that the derivatives $\frac{\partial g}{\partial t}$, $\frac{\partial g}{\partial x}$, and $\frac{\partial^2 g}{\partial x^2}$ of $g(t,x) := \mathbf{E}^{0,x}[h(X(t))]$ exist and are bounded in the corresponding spaces. Then $g(t,x)$, $t \geq 0$, $x \in \mathbb{R}$, is a solution of the infinite-dimensional forward Cauchy problem

$$\frac{\partial g}{\partial t}(t,x) = \frac{\partial g}{\partial x}(t,x)Ax + \frac{1}{2}Tr\left[B^*\frac{\partial^2 g}{\partial x^2}(t,x)BQ\right], \quad t \geq 0,$$

$$g(0,x) = h(x). \qquad (4.4.26)$$

Proof. To prove (4.4.26) we show that the left-hand side of Equation (4.4.26) is equal to $\partial[\mathbf{U}_t h_y(x)]/\partial t$ for $h_y(\cdot) = e^{i\langle \cdot, y\rangle}$, where $y \in H$, and the right-hand side is equal to $\mathbf{A}\,[\mathbf{U}_t h_y](x)$, where $\mathbf{A}$ is the infinitesimal generator of the semi-group $\{\mathbf{U}_t,\, t \geq 0\}$ and $\overline{\mathbf{A}}$ is its generator. Then the equality holds.

We start by calculating the infinitesimal generator of the semi-group. Using the equalities (4.2.18) and (4.4.23) obtained for $Q_t x$ and $[\mathbf{U}_t h_y](x)$ we have

$$
\begin{aligned}
[\mathbf{A}h_y](x) &:= \lim_{t \to 0} \frac{[\mathbf{U}_t h_y](x) - h_y(x)}{t} = \frac{\partial[\mathbf{U}_t h_y](x)}{\partial t}\Big|_{t=0} \\
&= \left[-\frac{1}{2}\langle U(t)BQB^*U^*(t)y, y\rangle + i\langle U'(t)x, y\rangle\right][\mathbf{U}_t h_y](x)\Big|_{t=0} \\
&= \left[-\frac{1}{2}\langle IBQB^*Iy, y\rangle + i\langle Ax, y\rangle\right]e^{i\langle x, y\rangle} \\
&= \left[-\frac{1}{2}\langle BQB^*y, y\rangle + i\langle Ax, y\rangle\right]h_y(x), \quad x \in dom\,A. \quad (4.4.27)
\end{aligned}
$$

Now we show that the equality (4.4.27) obtained for $x \in dom\,A$ can be written as

$$[\mathbf{A}h_y](x) = \frac{1}{2}Tr\left[(B^*\frac{\partial^2 h_y}{\partial x^2}(x)BQ\right] + \frac{\partial h_y}{\partial x}(x)Ax. \qquad (4.4.28)$$

For this purpose we need to calculate the Frechet derivatives $\partial h_y(x)/\partial x$ and $\partial^2 h_y(x)/\partial x^2$. We have the Taylor expansion $h_y(x) = e^{i\langle x, y\rangle} = 1 + i\langle x, y\rangle + o(x)$ as $x \to 0$. Hence

$$e^{i\langle x+\Delta x, y\rangle} - e^{i\langle x, y\rangle} = e^{i\langle x, y\rangle}i\langle \Delta x, y\rangle + o(\Delta x) \quad \text{as} \quad \Delta x \to 0$$

and by the definition of Frechet differentials we obtain

$$
\frac{\partial h_y(x)}{\partial x}\Delta x = \langle \Delta x, y\rangle e^{i\langle x, y\rangle},
$$

$$
\frac{\partial^2 h_y(x)}{\partial x^2}\Delta x = \frac{\partial}{\partial x}\left(\frac{\partial h_y(x)}{\partial x}\right)\Delta x = -y\langle \Delta x, y\rangle e^{i\langle x, y\rangle}.
$$

4.4. Extension of the Feynman–Kac theorem to Hilbert spaces   195

Now, using these expressions we show that for the infinitesimal generator $\mathbf{A}$ of the semi-group $\{\mathbf{U}_t,\, t \geq 0\}$ the equality (4.4.28) holds. Writing $Tr\left[BQB^*\frac{\partial^2 h_y(x)}{\partial x^2}\right]$ via the scalar product (i.e., identifying $\mathbb{H}$ with $\mathbb{H}^*$ for the sake of simplicity) we have

$$Tr\left[BQB^*\frac{\partial^2 h_y(x)}{\partial x^2}\right] = -\sum_{j=1}^{\infty}\langle BQB^*y\langle e_j, y\rangle e^{i\langle x, y\rangle}, e_j\rangle$$

$$= -\sum_{j=1}^{\infty}\langle\langle e_j, y\rangle e_j, BQB^*y e^{i\langle x, y\rangle}\rangle$$

$$= -\langle BQB^*y e^{i\langle x, y\rangle}, \sum_{j=1}^{\infty}\langle e_j, y\rangle e_j\rangle = -\langle BQB^*y, y\rangle e^{i\langle x, y\rangle}.$$

Hence, taking into account (4.4.27) and properties of traces, we obtain

$$[\mathbf{A}h_y](x) = \frac{1}{2}Tr\left[B^*\frac{\partial^2 h_y}{\partial x^2}(x)BQ\right] + \frac{\partial h_y}{\partial x}(x)Ax, \quad x \in dom\, A.$$

For further proof we note that $\overline{dom\, A} = H$ and use the fact that Z (the linear span of the elements $\{e^{i\langle x, y\rangle},\, x \in H\},\, y \in H$) is the core of the operator $\mathbf{A}$ in the space $L(H, d\mu),\, d\mu = dN_{0,Q_\infty}$ (see [24]). It follows that (4.4.28) can be extended to all functions from $L(H, d\mu)$, in particular, to the functions $g(t, x) = \mathbf{E}^{0,x}[h(X(t))] = [\mathbf{U}_t h](x),\, x \in H$:

$$\overline{\mathbf{A}}g(t, x) = \frac{1}{2}Tr\left[B^*\frac{\partial^2 g}{\partial x^2}(t, x)BQ\right] + \frac{\partial g}{\partial x}(t, x)Ax, \quad x \in dom\, A.$$

Since $u(t) = [\mathbf{U}_t h]$ solves the Cauchy problem $u'(t) = \overline{\mathbf{A}}u(t),\, t \geq 0,\, u(0) = h$, with the generator $\overline{\mathbf{A}}$ of the semi-group $\{\mathbf{U}_t,\, t \geq 0\}$ in the space $L(H; \mu)$, we have

$$\frac{\partial g}{\partial t}(t, x) = \frac{1}{2}Tr\left[B^*\frac{\partial^2 g}{\partial x^2}(t, x)BQ\right] + \frac{\partial g}{\partial x}(t, x)Ax, \quad g(0, x) = h(x),\, x \in dom\, A.$$

This is the deterministic infinite-dimensional forward Cauchy problem (4.4.26). For functions $h \in L(H; \mu)$ the equation, of course, is understood in a generalized sense.   □

Thus, using the Itô approach in Theorems 4.4.1–4.4.2, we have proved the relationship between the stochastic Cauchy problem (4.4.3) and the backward Cauchy problem for the deterministic partial differential equation (4.4.4). Using the semi-group techniques in Theorem 4.4.3 we have proved the relationship between the stochastic Cauchy problem (4.4.25) and the forward Cauchy problem (4.4.26). It is well known that by the change of variables $\tau = T - t$ the backward Cauchy problem for an abstract equation $dg(t)/dt = -\mathbf{A}g(t)$ can be transformed into the forward Cauchy problem for the equation $dg(\tau)/dt = \mathbf{A}g(\tau)$. This shows the relationship between the deterministic Cauchy problems obtained in Theorems 4.4.1, 4.4.2, and 4.4.3.

Chapter 5

Infinite-dimensional stochastic Cauchy problems with white noise processes in spaces of distributions

We continue to study infinite-dimensional stochastic Cauchy problems, now in spaces of distributions. These problems arise in numerous applications as mathematical models reflecting random influence of white noise type on systems that are under consideration. Among the problems, the important one is the Cauchy problem (P.1) in Hilbert spaces H, $\mathbb{H}$:

$$X'(t) = AX(t) + F(t, X) + B(t, X)\mathbb{W}(t), \quad t \in [0, T], \qquad X(0) = \zeta,$$

with an $\mathbb{H}$-valued white noise process $\mathbb{W}$, the generator A of a regularized semi-group in H, a nonlinear term $F : H \to H$, and $B \in \mathcal{L}(\mathbb{H}, H)$.

As noted in the Introduction, the problem is ill-posed due to the irregularity of the white noise and the unboundedness of solution operators generated by A. In Chapter 4 we overcame these obstacles by studying the problem in the integrated form with stochastic integrals wrt Wiener processes and by constructing regularized solutions. In the present chapter we study the problem in spaces of distributions. The spaces are chosen in dependence on the type of semi-group generated by A.

In Section 5.1, for the linear stochastic Cauchy problem in the case of A generating an integrated or convoluted semi-group, we construct generalized (wrt t) solutions in spaces of abstract distributions or ultra-distributions. In the case of operators generating R-semi-groups, especially differential operators $A = \mathbf{A}(i\partial/\partial x)$, we construct generalized (wrt t and x) solutions using Gelfand–Shilov spaces.

In Section 5.2 we construct generalized solutions to semi-linear stochastic Cauchy problems. We overcome the additional difficulties arising here connected with multiplication of distributions by the applicatio of the Colombeau technique extended to the case of Hilbert space valued distributions.

197

198   5. Stochastic Cauchy problems in spaces of distributions

5.1 Generalized solutions to linear stochastic Cauchy problems with generators of regularized semi-groups

In the present section we consider the linear stochastic problem with additive noise

$$X'(t) = AX(t) + B\mathbb{W}(t), \quad t \geq 0, \quad X(0) = \zeta, \tag{5.1.1}$$

in space of abstract distributions. We construct generalized (wrt t) solutions satisfying the equation

$$\langle \varphi, X' \rangle = A\langle \varphi, X \rangle + \langle \varphi, \delta \rangle \zeta + \langle \varphi, B\mathbb{W} \rangle, \quad \varphi \in \Phi, \tag{5.1.2}$$

for the case of "regularized in t" integrated and convoluted semi-groups. Here the test function space Φ is taken in dependence on the type of the semi-group generated by A, and white noise distribution $\mathbb{W} \in \Phi'(\mathbb{H})$ is defined as a generalized derivative of a Wiener process imbedded in $\Phi'(\mathbb{H})$. In the case of differential operators A generating an R-semi-groups ("regularized in x"), we construct solutions generalized wrt t and x.

5.1.1 Setting the problem in spaces of abstract distributions. Generalized solutions in the case of integrated and convoluted semi-groups

In order to give the statement of the stochastic Cauchy problem in spaces of abstract distributions and define white noise in these spaces, recall some notations from the theory of abstract distributions presented in Sections 2.1 and 2.2.

Let H be a Banach space, in particular a Hilbert space. We denote by $\mathcal{D}'(H)$ the space of H-valued distributions over the space of Schwartz test functions $\mathcal{D}$ and by $\mathcal{S}'(H)$ the space of H-valued tempered distributions. Unlike distributions over $\mathbb{R}$ or $\mathbb{C}$, such distributions are called abstract.

In this subsection Hilbert space valued distributions are of special interest. The distribution spaces $\mathcal{D}'(H)$, $\mathcal{S}'(H)$, $\mathcal{S}'_\omega(H)$, and $\mathcal{D}'_{\{M_q\}}(H)$ will be taken as $\Phi'(H)$ for solutions to (5.1.2) with generators of integrated and convoluted semi-groups.

Let $(\Omega, \mathcal{F}, P)$ be a probability space and $\{W(t), t \geq 0\}$ be a Q-Wiener process with values in a Hilbert space $\mathbb{H}$ (Definition 4.1.7). We have

$$W(t) = W(t, \omega), \ \omega \in \Omega; \quad W(t, \omega) \in \mathbb{H}, \ t \geq 0, \ P_{\text{a.s.}}; \quad W(t, \cdot) \in L^2(\Omega; \mathbb{H}).$$

A Q-white noise $\mathbb{W}$ in spaces of abstract distributions is defined as the generalized t-derivative of W:

$$\langle \varphi, \mathbb{W} \rangle := -\langle \varphi', W \rangle = -\int_0^\infty W(t)\varphi'(t)\,dt, \quad \varphi \in \Phi, \tag{5.1.3}$$

5.1. Generalized solutions to linear stochastic Cauchy problems 199

where W is regarded as a (regular) element of $\Phi_0'(\mathbb{H})$ $P_{a.s.}$ and of $\Phi_0'(L^2(\Omega;\mathbb{H}))$. The integral in (5.1.3) is understood as the Bochner integral of a function with values either in $\mathbb{H}$ or in $L^2(\Omega;\mathbb{H})$. Thus we have

$$\mathbb{W} = \mathbb{W}(\cdot,\omega) \in \Phi_0'(\mathbb{H}) \ \text{ for } \ \omega \ P_{\mathbf{a}.\mathbf{s}.}, \quad \text{and} \quad \mathbb{W} \in \Phi_0'(L^2(\Omega;\mathbb{H})).$$

In Section 4.1 we noticed that an $\mathbb{H}$-valued cylindrical Wiener process W can be considered as an $\mathbb{H}_1$-valued Q_1-Wiener process; therefore, without loss of generality, we can confine ourselves to Q-Wiener processes.

To explain the setting (5.1.2) for the generalized Cauchy problem, we will use the general idea of reducing boundary-value problems to equations with δ-functions and their derivatives multiplied by boundary (initial) data in spaces of distributions (see, e.g., [30, 79]). Within this approach we consider the generalized wrt t stochastic Cauchy problem (5.1.1) with $\mathbb{W} = W'$ and A generating an integrated (convoluted) semi-group as (5.1.2) taking Φ to be a space of (ultra)differentiable functions, i.e.,

$$\langle\varphi,X'\rangle = A\langle\varphi,X\rangle + \langle\varphi,\delta\rangle\zeta + \langle\varphi,B\mathbb{W}\rangle, \quad \varphi \in \mathcal{D} \quad (\varphi \in \mathcal{D}_{\{M_q\}}). \quad (5.1.4)$$

It holds for ω a.s., that is, $P_{\mathbf{a}.\mathbf{s}.}$, and can be considered as a stochastic extension of the generalized Cauchy problem (2.1.7) with initial data ζ.

The equality (5.1.4) can be obtained if we formally consider X and $\mathbb{W}$ in (5.1.1) as functions, multiply the equation by a test function φ, and integrate from zero to infinity:

$$\int_0^\infty X'(t)\varphi(t)\,dt = -\varphi(0)\zeta - \int_0^\infty X(t)\varphi'(t)\,dt$$
$$= \int_0^\infty AX(t)\varphi(t)\,dt + \int_0^\infty B\mathbb{W}(t)\varphi(t)\,dt.$$

Taking into account the equalities $\langle\varphi,X'\rangle := -\langle\varphi',X\rangle$ and $\langle\varphi,\delta\rangle\zeta = \varphi(0)\zeta$, we get to (5.1.4). Here, in addition to stochastic inhomogeneity, Equation (5.1.4) has the term $\delta\zeta$ due to the jump of the solution from zero at $t < 0$ up to the initial data ζ at $t = 0$.

Following results from Section 2.1 we can write (5.1.4) as the equation in convolutions:

$$\mathbf{P} * X = \delta \otimes \zeta + B\mathbb{W}, \quad \mathbf{P} := \delta' \otimes I - \delta \otimes A \in \mathcal{D}_0'\big(\mathcal{L}([dom\,A],H)\big), \quad (5.1.5)$$

where

$$\langle\varphi,\delta' \otimes I\rangle := \langle\varphi,\delta'\rangle I = \varphi'(0), \quad \langle\varphi,\delta \otimes A\rangle := \langle\varphi,\delta\rangle A = \varphi(0)A, [1]$$

and $[dom\,A]$ is the domain of A with the graph norm $\|x\|_{[dom\,A]} = \|x\| + \|Ax\|$.

The convolution of distributions here and below will be understood in the following sense (consistent with definitions in Section 2.1).

---

[1] For $u \in \mathcal{D}'$, $h \in H$, we denote by $u \otimes h$ the distribution from $\mathcal{D}'(H)$ defined by the equality $\langle\theta,u \otimes h\rangle := \langle\theta,u\rangle h, \theta \in \mathcal{D}$.

200   5. Stochastic Cauchy problems in spaces of distributions

Definition 5.1.1 *Let $\mathcal{X}$, $\mathcal{Y}$, and $\mathcal{Z}$ be Banach spaces such that a bilinear operation $(u, v) \mapsto uv \in \mathcal{Z}$ is defined on $\mathcal{X} \times \mathcal{Y}$. Then, for any distributions $G \in \mathcal{D}'_0(\mathcal{X})$ and $F \in \mathcal{D}'_0(\mathcal{Y})$, the convolution $G * F \in \mathcal{D}'_0(\mathcal{Z})$ is defined by the equality*

$$\langle \varphi, G*F \rangle := \langle \varphi, (g*f)^{(n+m)} \rangle = (-1)^{(n+m)} \int_0^\infty (g*f)(t)\varphi^{(n+m)}(t)\, dt, \quad \varphi \in \mathcal{D},$$

where $g : \mathbb{R} \mapsto \mathcal{X}$ and $f : \mathbb{R} \mapsto \mathcal{Y}$ are continuous functions such that

$$\langle \varphi, G \rangle = (-1)^n \int_0^\infty g(t)\varphi^{(n)}(t)\, dt, \qquad \langle \varphi, F \rangle = (-1)^m \int_0^\infty f(t)\varphi^{(m)}(t)\, dt,$$

$(g*f)(t) := \int_0^t g(t-s)f(s)\, ds$ (g, f, n, m here depend on G, F, and φ). If G is a regular distribution, i.e., $\langle \varphi, G \rangle = \int_0^\infty \varphi(t)G(t)\, dt$, the following equality holds:

$$\langle \varphi, G * F \rangle = \int_0^\infty G(s) \langle \varphi(s + \cdot), F(\cdot) \rangle\, ds. \tag{5.1.6}$$

Thus we will study the stochastic Cauchy problem (5.1.1) in spaces of abstract distributions in the form (5.1.4) or in the equivalent form (5.1.5). To obtain a *generalized solution to* (5.1.1), i.e., a solution to the convolution equation (5.1.5), we need a distribution which is convolution inverse to $\mathbf{P}$.

A distribution $G \in \mathcal{D}'_0(\mathcal{L}(H, [dom\, A]))$ is called *convolution inverse* to $\mathbf{P} \in \mathcal{D}'_0(\mathcal{L}([dom\, A], H))$ if

$$G * \mathbf{P} = \delta \otimes I_{[dom\, A]}, \qquad \mathbf{P} * G = \delta \otimes I_H,$$

where $I_{[dom\, A]}$ and I_H are unit operators in $[dom\, A]$ and H, respectively. Using these equalities and the fact that the δ-function plays the role of unity wrt convolution, we obtain that the unique solution to the equation $\mathbf{P} * X = \mathbb{F}$ with an inhomogeneity term $\mathbb{F} \in \mathcal{D}'_0(H)$ has the form $X = G * \mathbb{F}$. It follows that, for $\zeta \in L^2(\Omega; H)$,

$$X = G * \delta\zeta + G * B\mathbb{W}, \tag{5.1.7}$$

where $X = X(\cdot, \omega) \in \mathcal{D}'_0([dom A])$ $P_{\text{a.s.}}$ and $X = X(\cdot, \cdot) \in \mathcal{D}'_0(L^2(\Omega; [dom A]))$ is the unique solution to the Cauchy problem (5.1.5) with stochastic inhomogeneity $\mathbb{F} = \delta\zeta + B\mathbb{W} \in \mathcal{D}'_0(L^2(\Omega; H))$.

Let us show how to construct G, the convolution inverse to $\mathbf{P}$. We begin with the case of A generating a C_0-semi-group or an integrated semi-group and arrive at the construction of the solution to (5.1.5) in the space of abstract distributions $\mathcal{D}'_0(L^2(\Omega; [dom\, A]))$ for such generators.

Theorem 5.1.1 *Let H and $\mathbb{H}$ be Hilbert spaces, A be the generator of a C_0-semi-group or of an n-times integrated semi-group in H, $\zeta \in L^2(\Omega; H)$, and $\mathbb{W}$ be an $\mathbb{H}$-valued Q-white noise defined by (5.1.3). Then there exists a unique generalized solution $X \in \mathcal{D}'_0(L^2(\Omega; [dom\, A]))$ of the Cauchy problem (5.1.1), i.e., a unique solution of (5.1.5).*

5.1. Generalized solutions to linear stochastic Cauchy problems 201

Proof. Begin with the case when A is the generator of a C_0-semi-group $\{S(t),\, t \geq 0\}$. Define an operator-valued distribution G by

$$\langle \varphi, G \rangle = \int_0^\infty \varphi(t) S(t)\, dt.$$

Let us prove that it is the convolution inverse to $\mathbf{P}$ and hence $X = \mathbf{S}\zeta := \mathbf{S} \otimes \zeta$ is a solution to (5.1.5) if $B = 0$. Here $\mathbf{S} \in \mathcal{D}'_0(H)$ is the H-valued regular distribution defined by the semi-group $\{S(t),\, t \geq 0\}$ continued by zero for $t < 0$.

By the definition of C_0-semi-groups, the operators $S(t),\, t \geq 0$, are bounded in H and strongly continuous wrt $t \geq 0$ solution operators to the homogeneous Cauchy problem $X'(t) = AX(t),\, t \geq 0$, with initial data $X(0) = \zeta$: $X(t) = S(t)\zeta,\, \zeta \in dom A$. From the properties of the solution operators the equalities follow

$$\langle \varphi, \mathbf{S}\zeta \rangle = \int_0^\infty \varphi(t) S(t)\zeta\, dt, \qquad \varphi \in \mathcal{D}, \quad \zeta \in H,$$

and

$$A\langle \varphi, \mathbf{S}\zeta \rangle = -\varphi(0)\zeta - \int_0^\infty \varphi'(t) S(t)\zeta\, dt, \qquad \varphi \in \mathcal{D}, \quad \zeta \in dom\, A. \qquad (5.1.8)$$

Let us show that using closedness of A and density of $dom\, A$ in H we can extend (5.1.8) to $\zeta \in H$. Let $\zeta_n \in dom\, A$ and $\zeta_n \to \zeta$, then

$$\lim_{n \to \infty} S(t)\zeta_n = S(t)\zeta, \qquad \lim_{n \to \infty} \langle \varphi, \mathbf{S}\zeta_n \rangle = \langle \varphi, \mathbf{S}\zeta \rangle,$$

and by (5.1.8), the sequence $A\langle \varphi, \mathbf{S}\zeta_n \rangle$ is convergent. Since A is closed, it follows that $\langle \varphi, \mathbf{S}\zeta \rangle \in dom\, A$ for each $\zeta \in H$ and (5.1.8) holds for $\zeta \in H$.

Now we show that $\mathbf{S}\zeta,\, \zeta \in L^2(\Omega; H)$ is a generalized solution to the homogeneous Cauchy problem corresponding to (5.1.5). Since $\langle \varphi, \mathbf{S}\zeta \rangle \in dom\, A$ for each $\zeta \in H$, we can take convolution of the operator-distribution $\mathbf{P} \in \mathcal{D}'_0(\mathcal{L}([dom\, A], H))$ with the distribution $\mathbf{S}\zeta$. Using the equality (5.1.6), we obtain

$$\langle \varphi, \mathbf{P} * \mathbf{S}\zeta \rangle = \int_0^\infty \langle \varphi(s + \cdot), \mathbf{P}(\cdot) \rangle S(s)\zeta\, ds = -\langle \varphi'(s), \mathbf{S}\zeta \rangle - A\langle \varphi(s), \mathbf{S}\zeta \rangle$$

$$= -\int_0^\infty \varphi'(s) S(s)\zeta\, ds + \varphi(0)\zeta + \int_0^\infty \varphi'(s) S(s)\zeta\, ds = \varphi(0)\zeta = \langle \varphi, \delta \otimes \zeta \rangle.$$

Thus $\langle \varphi, \mathbf{P} * \mathbf{S}\zeta \rangle = \langle \varphi, \delta \otimes \zeta \rangle$. Hence $\mathbf{S}$ is the convolution inverse to $\mathbf{P}$. In this case the solution (5.1.7) can be written as

$$\langle \varphi, X \rangle = \int_0^\infty \varphi(t) S(t)\zeta\, dt - \int_0^\infty \varphi'(t) \int_0^t S(t - s) BW(s)\, ds\, dt,$$

$$P_{\text{a.s.}}, \quad \varphi \in \mathcal{D}, \quad (5.1.9)$$

202   5. *Stochastic Cauchy problems in spaces of distributions*

where the Q-Wiener process W extended by zero for $t < 0$ is a regular distribution in $\mathcal{D}_0'(\mathbb{H}) \cap \mathcal{D}_0'(L^2(\Omega; \mathbb{H}))$ and $X \in \mathcal{D}_0'([dom\ A]) \cap \mathcal{D}_0'(L^2(\Omega; [dom\ A]))$. Rigorously that means $X \in \mathcal{D}_0'([dom\ A])$ $P_{\text{a.s.}}$ and $X \in \mathcal{D}_0'(L^2(\Omega; [dom\ A]))$.

If A is the generator of an n-times integrated semi-group $\{S_n(t),\ t \geq 0\}$, then the solution can be obtained by differentiating n times the regular distribution $\mathbf{S}_n(\cdot)$ defined as $S_n(t)$ for $t \geq 0$ and as zero for $t < 0$. Hence the convolution inverse to the distribution $\mathbf{P}$ is the operator-valued distribution G defined by

$$\langle \varphi, Gx \rangle = (-1)^n \int_0^\infty \varphi^{(n)}(t) S_n(t) x\, dt, \quad \varphi \in \mathcal{D}, \quad x \in H.$$

Hence $X = G\zeta + G * B\mathbb{W} \in \mathcal{D}_0'([dom\ A]) \cap \mathcal{D}_0'(L^2(\Omega; [dom\ A]))$ is the unique generalized solution of the problem (5.1.1), where A is the generator of an n-times integrated semi-group S_n and it can be written as

$$\langle \varphi, X \rangle = (-1)^n \left[\int_0^\infty \varphi^{(n)}(t) S_n(t) \zeta\, dt - \int_0^\infty \varphi^{(n+1)}(t)\, dt \int_0^t S_n(t-s) BW(s)\, ds \right]$$

for $\varphi \in \mathcal{D}$. $\qquad\qquad\qquad\qquad\qquad\qquad\qquad\qquad\qquad\qquad\qquad\qquad\qquad$ $\square$

Now we consider a generalized solution to the Cauchy problem (5.1.1), where A is the generator of a K-convoluted semi-group S_K.

As shown above, in order to obtain a generalized solution to (5.1.1), or equivalently a solution to (5.1.5), it is necessary to have G, the convolution inverse to $\mathbf{P}$. We will use as G the solution operators for the Cauchy problem with the generator of a K-convoluted semi-group. Unlike the case of n-times integrated semi-groups, where $G = \mathbf{S}^{(n)}$, an infinite order differential operator is needed to define the solution operator via the K-convoluted semi-group. Such an operator can be well defined only on spaces of test functions narrower than $\mathcal{D}$, namely, on the spaces of ultra-differentiable test functions. Thus the solutions become elements of spaces which are wider than the spaces of abstract distributions, namely, the spaces of abstract ultra-distributions.

In Section 2.2 we considered in detail generalized solutions to the homogeneous Cauchy problem on spaces of Roumier ultra-differentiable test functions $\mathcal{D}^{\{M_q\}}$. Here we consider another class of ultra-differentiable test functions, Beurling class $\mathcal{D}_{\{M_q\}}$. It is defined via estimates for their derivatives that depend on a sequence of positive numbers $\{M_q\}$ under conditions (M.1)–(M3) (see Section 3.3):

$$\mathcal{D}_{\{M_q\}} = ind \lim_{[-n,n] \subset \mathbb{R}} proj \lim_{h \to 0} \mathcal{D}_{\{M_q\},h,n},$$

where $\mathcal{D}_{\{M_q\},h,n}$ is the space of functions $\varphi \in C^\infty(\mathbb{R})$ with compact supports $[-n, n]$ satisfying the inequalities $\|\varphi^{(q)}\|_{C[-n,n]} \leq C M_q h^q$ with the norm

$$\|\varphi\|_{\{M_q\},h,n} = \sup_q \left(\frac{\|\varphi^{(q)}\|_{C[-n,n]}}{M_q h^q} \right).$$

5.1. Generalized solutions to linear stochastic Cauchy problems 203

The corresponding space of abstract ultra-distributions is the space $\mathcal{D}'_{\{M_q\}}(H) := \mathcal{L}(\mathcal{D}_{\{M_q\}}, H)$.

Thus we consider the stochastic Cauchy problem (5.1.1) with A generating a K-convoluted semi-group in spaces of abstract ultra-distributions. In this case the equivalent problems (5.1.4) and (5.1.5) will be considered for test functions $\varphi \in \mathcal{D}_{\{M_q\}}$.

Theorem 5.1.2 *Let H and $\mathbb{H}$ be Hilbert spaces and A be the generator of a K-convoluted semi-group S_K in H. Let $\zeta \in L^2(\Omega; H)$, $\mathbb{W}$ be an $\mathbb{H}$-valued Q-white noise defined in $\mathcal{D}'_{\{M_q\}}(\mathbb{H})$, and $B \in \mathcal{L}(\mathbb{H}, H)$. Then the Cauchy problem*

$$\langle \varphi, \boldsymbol{P} * X \rangle = \langle \varphi, \delta \otimes \zeta + B\mathbb{W} \rangle, \qquad \varphi \in \mathcal{D}_{\{M_q\}}, \tag{5.1.10}$$

has a unique solution $X \in \mathcal{D}'_{\{M_q\},0}([\operatorname{dom} A]) \cap \mathcal{D}'_{\{M_q\},0}(L^2(\Omega; [\operatorname{dom} A]))$.

Proof. In the case considered the unique solution $G\zeta$ of the homogeneous Cauchy problem $\langle \varphi, \boldsymbol{P} * G\zeta \rangle = \langle \varphi, \delta \otimes \zeta \rangle$, $\varphi \in \mathcal{D}_{\{M_q\}}$, related to (5.1.10), is defined by the equality

$$\langle \varphi, G\zeta \rangle := \left\langle \varphi, P_{ult}\left(\frac{d}{dt}\right) \mathbf{S}_K \zeta \right\rangle = \left\langle P^*_{ult}\left(\frac{d}{dt}\right) \varphi, \mathbf{S}_K \zeta \right\rangle, \quad \varphi \in \mathcal{D}_{\{M_q\}}. \tag{5.1.11}$$

Here $\mathbf{S}_K$ is the regular ultra-distribution in $\mathcal{D}'_{\{M_q\},0}(\mathcal{L}(H, [\operatorname{dom} A]))$ defined as the semi-group S_K continued by zero as $t < 0$. By the properties of K-convoluted semi-groups (see Section 2.2) the ultra-differential operator $P_{ult}\left(\frac{d}{dt}\right) = \sum_{i=1}^{\infty} \alpha_i \frac{d^i}{dt^i}$ is defined via a convolution inverse to K:

$$\langle \varphi, P_{ult}(\delta) * K \rangle = \left\langle \varphi, \sum_{i=1}^{\infty} \alpha_i \delta^{(i)} * K \right\rangle = \langle \varphi, \delta \rangle, \quad \varphi \in \mathcal{D}_{\{M_q\}}. \tag{5.1.12}$$

(Compare with n-times integrated semi-groups, where

$$K(t) = \frac{t^{n-1}}{(n-1)!}, \qquad P_{ult}\left(\frac{d}{dt}\right) = \frac{d^n}{dt^n} \quad \text{and} \quad \delta^{(n)} * \frac{t^{n-1}}{(n-1)!} = \delta.)$$

In the K-convoluted case the ultra-distribution G defined by (5.1.11) belongs to the space $\mathcal{D}'_{\{M_q\},0}(\mathcal{L}(H, [\operatorname{dom} A]))$ and

$$G\zeta = P_{ult}\left(\frac{d}{dt}\right) \mathbf{S}_K \zeta \in \mathcal{D}'_{\{M_q\},0}([\operatorname{dom} A])$$

is the solution to the corresponding homogeneous Cauchy problem.

Now let us consider the stochastic part of the solution, i.e., the generalized stochastic convolution $G * B\mathbb{W}$:

$$\langle \varphi, G * B\mathbb{W} \rangle := -\int_0^\infty S_K(t) \left\langle P^*_{ult}\left(\frac{d}{dt}\right) \varphi'(t + \cdot), B\mathbb{W}(\cdot) \right\rangle dt, \quad \varphi \in \mathcal{D}_{\{M_q\}}.$$

204   5. Stochastic Cauchy problems in spaces of distributions

Since $G \in \mathcal{D}'_{\{M_q\},0}(\mathcal{L}(H, [dom\, A]))$ and $BW \in \mathcal{D}'_{\{M_q\},0}(H) \cap \mathcal{D}'_{\{M_q\},0}(L^2(\Omega; H))$, the stochastic convolution $G * BW$ is well defined. It belongs to the space $\mathcal{D}'_{\{M_q\},0}(L^2(\Omega; [dom\, A]))$ and for $X = G\zeta + G * BW$ we have the following equalities:

$$
\begin{aligned}
\langle \varphi, X \rangle &= \left\langle P^*_{ult}\left(\frac{d}{dt}\right)\varphi, \mathbf{S}_K\zeta \right\rangle \\
&\quad - \int_0^\infty S_K(t) \left\langle P^*_{ult}\left(\frac{d}{dt}\right)\varphi'(t+s), BW(s) \right\rangle dt \\
&= \int_0^\infty P^*_{ult}\left(\frac{d}{dt}\right)\varphi(t) S_K(t)\zeta\, dt \\
&\quad + \int_0^\infty P^*_{ult}\left(\frac{d}{dt}\right)\varphi(t)\, dt \int_0^t S_K(t-s)B\, dW(s) \\
&= \langle \varphi, \mathbf{U}_K\zeta + \mathbf{U}_K * BW \rangle, \qquad \varphi \in \mathcal{D}_{\{M_q\}}, \tag{5.1.13}
\end{aligned}
$$

where $\mathbf{U}_K = P_{ult}\left(\frac{d}{dt}\right)\mathbf{S}_K$ is the distribution of solution operators corresponding to the K-convoluted semi-group $\{S_K(t), t \geq 0\}$. Therefore, $X = G\zeta + G * BW$ is a solution of the stochastic Cauchy problem (5.1.10). □

5.1.2   Connections between weak and generalized solutions

Now we compare the generalized solutions obtained in this section for the linear stochastic Cauchy problem (5.1.2), where A is the generator of a C_0-semi-group (as well an integrated and a convoluted one), with weak and regularized solutions obtained in Section 4.2 for the Itô integrated problem in Hilbert spaces.

We prove the relationship between these solutions under the same conditions on the initial data of the problem [84].

Theorem 5.1.3 *Let A be the generator of a C_0-semi-group S in a Hilbert space H and ζ be an H-valued $\mathcal{F}_0$-measurable random variable. Then a weak solution to the Itô integral Cauchy problem defined as a solution to the Cauchy problem (4.2.3),*

$$
\langle X(t), y \rangle = \langle \zeta, y \rangle + \int_0^t \langle X(s), A^*y \rangle\, ds + \langle BW(t), y \rangle \quad P_{a.s.},
$$

$$
t \geq 0, \quad y \in dom\, A^*, \tag{5.1.14}
$$

and continued by zero for $t < 0$, is a solution to the generalized Cauchy problem (5.1.4), where Q-white noise $\mathbb{W}$ is the generalized derivative of the Q-Wiener process $\{W(t), t \geq 0\}$ continued by zero for $t < 0$. Conversely, the generalized solution defined by (5.1.9) is a solution to the Cauchy problem (5.1.14).

5.1. Generalized solutions to linear stochastic Cauchy problems 205

Proof. Let us verify that the solution of the Cauchy problem (5.1.14) defined for each H-valued $\mathcal{F}_0$-measurable random variable ζ as the H-valued process

$$X(t) = S(t)\zeta + \int_0^t S(t-s)B\,dW(s), \qquad t \geq 0, \tag{5.1.15}$$

continued by zero as $t < 0$, satisfies (5.1.4). For this purpose we first prove the important equality between the abstract Itô and Bochner integrals, which is essentially based on the abstract formula of integration by parts for the Itô integrals,

$$-\int_0^\infty W(t)\varphi'(t)\,dt = \int_0^\infty \varphi(t)\,dW(t), \qquad \varphi \in \mathcal{D}, \tag{5.1.16}$$

and is closely related to (5.1.3), which is the definition of the Q-white noise distribution $\mathbb{W}$ in spaces of abstract distributions

$$\langle \varphi, \mathbb{W} \rangle := -\int_0^\infty W(t)\varphi'(t)\,dt, \qquad \varphi \in \mathcal{D}.$$

By Definition 4.1.9 of the abstract (H-valued) Itô integral, it is a limit in the space $L^2(\Omega; H)$ of specially constructed integral sums:

$$\int_0^\infty \varphi(t)\,dW(t) = \lim_{n \to \infty} \sum_{i=0}^{n-1} \varphi(t_i)\Delta W(t_i).$$

By the properties of these sums we have

$$\int_0^\infty \varphi(t)\,dW(t) = \lim_{n \to \infty} \sum_{i=0}^{n-1} \Delta\big(\varphi(t_i)W(t_i)\big) - \lim_{n \to \infty} \sum_{i=0}^{n-1} W(t_{i+1})\Delta\varphi(t_i)$$

$$= \int_0^\infty d\big(\varphi(t)W(t)\big) - \int_0^\infty W(t)\,d\varphi(t) = -\int_0^\infty W(t)\varphi'(t)\,dt.$$

Now multiply by $\varphi \in \mathcal{D}$ the equality (5.1.15), which defines a weak solution, and integrate both sides of the equality wrt t from zero to infinity. Using change of order of integration (Theorem 4.1.3) and then applying the equality (5.1.16), we obtain the following equalities for X:

$$\langle \varphi, X \rangle = \int_0^\infty \varphi(t)S(t)\zeta\,dt + \int_0^\infty \varphi(t)\int_0^t S(t-s)B\,dW(s)\,dt$$

$$= \langle \varphi, \mathbf{S}\zeta \rangle - \Big\langle \varphi', \int_0^t S(t-s)B\,W(s)ds \Big\rangle\ P_{a.s.}. \tag{5.1.17}$$

The equalities (5.1.17) can be written in the form

$$\langle \varphi, X \rangle = \langle \varphi, \mathbf{S}\,\zeta \rangle - \langle \varphi', \mathbf{S} * B\,W \rangle = \langle \varphi, \mathbf{S}\,\zeta \rangle - \langle \varphi, \mathbf{S} * B\,\mathbb{W} \rangle, \tag{5.1.18}$$

206   5. Stochastic Cauchy problems in spaces of distributions

where $\mathbf{S} \in \mathcal{D}_0'(H)$ is the H-valued regular distribution defined by the semi-group of solution operators $\{S(t), t \geq 0\}$ continued by zero for $t < 0$. The equality (5.1.18) means that the process (5.1.15) continued by zero for $t < 0$ and being a weak solution of (5.1.4) coincides with its generalized solution.

Conversely, having a C_0-semi-group S and corresponding distribution $\mathbf{S}$, we move from the bottom up in the above proof, i.e., from the equality (5.1.18) to (5.1.17). We obtain that generalized solution $X = \mathbf{S}\zeta + \mathbf{S} * B\mathbb{W}$ of the problem (5.1.4) with the generator of the semi-group $\{S(t), t \geq 0\}$ coincides with (5.1.15), the solution of (5.1.14). $\qquad\square$

The analysis of the relations between weak and generalized solutions shows that in the case of the generator of a C_0-semi-group $\{S(t), t \geq 0\}$ the sum of two terms $S(t)\zeta + \mathbb{W}_A(t)$, $t \geq 0$, is a weak solution for each H-valued $\mathcal{F}_0$-measurable random variable ζ, due to the fact that there is no need to apply A to $S(t)\zeta$ or to $\mathbb{W}_A(t)$. Instead, A^* is applied to the elements $y \in dom\, A^*$.

On the other hand, $S(t)\zeta + \mathbb{W}_A(t)$, $t \geq 0$, is the generalized solution due to the equality (5.1.8), which implies that the action of A is "relaxed" by the test functions φ. More specifically, according to the properties of the semi-group operators $\{S(t), t \geq 0\}$, the action of the generator is transformed into the t-differentiation, which is in turn transferred to (infinitely differentiable) test functions φ due to the properties of generalized differentiation.

In the case of n-times integrated or K-convoluted semi-groups we show that generalized solutions coincide with the nth derivative of the integrated solution or with the ultra-differential derivative of the K-convoluted solution, respectively (under the same conditions on ζ and W, of course).

Theorem 5.1.4 *Let A be the generator of an n-times integrated semi-group $\{S_n(t), t \geq 0\}$. Then the nth generalized derivative of a weak n-times integrated solution (continued by zero for $t < 0$) is a solution of the generalized Cauchy problem (5.1.4). Conversely, a generalized solution is the nth generalized derivative of a weak n-times integrated solution to the Cauchy problem (4.2.1).*

Proof. As proved in the previous chapter, the process

$$X(t) = S_n(t)\zeta + \int_0^t S_n(t - s)B\,dW(s), \qquad t \geq 0,$$

is a weak n-times integrated solution to the Cauchy problem (4.2.1). Similar to (5.1.17), we have the following equalities for X:

$$\langle \varphi, X^{(n)} \rangle = (-1)^n \Bigg[\int_0^\infty \varphi^{(n)}(s) S_n(s)\zeta\,ds$$

$$+ \int_0^\infty \varphi^{(n)}(t)\,dt \int_0^t S_n(t - s)B\,dW(s) \Bigg] = (-1)^n \Bigg[\int_0^\infty \varphi^{(n)}(t) S_n(t)\zeta\,dt$$

$$- \int_0^\infty \varphi^{(n+1)}(t)\,dt \int_0^t S_n(t - s)BW(s)\,ds \Bigg]. \quad (5.1.19)$$

5.1. Generalized solutions to linear stochastic Cauchy problems 207

Hence, by (5.1.16) and by the properties of n-times integrated solutions, for the regular distribution defined as the X continued by zero for $t < 0$ we obtain that $X^{(n)}$ is a solution to the problem (5.1.4). From the equalities (5.1.19) and the properties of convolution, the converse statement follows. $\qquad\square$

Theorem 5.1.5 *Let A be the generator of a K-convoluted semi-group $\{S_K(t),\ t \geq 0\}$. Then $P_{ult}\left(\frac{d}{dt}\right) X(\cdot)$, where $X(t),\ t \geq 0$, is a weak K-convoluted solution of the Cauchy problem (4.2.1) continued by zero for $t < 0$, is a generalized solution of (5.1.4). Conversely, the generalized solution of (5.1.4) is the result of the action of ultra-differential operator $P_{ult}\left(\frac{d}{dt}\right)$ on a weak K-convoluted solution.*

Proof. As proved in Section 4.2, a weak K-convoluted solution to the Cauchy problem (4.2.1), i.e., a solution to the problem (4.2.25), has the form

$$X(t) = S_K(t)\zeta + \int_0^t S_K(t-s)B\,dW(s), \quad t \geq 0.$$

Applying to the process X continued by zero for $t < 0$ the ultra-differential operator $P_{ult}\left(\frac{d}{dt}\right)$ defined by the equality (5.1.12), we obtain

$$\langle \varphi, P_{ult}\left(\frac{d}{dt}\right) X \rangle = \langle P^*_{ult}\left(\frac{d}{dt}\right)\varphi(t), X(t) \rangle$$

$$= \int_0^t P^*_{ult}\left(\frac{d}{dt}\right)\varphi(t) S_K(t)\zeta\,dt + \int_0^t P^*_{ult}\left(\frac{d}{dt}\right)\varphi(t)\,dt \int_0^t S_K(t-s)B\,dW(s).$$
$$(5.1.20)$$

Hence, by Theorem 5.1.2, $P_{ult}\left(\frac{d}{dt}\right) X$ is a solution to the problem (5.1.10).

Conversely, taking into account (5.1.11) and moving upwards in the equalities (5.1.20), we obtain that a generalized solution to the problem (5.1.10) with A being the generator of a K-convoluted semi-group $\{S_K(t),\ t \geq 0\}$ is equal to

$$P_{ult}\left(\frac{d}{dt}\right)\mathbf{S}_K\zeta + P_{ult}\left(\frac{d}{dt}\right)\mathbf{S}_K * B\,\mathbb{W},$$

where the regular ultra-distribution $\mathbf{S}_K$ coincides with $S_K(t)$ as $t \geq 0$ and is equal to zero as $t < 0$ and the ultra-differential operator $P_{ult}\left(\frac{d}{dt}\right)$ is defined by the equality (5.1.12). $\qquad\square$

5.1.3 Generalized solutions in the case of R-semi-groups

Now let the operator A in the stochastic Cauchy problem (5.1.1) be the generator of an R-semi-group $\{S_R(t),\ t \geq 0\}$. It follows from the results obtained in this section that the problem with such an A has the generalized solution, similar to (5.1.9), given by the formula

$$\langle \varphi, X \rangle = \langle \varphi, R^{-1}\mathbf{S}_R\zeta \rangle - \int_0^\infty \varphi'(t) \int_0^t R^{-1}S_R(t-s)BW(s)\,ds\,dt, \quad (5.1.21)$$

208   5. *Stochastic Cauchy problems in spaces of distributions*

where $\mathbf{S}_R(\cdot)$ is defined as $S_R(t)$ for $t \geq 0$ and zero for $t < 0$. However, (5.1.21) holds under a rather restrictive condition on $\{R^{-1}S_R(t)\}$ to be bounded operators, which is not typical for the case of R-semi-groups. In this case, in addition to the regularization by test functions $\varphi = \varphi(t), t \in \mathbb{R}$, we need a regularization in spatial variables of A, which generates an R-semi-group.

In this subsection we study generalized solutions for (5.1.1) with abstract operators A and differential operators $A = \mathbf{A}(i\partial/\partial x)$ generating R-semi-groups.

We begin the study with the case of self-adjoint operators in Hilbert spaces. We show that for (5.1.1) with the generator of an R-semi-group, a solution can be obtained in appropriate spaces of distributions, which we denote by H_{-k} and $H_{-\infty}$, wherein the operator R^{-1} is bounded.

The Ivanov spaces H_{-k}, $H_{-\infty}$ are given as a generalization of the spaces of Schwartz, Sobolev, Zemanyan, Pilipović type (see, e.g., [79, 98, 115]), which were introduced in such a way that different differential operators are defined in these spaces. The spaces H_{-k}, $H_{-\infty}$ are constructed in such a way that an unbounded operator $\mathcal{P}$, unnecessarily differential, is used in their definition [46]. For our purpose of solving the Cauchy problem with the generator of an R-semi-group, this operator is $\mathcal{P} := R^{-1}$.

Definition 5.1.2 *Let $\mathcal{P}$ be a self-adjoint (unbounded) operator in a Hilbert space H with eigenvectors $\{e_k\}$ forming an orthonormal basis in H and corresponding eigenvalues $|\mu_1| \leq |\mu_2| \leq \dots$. We define a sequence of Hilbert spaces H_k, $k = 1, 2, \dots$, and a countably normed space H_∞ as follows:*

$$H_k := \{\varphi \in dom\, \mathcal{P}^k, \ \|\varphi\|_k = \sum_{i=0}^{k} \|\mathcal{P}^i u\|_H\}, \quad H_\infty := \{\varphi \in \bigcap_{k=0}^{\infty} dom\, \mathcal{P}^k\}.$$

The spaces H_{-k}, $k = 0, 1, 2, \dots$, and $H_{-\infty}$ are defined as adjoint to the introduced spaces H_k, H_∞, respectively.

An equivalent definition of such spaces can be given via the behavior of Fourier coefficients of elements $\varphi = \sum_1^\infty \varphi_j e_j$ from H and formal sums $f = \sum_1^\infty f_j e_j$:

$$\varphi \in H_k \iff \sum_1^\infty |\varphi_j|^2 |\mu_j|^{2k} < \infty, \qquad f \in H_{-k} \iff \sum_1^\infty \frac{|f_j|^2}{(1 + |\mu_j|)^{2k}} < \infty.$$

In these spaces for ill-posed homogeneous (deterministic) Cauchy problems with self-adjoint operators A the following result holds.

Theorem 5.1.6 [46] *Let A be a self-adjoint (unbounded) operator in a Hilbert space H with eigenvectors $\{e_k\}$ forming an orthonormal basis in H corresponding to eigenvalues $|\lambda_1| \leq |\lambda_2| \leq \dots$. Let $\mathcal{P} := e^{A\tau}$, $\tau > 0$, with eigenvalues $\mu_j = e^{\lambda_j \tau}$. Then for any $\zeta = \sum_{j \in \mathbb{N}} \zeta_j e_j \in H_{-k}$ there exists the unique solution to the Cauchy problem*

$$u'(t) = Au(t), \quad t \in [0, T], \ T < \tau, \qquad u(0) = \zeta,$$

5.1. Generalized solutions to linear stochastic Cauchy problems　　209

defined by

$$u(t) = \sum_{j \in \mathbb{N}} e^{\lambda_j t} \zeta_j e_j \in H_{-(k+1)}.$$

The solution is stable with respect to $\zeta \in H_{-k}$. If $\zeta \in H_{-\infty}$, then $u(t) \in H_{-\infty}$.

It is easy to check that, similarly to numerous examples of R-semi-groups generated by operators presented in Sections 3.1 and 4.2, the operator A under the conditions of Theorem 5.1.6 generates the R-semi-group in H defined as $\{S_R(t) = e^{A(t-\tau)}, t \in [0, \tau)\}$ with $R = e^{-A\tau}$. In this case the solution operators $R^{-1}S_R(t)$ are bounded as operators from H_{-k} to $H_{-(k+1)}$.

Considering solutions to the stochastic Cauchy problem (5.1.1) in these Hilbert spaces we have the following result on the generalized solution defined by the formula (5.1.21) in the space $\mathcal{D}'(H_{-(k+1)})$.

Theorem 5.1.7 *Let A be the generator of an R-semi-group, where R is a self-adjoint bounded operator in a Hilbert space H with eigenvectors $\{e_k\}$ forming an orthonormal basis in H corresponding to eigenvalues μ_i: $|\mu_1^{-1}| \leq |\mu_2^{-1}| \leq \dots$. Let $\mathcal{P} := R^{-1}$ and $\mathbb{W}$ be a Q-white noise defined as the generalized derivative of an H-valued Q-Wiener process. Then $X \in \mathcal{D}_0'(L^2(\Omega; [dom\, A]_{k+1}))$ defined by (5.1.21), where $[dom\, A]_{k+1}$ is dom A endowed with the graph-norm in $H_{-(k+1)}$ and $B \in \mathcal{L}(H)$, is the unique solution to the stochastic Cauchy problem (5.1.1) for each $\zeta \in L^2(\Omega; H_{-k})$. In particular, for $\zeta \in L^2(\Omega; H)$, the solution $X \in \mathcal{D}_0'(L^2(\Omega; [dom\, A]_1))$.*

The solution $X \in \mathcal{D}_0'(L^2(\Omega; [dom\, A]_{k+1}))$ in fact can be considered as the solution generalized wrt t and "spatial" variables of the elements of H.

Now we consider generalized solutions to the important class of equations with differential operators $A = \mathbf{A}(i\partial/\partial x)$, which, generally, are not self-adjoint. As shown in Section 4.2, the operators generate different R-semi-groups in the Hilbert space $H = L_m^2(\mathbb{R}^n)$.

Let $(\Omega, \mathcal{F}, P)$ be a probability space. We consider the stochastic Cauchy problem for the system of partial differential equations

$$\frac{\partial X(t, x, \omega)}{\partial t} = \mathbf{A}\left(i\frac{\partial}{\partial x}\right) X(t, x, \omega) + B\mathbb{W}(t, x, \omega),$$

$$t \geq 0, \ x \in \mathbb{R}^n, \ P_{\text{a.s.}}, \quad X(0, x, \omega) = \zeta(x, \omega).$$

Further, we usually omit ω and write the problem as

$$\frac{\partial X(t, x)}{\partial t} = \mathbf{A}\left(i\frac{\partial}{\partial x}\right) X(t, x) + B\,\mathbb{W}(t, x),$$

$$t \geq 0, \ x \in \mathbb{R}^n, \quad X(0, x) = \zeta(x), \quad (5.1.22)$$

still implying the equalities hold $P_{\text{a.s.}}$. The operator of the equation is a matrix

210   *5. Stochastic Cauchy problems in spaces of distributions*

operator producing systems of different types in the Gelfand–Shilov classification (Definition 2.3.1) and generating corresponding R-semi-groups (Theorem 4.2.6).

Due to the singularity of white noise and the unboundedness of solution operators generated by A, we consider (5.1.22) as a generalized problem with $\mathbb{W}$ defined as the t-derivative of a Wiener process W in the sense of distributions (wrt t) and with values in an appropriate topological space Ψ' (wrt x). Thus we will consider the problem (5.1.22) as a generalized one wrt both t and x.

We start the construction with generalized wrt x solutions, constructing them on the basis of R-semi-groups generated by $\mathbf{A}\,(i\partial/\partial x)$. It is proved in Theorem 4.2.6 that the family of convolution operators

$$[S(t)f](x) := G_R(t, x) * f(x), \qquad t \in [0, \tau), \quad x \in \mathbb{R}^n, \tag{5.1.23}$$

where

$$G_R(t, x) = \frac{1}{(2\pi)^n} \int_{\mathbb{R}^n} e^{i\sigma x} K(\sigma) e^{t\mathbf{A}(\sigma)}\, d\sigma, \qquad t \in [0, \tau), \quad x \in \mathbb{R}^n,$$

forms an R-semi-group $\{S(t), t \in [0, \tau)\}$ in $L^2_m(\mathbb{R}^n)$ with the generator $\mathbf{A}\,(i\partial/\partial x)$ and

$$Rf(x) = \frac{1}{(2\pi)^n} \int_{\mathbb{R}^n} e^{i\sigma x} K(\sigma) \widetilde{f}(\sigma)\, d\sigma, \qquad x \in \mathbb{R}^n. \tag{5.1.24}$$

The semi-group is local if $\tau < \infty$. The regularizing operators R and corresponding functions $K(\sigma), \sigma \in \mathbb{R}^n$, are defined in dependence on the type of the system in the Gelfand–Shilov classification (see Section 2.3).

Since the operator A of the system generates only a regularized semi-group (5.1.23), we cannot construct a solution to the Cauchy problem (5.1.22) in the form

$$X(t) = U(t)\zeta + \int_0^t U(t - s)B\, dW(s), \qquad t \geq 0$$

as we did in the case of a C_0-semi-group $\{U(t), t \geq 0\}$ while constructing the solution to the Itô integrated Cauchy problem:

$$dX(t, x) = \mathbf{A}\left(i\frac{\partial}{\partial x}\right) X(t, x)\, dt + B\, dW(t, x),$$

$$t \in [0, T], \quad x \in \mathbb{R}^n, \quad X(0, x) = \zeta(x).$$

In the case of an R-semi-group $\{S(t), t \in [0, \tau)\}, \tau > T$, the solution operators $U(t)$ of the corresponding homogeneous Cauchy problem are not bounded. They are defined in the spaces of distributions Ψ', which are related to the classes of the systems via the R-semi-groups (5.1.23) as follows (see [71] and Section 2.3):

$$\langle \psi, U(t)\zeta \rangle = \langle \psi, R^{-1}S(t)\zeta \rangle = \langle \left(R^{-1}\right)^* \psi, S(t)\zeta \rangle, \qquad \psi \in \Psi.$$

5.1. Generalized solutions to linear stochastic Cauchy problems 211

The generalized wrt x process $X(t) = X(t, \cdot)$ defined by

$$\langle \psi(\cdot), X(t, \cdot) \rangle := \langle \left(R^{-1} \right)^* \psi(\cdot), S(t)\zeta(\cdot) \rangle + \langle \left(R^{-1} \right)^* \psi(\cdot), \int_0^t S(t-s)B\, dW(s, \cdot) \rangle$$

for $\psi \in \Psi$ is a solution of the generalized Cauchy problem

$$\langle \psi, X(t) \rangle - \langle \psi, \zeta \rangle = \int_0^t \langle \mathbf{A}^*\psi, X(s) \rangle ds + \langle \psi, \int_0^t B\, dW(s) \rangle, \quad t \in [0, T]. \quad (5.1.25)$$

The problem holds $P_{\text{a.s.}}$.

Now using the results of Section 2.3 we show how to choose the spaces Ψ in order to construct generalized solutions for different classes of systems, and we do it for the (important in applications) Cauchy problem with initial data $\zeta \in L_m^2(\mathbb{R})$.[2]

For this we use the connection of the homogeneous Cauchy problem corresponding to (5.1.22) with its Fourier transformed one:

$$\frac{\partial \widetilde{u}(t, \sigma)}{\partial t} = \mathbf{A}(\sigma)\widetilde{u}(t, \sigma), \quad t \in [0, T], \; \sigma \in \mathbb{R}, \quad \widetilde{u}(t, \sigma) = e^{t\mathbf{A}(\sigma)}\widetilde{\zeta}(\sigma), \quad (5.1.26)$$

where

$$\langle \widetilde{\psi}(\sigma), \widetilde{u}(t, \sigma) \rangle = \langle \widetilde{\psi}(\sigma), e^{t\mathbf{A}(\sigma)}\widetilde{\zeta}(\sigma) \rangle = \langle e^{t\mathbf{A}^*(\sigma)}\widetilde{\psi}(\sigma), \widetilde{\zeta}(\sigma) \rangle, \qquad \widetilde{\psi} \in \widetilde{\Psi},$$

and recall the estimates for $e^{t\mathbf{A}(\cdot)}$ in the Gelfand–Shilov classification.

If p is the maximal order of the differential operators $A_{jk}\left(i\frac{\partial}{\partial x}\right)$, then the solution operators of (5.1.26) satisfy the estimate

$$e^{t\Lambda(s)} \leq \left\| e^{t\mathbf{A}(s)} \right\|_m \leq C(1+|s|)^{p(m-1)} e^{t\Lambda(s)}, \quad t \geq 0, \; s = \sigma + i\tau \in \mathbb{C}. \quad (5.1.27)$$

In the classification, these estimates are done more accurately:
- for a Petrovsky correct system $\left\| e^{t\mathbf{A}(\sigma)} \right\|_m$ is polynomially bounded;
- for a conditionally correct system

$$\left\| e^{t\mathbf{A}(\sigma)} \right\|_m \leq Ce^{a_0 t|\sigma|^h}, \quad t \geq 0, \quad \sigma \in \mathbb{R}, \quad (5.1.28)$$

where $C > 0, a_0 > 0$, and $0 < h < 1$;
- for an incorrect system

$$\left\| e^{t\mathbf{A}(\sigma)} \right\|_m \leq Ce^{b_1 t \cdot |\sigma|^{p_0}}, \quad t \geq 0, \quad \sigma \in \mathbb{R}, \quad (5.1.29)$$

where $C > 0, b_1 > 0$, and p_0 is the exact order of the system.

---

[2] It is necessary to warn the reader that the designation of the space Ψ ($\widetilde{\Psi}$, Ψ') used here differs from the designations in Section 2.3, where generalized solutions are constructed in Φ'. That is because in this section test functions $\varphi \in \Phi$ have been used as functions $\varphi = \varphi(t)$.

212   5. Stochastic Cauchy problems in spaces of distributions

In Theorem 2.3.4 for the classes of systems (5.1.26), spaces $\widetilde{\Psi}$ such that $e^{t\mathbf{A}(\cdot)}$ are multiplication operators from $L_m^2(\mathbb{R}) = L^2(\mathbb{R}) \times \cdots \times L^2(\mathbb{R})$ to $\widetilde{\Psi}'$, are obtained:

  • for a Petrovsky correct system, $e^{t\mathbf{A}(\cdot)}$ is a bounded multiplication operator acting from $L_m^2(\mathbb{R})$ to $\mathcal{S}_m' = \mathcal{S}' \times \cdots \times \mathcal{S}'$;

  • for a conditionally correct system, $e^{t\mathbf{A}(\cdot)}$ is a bounded multiplication operator from $L_m^2(\mathbb{R})$ to $(\mathcal{S}_{\alpha,A})_m'$ with $\alpha = 1/h$, and $1/(h\,e\,A^h) > a_0$, where the constants a_0, h are from (5.1.28);

  • for an incorrect system, $e^{t\mathbf{A}(\cdot)}$ is a bounded multiplication operator from $L_m^2(\mathbb{R})$ to $(\mathcal{S}_{\alpha,A})_m'$ with $\alpha = 1/p_0$, $1/(p_0\,e\,A^{p_0}) > b_1$, where b_1, p_0 are from (5.1.29).

Using the spaces $\widetilde{\Psi}'$ obtained, taking into account the relations between the spaces $\widetilde{\Psi}'$ and Ψ', and choosing the regularizing function K in (5.1.24) in such a way that $K^{-1}(\sigma)$ grows no faster than $\left\|e^{t\mathbf{A}(\sigma)}\right\|_m$ multiplied by any polynomial, we arrive at the following result.

Theorem 5.1.8 *Let the matrix-function $e^{t\mathbf{A}(\cdot)}$ satisfy the estimate (5.1.27). Then for the Cauchy problem (5.1.25) with $\zeta \in L_m^2(\mathbb{R})$ and $\mathbf{A}\,(i\partial/\partial x)$ generating a Petrovsky correct system, there exists a unique solution $X(t, \cdot) \in \mathcal{S}_m'$; for a conditionally correct system, $X(t, \cdot) \in \left(\mathcal{S}^{\alpha,A}\right)_m'$ with $\alpha = \frac{1}{h}$ and $\frac{1}{h\,e\,A^h} > a_0$; for an incorrect system, $X(t, \cdot) \in \left(\mathcal{S}^{\alpha,A}\right)_m'$ with $\alpha = \frac{1}{p_0}$ and $\frac{1}{p_0\,e\,A^{p_0}} > b_1$.*

Now we consider the *generalized wrt t and x solutions*. We start with the properties of spaces $\mathcal{D}'(\Psi')$, where we are going to construct the solutions.

Define $\mathcal{D}'(\Psi')$ as the space $\mathcal{L}(\mathcal{D}, \Psi')$ of linear continuous operators from $\mathcal{D}$ to Ψ'. Here Ψ is a locally convex space, and Ψ' is its adjoint with weak topology, i.e., topology corresponding to the convergence of a sequence on each element of Ψ. We assume $\mathcal{D}'(\Psi')$ to be equipped with the topology of the uniform convergence on bounded subsets of $\mathcal{D}$ (strong topology).

It is a well-known property of the Schwartz space $\mathcal{D}' = \mathcal{L}(\mathcal{D}, \mathbb{R})$ that for any compact $\Upsilon \subset \mathbb{R}$ and any $f \in \mathcal{D}'$, there exist $p \in \mathbb{N}_0$ and $C > 0$ such that $|f(\varphi)| \leq C\|\varphi\|_p$, $\varphi \in \mathcal{D}_\Upsilon$, where

$$\|\varphi\|_p = \sup_{k,n \leq p} \sup_{x \in \Upsilon} |x^k \varphi^{(n)}(x)|.$$

This reflects the structure of the space

$$\mathcal{D}' = \bigcap_\Upsilon \bigcup_p \mathcal{D}_{\Upsilon,p}'.$$

We prove a similar statement for $\mathcal{D}'(\Psi')$.

Proposition 5.1.1 *Let $\Upsilon \subset \mathbb{R}$ be a compact set. For any $f \in \mathcal{D}'(\Psi')$ there exists such $p \in \mathbb{N}_0$ that for any bounded set $\mathcal{B} \subset \mathcal{D}_{\Upsilon,p}$, the set $f(\mathcal{B})$ is bounded in Ψ'.*

5.1. Generalized solutions to linear stochastic Cauchy problems 213

Proof. Suppose the opposite: for each $p \in \mathbb{N}_0$, there exists such a set $\mathcal{B}_p$ bounded in $\mathcal{D}_{\Upsilon,p}$ that $f(\mathcal{B}_p)$ is not bounded in Ψ', i.e., there exists a weak neighborhood V in Ψ' such that $\lambda f(\mathcal{B}_p) \not\subset V$ for any $\lambda > 0$. Thus we obtain

$$\forall p \in \mathbb{N}_0 \; \exists \mathcal{B}_p \subset \mathcal{D}_{\Upsilon,p}, \; \exists V \subset \Psi' \; : \; \left(\forall \lambda > 0 \; \exists \psi_\lambda \in \mathcal{B}_p \; : \; \lambda f(\psi_\lambda) \notin V \right).$$

Take $\lambda = \frac{1}{n} \to 0$, then $\frac{1}{n} \psi_n \to 0$ in $\mathcal{D}_{\Upsilon,p}$ since ψ_n belong to a bounded set. Therefore $\frac{1}{n} f(\psi_n) = f(\psi_n/n) \to 0$ in Ψ' weakly, that is, on each element of Ψ. But we have a weak neighborhood V in which there are no elements of this sequence. This contradiction ends the proof. $\qquad \square$

Due to Proposition 5.1.1, the following structure theorem is true in the space introduced. It is similar to the one considered in [30].

Theorem 5.1.9 *Let $F \in \mathcal{D}'(\Psi')$ and $\gamma \subset \mathbb{R}$ be an open bounded set. Then there exist a continuous function $f : \mathbb{R} \to \Psi'$ and $m \in \mathbb{N}_0$ such that for any $\varphi \in \mathcal{D}$ with $\mathrm{supp}\, \varphi \subset \gamma$*

$$\langle \varphi(t), F \rangle = \langle \varphi(t), f^{(m)}(t, \cdot) \rangle \in \Psi'.$$

Proof. Let $\Upsilon = \overline{\gamma}$ and $\varepsilon > 0$. Denote $\Upsilon_\varepsilon = \{ t \in \mathbb{R} : \rho(t, \Upsilon) \leq \varepsilon \}$ and let p be the constant provided by Proposition 5.1.1 for Υ_ε and F. We extend F to $\mathcal{D}_{\Upsilon_\varepsilon,p}$ by continuity and save the same notation. Then for any bounded set $\mathcal{B} \subset \mathcal{D}_{\Upsilon_\varepsilon,p}$ the set $F(\mathcal{B})$ is bounded in Ψ'. To prove this, let

$$\eta(t) = \begin{cases} \frac{t^{p+1}}{(p+1)!}, & t \geq 0, \\ 0, & t < 0. \end{cases}$$

The function η is obviously p times continuously differentiable. Take $\chi \in \mathcal{D}$ with support in Υ_ε and consider $\lambda_t(s) := \chi(s)\eta(t - s)$, $t, s \in \mathbb{R}$. The function $\lambda_t(\cdot)$ is p times continuously differentiable and $\mathrm{supp}\, \lambda_t \subseteq \Upsilon_\varepsilon$, so $\lambda_t \in \mathcal{D}_{\Upsilon_\varepsilon,p}$ for each $t \in \mathbb{R}$. In addition, the function is continuous wrt $t \in \mathbb{R}$; therefore, $f(t) := F(\lambda_t)$, $t \in \mathbb{R}$, is a well-defined continuous function with values in Ψ'. This function f defines a regular functional with values in Ψ':

$$\int \varphi(t) f(t)\, dt = \int \varphi(t) F(\lambda_t)\, dt, \qquad \varphi \in \mathcal{D}.$$

Considering this integral as a limit of Riemann sums, we arrive at the equality

$$\int \varphi(t) F(\lambda_t)\, dt = F\left(\int \varphi(t) \lambda_t\, dt \right),$$

where

$$\int \varphi(t) \lambda_t\, dt = \int \varphi(t) \chi(s) \eta(t - s)\, dt = \chi(s) \int \varphi(t) \eta(t - s)\, dt$$

214   *5. Stochastic Cauchy problems in spaces of distributions*

is in $\mathcal{D}_{\Upsilon_\varepsilon, p}$ as a function of s.

To complete the proof we choose $\chi(s) = 1$ for $s \in \Upsilon$, take $\varphi \in \mathcal{D}$ with $supp\,\varphi \subset \gamma$, and find the generalized derivative of f of order $p + 2$:

$$
\begin{aligned}
f^{(p+2)}(\varphi) &:= (-1)^{p+2} f(\varphi^{(p+2)}) = (-1)^{p+2} \int \varphi^{(p+2)}(t) f(t)\, dt \\
&= (-1)^{p+2} U \left(\chi(s) \int \varphi^{(p+2)}(t) \eta(t - s)\, dt \right).
\end{aligned}
$$

Integrating by parts $p + 1$ times, we obtain

$$
\begin{aligned}
\int \varphi^{(p+2)}(t) \eta(t - s)\, dt &= (-1)^{p+1} \int \varphi'(t) \eta^{(p+1)}(t - s)\, dt \\
&= (-1)^{p+1} \int_{t \geq s} \varphi'(t)\, dt = (-1)^{p+2} \varphi(s).
\end{aligned}
$$

Then, since $\chi \equiv 1$ in γ, we obtain

$$
f^{(p+2)}(\varphi) = (-1)^{p+2} F \left(\chi(s)(-1)^{p+2} \varphi(s) \right) = F(\varphi).
$$

$\square$

Now we return to the original problem (5.1.22). We consider $\mathbb{W}$ defined by (5.1.3), assuming the equality holds $P_{\text{a.s.}}$ for each $x \in \mathbb{R}^n$. Therefore, the problem (5.1.22) is understood in the generalized sense too:

$$
\begin{aligned}
\langle \varphi(t), X_t'(t, x) \rangle &= \langle \varphi(t), \mathbf{A} \left(i \frac{\partial}{\partial x} \right) X(t, x) \rangle \\
&\quad + \varphi(0)\zeta(x) + B \langle \varphi(t), W_t'(t, x) \rangle\ P_{\text{a.s.}}, \quad \varphi \in \mathcal{D}. \quad (5.1.30)
\end{aligned}
$$

Since the differential operator $\mathbf{A} \left(i \frac{\partial}{\partial x} \right)$ generates in $L_m^2(\mathbb{R}^n)$ the R-semigroup S defined by (5.1.23), the unique solution of the homogeneous Cauchy problem corresponding to (5.1.22) exists for any $\zeta \in R \left(dom\, \mathbf{A} \left(i \frac{\partial}{\partial x} \right) \right)$ and can be found as follows:

$$
\begin{aligned}
R^{-1} S(t, x) \zeta(x) = R^{-1} [G_R(t, x) * \zeta(x)] &= \mathcal{F}^{-1} \left[\frac{K(\sigma) e^{t A(\sigma)} \widetilde{\zeta}(\sigma)}{K(\sigma)} \right] \\
&= \mathcal{F}^{-1} \left[e^{t A(\sigma)} \widetilde{\zeta}(\sigma) \right] =: G_t(x) * \zeta(x).
\end{aligned}
$$

Here the Green function $G_t(\cdot)$ is a generalized function wrt $x \in \mathbb{R}^n$ in an appropriate space Ψ' depending on the properties of the system and its convolution with ζ is well-defined on the set $R \left(dom\, \mathbf{A} \left(i\partial/\partial x \right) \right)$. Nevertheless, to obtain a solution of (5.1.22) we need to define the convolution of $G_t(\cdot)$ with the stochastic inhomogeneity $B\mathbb{W}$. This inhomogeneity, being an $L_m^2(\mathbb{R}^n)$-valued white noise, generally does not belong to the set indicated. Therefore, we need to construct the convolution well-defined on the whole $L_m^2(\mathbb{R}^n)$. This forces us

5.1. Generalized solutions to linear stochastic Cauchy problems 215

to consider the problem (5.1.30) generalized wrt t in a generalized sense wrt $x \in \mathbb{R}^n$, i.e., in the space Ψ'. Thus we arrive at the next generalized Cauchy problem. For each $\varphi \in \mathcal{D}$, $\psi \in \Psi$

$$\langle \psi(x), \langle \varphi(t), X_t'(t, x) \rangle \rangle = \langle \psi(x), \mathbf{A}\, (i\partial/\partial x)\, \langle \varphi(t), X(t, x) \rangle \rangle$$

$$+ \varphi(0)\langle \psi(x), \zeta(x) \rangle + \langle \psi(x), B\langle \varphi(t), W_t'(t, x) \rangle \rangle \quad P_{\text{a.s.}}. \quad (5.1.31)$$

Here the notation $X(t, x)$ means that the distribution $X(\cdot, \cdot)$ acts on $\varphi(t)$ wrt the first argument and on $\psi(x)$ wrt the second one.

Theorem 5.1.10 *Let $\mathbb{W}$ be a Q-white noise. Let $\mathbf{A}\, (i\partial/\partial x)$ generate an R-semi-group $\{S(t, \cdot),\, t \in [0, \infty)\}$ in $L_m^2(\mathbb{R}^n)$ and $R^{-1}S(t, \cdot) : L_m^2(\mathbb{R}^n) \to \Psi'$ be a bounded operator for each $t \in [0, \infty)$. Then*

$$X(t, x) = R^{-1}\mathbf{S}(t, x)\zeta(x) + R^{-1}\mathbf{S}(t, x) * B\mathbb{W}(t, x), \quad \zeta \in L_m^2(\mathbb{R}^n), \quad (5.1.32)$$

where $\mathbf{S}$ is the semi-group S continued by zero for $t < 0$, is the solution to (5.1.31) in $\mathcal{D}'(\Psi')$.

Proof. We begin with the first term of the prospective solution (5.1.32). Note that when we write the arguments of the distributions in (5.1.32), we mean that the distribution acts on the test function of the corresponding argument. The properties of R-semi-groups and boundness of $R^{-1}\mathbf{S}(t, \cdot)$ as an operator acting from $L_m^2(\mathbb{R}^n)$ into Ψ' imply that

$$\langle \psi(x), R^{-1}\mathbf{S}(t, x)\zeta(x) \rangle = \langle \psi(x), G_t(x) * \zeta(x) \rangle, \quad \psi \in \Psi,$$

is a solution of the homogeneous Cauchy problem, corresponding to (5.1.22). In particular, it is a continuous function of t, hence we can consider this function as a regular distribution over $\mathcal{D}$:

$$\langle \varphi(t), \langle \psi(x), R^{-1}\mathbf{S}(t, x)\zeta(x) \rangle \rangle = \int \varphi(t)\langle \psi(x), G_t(x) * \zeta(x) \rangle\, dt$$

$$= \int \varphi(t)\langle \psi(x), R^{-1}\mathbf{S}(t, x)\zeta(x) \rangle\, dt, \quad \varphi \in \mathcal{D}.$$

Replacing the last integral with the integral sums, we obtain

$$\sum \varphi(t_i)\langle \psi(x), R^{-1}\mathbf{S}(t_i, x)\zeta(x) \rangle \Delta t_i = \langle \psi(x), \sum \varphi(t_i)R^{-1}\mathbf{S}(t_i, x)\zeta(x)\Delta t_i \rangle.$$

Since the left-hand side of the equality converges, so does the right one. Passing to the limit, we obtain the following representation of $R^{-1}\mathbf{S}(t, x)\zeta(x)$ in $\mathcal{D}'(\Psi')$:

$$\langle \psi(x), \langle \varphi(t), R^{-1}\mathbf{S}(t, x)\zeta(x) \rangle \rangle = \langle \varphi(t), \langle \psi(x), R^{-1}\mathbf{S}(t, x)\zeta(x) \rangle \rangle$$

$$= \int \varphi(t)\langle \psi(x), G_t(x) * \zeta(x) \rangle\, dt, \quad \varphi \in \mathcal{D},\ \psi \in \Psi. \quad (5.1.33)$$

216   5. Stochastic Cauchy problems in spaces of distributions

Now define the second term in (5.1.32). A convolution of distributions on $\mathcal{D}$ is defined as a convolution of their primitives. Hence

$$\langle \varphi(t), R^{-1}\mathbf{S}(t,x) * BW_t'(t,x) \rangle = -\langle \varphi'(t), R^{-1}\mathbf{S}(t,x) * BW(t,x) \rangle$$

$$= -\langle \varphi'(t), \int_0^t R^{-1}\mathbf{S}(t-h,x)BW(h,x)\,dh \rangle, \quad \varphi \in \mathcal{D}, \quad (5.1.34)$$

where $R^{-1}\mathbf{S}(t-h,\cdot)BW(h,\cdot) \in \Psi'$. Therefore, it can be considered only on $\psi \in \Psi$. We obtain from (5.1.33) and (5.1.34)

$$\langle \varphi(t), \langle \psi(x), R^{-1}\mathbf{S}(t,x) * BW_t'(t,x) \rangle \rangle$$

$$= -\langle \varphi'(t), \int_0^t \langle \psi(x), R^{-1}\mathbf{S}(t-h,x)BW(h,x) \rangle\,dh \rangle.$$

As above, $\langle \psi(x), R^{-1}\mathbf{S}(t-h,x)BW(h,x) \rangle$ is a continuous function wrt t, hence it defines a regular functional on $\mathcal{D}$. Using the definition of a generalized derivative and approximating the integrals by Riemann sums due to the linearity property of the functionals considered, we obtain

$$\langle \varphi(t), \langle \psi(x), R^{-1}\mathbf{S}(t,x) * BW_t'(t,x) \rangle \rangle$$

$$= -\int_{\mathbb{R}} \varphi'(t) \int_0^t \langle \psi(x), R^{-1}\mathbf{S}(t-h,x)BW(h,x) \rangle\,dh\,dt$$

$$= \langle \psi(x), \langle -\varphi'(t), \int_0^t R^{-1}\mathbf{S}(t-h,x)BW(h,x)\,dh \rangle \rangle$$

$$= \langle \psi(x), \langle -\varphi'(t), R^{-1}\mathbf{S}(t,x) * BW(t,x) \rangle \rangle$$

$$= \langle \psi(x), \langle \varphi(t), R^{-1}\mathbf{S}(t,x) * BW_t'(t,x) \rangle \rangle.$$

It follows the next representation of $R^{-1}\mathbf{S}(t,x) * BW_t'(t,x)$ in $\mathcal{D}'(\Psi')$:

$$\langle \psi(x), \langle \varphi(t), R^{-1}\mathbf{S}(t,x) * BW_t'(t,x) \rangle \rangle$$

$$= \langle \varphi(t), \langle \psi(x), R^{-1}\mathbf{S}(t,x) * BW_t'(t,x) \rangle \rangle$$

$$= -\langle \varphi'(t), \int_0^t \langle \psi(x), G_{t-h}(x) * BW(h,x) \rangle\,dh \rangle. \quad (5.1.35)$$

Now we verify that the generalized function (5.1.32) satisfies (5.1.31):

$$\langle \psi(x), \langle \varphi(t), X_t'(t,x) \rangle \rangle = -\langle \psi(x), \langle \varphi'(t), X(t,x) \rangle \rangle$$

$$= -\langle \psi(x), \langle \varphi'(t), R^{-1}\mathbf{S}(t,x)\zeta(x) + R^{-1}\mathbf{S}(t,x) * BW_t'(t,x) \rangle \rangle$$

$$= -\langle \psi(x), \langle \varphi'(t), R^{-1}\mathbf{S}(t,x)\zeta(x) \rangle \rangle$$

$$- \langle \psi(x), \langle \varphi'(t), R^{-1}\mathbf{S}(t,x) * BW_t'(t,x) \rangle \rangle. \quad (5.1.36)$$

5.1. Generalized solutions to linear stochastic Cauchy problems 217

Due to (5.1.33) and the properties of solutions of the homogeneous Cauchy problem, for the first term in the right-hand side we have

$$-\langle \psi(x), \langle \varphi'(t), R^{-1}\mathbf{S}(t,x)\zeta(x)\rangle\rangle = -\langle \varphi'(t), \langle \psi(x), G_t(x) * \zeta(x)\rangle\rangle$$

$$= \langle \varphi(t), \frac{d}{dt}\langle \psi(x), G_t(x) * \zeta(x)\rangle\rangle + \varphi(0)\langle \psi(x), G_0(x) * \zeta(x)\rangle$$

$$= \langle \psi(x), \mathbf{A}\,(i\partial/\partial x)\,\langle \varphi(t), G_t(x) * \zeta(x)\rangle\rangle + \varphi(0)\langle \psi(x), \zeta(x)\rangle$$

$$= \langle \psi(x), \mathbf{A}\,(i\partial/\partial x)\,\langle \varphi(t), R^{-1}\mathbf{S}(t,x)\zeta(x)\rangle\rangle + \varphi(0)\langle \psi(x), \zeta(x)\rangle.$$

Using (5.1.35), the definition of a generalized derivative, the properties of solutions of the homogeneous Cauchy problem, the properties of R-semi-groups and convolutions, and replacing the integral by Riemann sums due to linearity of the functional for the second term of (5.1.36), we get

$$-\langle \psi(x), \langle \varphi'(t), R^{-1}\mathbf{S}(t,x) * BW_t'(t,x)\rangle\rangle$$

$$= \langle \varphi''(t), \int_0^t \langle \psi(x), R^{-1}\mathbf{S}(t-h,x)BW(h,x)\rangle\,dh\rangle$$

$$= -\langle \varphi'(t), \langle \psi(x), R^{-1}\mathbf{S}(0,x)BW(t,x)\rangle\rangle$$

$$\qquad - \langle \varphi'(t), \int_0^t \frac{d}{dt}\langle \psi(x), G_{t-h}(x) * BW(h,x)\rangle\,dh\rangle$$

$$= \langle \psi(x), \langle \varphi(t), BW_t'(t,x)\rangle\rangle$$

$$\qquad + \langle \psi(x), \mathbf{A}\,(i\partial/\partial x)\,\langle \varphi(t), R^{-1}\mathbf{S}(t,x) * BW_t'(t,x)\rangle\rangle.$$

Connecting both parts of (5.1.36), we obtain

$$\langle \psi(x), \langle \varphi(t), X_t'(t,x)\rangle\rangle = \langle \psi(x), \mathbf{A}\,(i\partial/\partial x)\,\langle \varphi(t), R^{-1}\mathbf{S}(t,x)\zeta(x)\rangle\rangle$$

$$\qquad + \varphi(0)\langle \psi(x), \zeta(x)\rangle + \langle \psi(x), \langle \varphi(t), BW_t'(t,x)\rangle\rangle$$

$$\qquad + \langle \psi(x), \mathbf{A}\,(i\partial/\partial x)\,\langle \varphi(t), R^{-1}\mathbf{S}(t,x) * BW_t'(t,x)\rangle\rangle$$

$$= \langle \psi(x), \mathbf{A}\,(i\partial/\partial x)\,\langle \varphi(t), X(t,x)\rangle\rangle + \varphi(0)\langle \psi(x), \zeta(x)\rangle$$

$$\qquad + \langle \psi(x), \langle \varphi(t), BW_t'(t,x)\rangle\rangle,$$

which completes the proof. $\qquad\square$

In conclusion, we note that the spaces Ψ' defined in Theorem 5.1.8 for each class of the Gelfand–Shilov systems provide the operators of convolution with Green function G_t

$$R^{-1}S(t,\cdot) = G_t(\cdot) * \quad : \quad L_m^2(\mathbb{R}^n) \quad \to \quad \Psi'$$

to be bounded for $t \geq 0$. For simplicity we considered the case of an R-semi-group $\{S(t), t \geq 0\}$. We can consider a local R-semi-group $\{S(t), t \in [0,\tau)\}$, but in this case a term $\varphi(T)$ with $T < \tau$ has to appear in (5.1.31), or ψ has to be taken with supp $\psi \subset [0,\tau)$.

218   5. Stochastic Cauchy problems in spaces of distributions

5.2   Quasi-linear stochastic Cauchy problem in abstract Colombeau spaces

In the present section we continue to study stochastic problems in spaces of distributions and consider generalized solutions to the quasi-linear stochastic problem,

$$X'(t) = AX(t) + F(X) + B\mathbb{W}(t), \quad t \geq 0, \quad X(0) = \zeta, \tag{5.2.1}$$

where A is the generator of a C_0-semi-group or an integrated semi-group in a Hilbert space H, F is a non-linear mapping from H to H, the operator B is linear and bounded from a Hilbert space $\mathbb{H}$ to H, the initial data ζ is an H-valued random variable, and $\mathbb{W} = \{\mathbb{W}(t), t \geq 0\}$ is a (generalized) stochastic process of white noise type with values in $\mathbb{H}$.

The known irregularity of the white noise makes it necessary to define the white noise $\mathbb{W}$ in such a way that the problem (5.2.1) makes sense.

The approach, which reduces (5.2.1) to the corresponding integral problem, where the white noise term is replaced by the Itô integral wrt a Wiener process W, was studied in Section 4.3. For the integrated Cauchy problem with a generator of a strongly continuous semi-group of solution operators and with Lipschitz non-linearity satisfying some growth conditions, we obtained mild and weak solutions and for the problem with a generator of an R-semi-group, we obtained R-solutions. Within this approach the questions of whether the solutions obtained are t-differentiable, whether they satisfy the problem (5.2.1), and whether these techniques can be applied for A generating integrated and convoluted semi-groups remain open.

Another approach is to consider the stochastic Cauchy problem in spaces of abstract distributions. It was realized in the previous section for the linear case of (5.2.1). We are going to use and generalize this approach considering the general problem (5.2.1), but here, due to the non-linear term F in the equation, the problem of a product of distributions arises. The novelty proposed here is to define an abstract stochastic Colombeau algebra $\mathcal{G}(\Omega, H_a)$ and extend the distribution approach to the algebra.

5.2.1   Statement of the problem. Definition of abstract Colombeau algebras

Let H_a be an (associative and commutative) algebra in a Hilbert space H. In the particular case $H = L^2(\mathbb{R})$ it might be the subspace of continuous or k times differentiable functions in $L^2(\mathbb{R})$ closed under the corresponding topology of $C^k(\mathbb{R})$, $k \in \mathbb{N}_0$.

We consider the Cauchy problem (5.2.1) in the abstract stochastic Colombeau algebra $\mathcal{G}(\Omega, H_a)$, supposing that F is an infinitely differentiable mapping, $B \in \mathcal{L}(\mathbb{H}, H_a)$, and $H_a \subset dom\, A$. We define the white noise $\mathbb{W}$

5.2. Quasi-linear stochastic Cauchy problem in abstract Colombeau spaces 219

as an element of $\mathcal{D}'_0(\mathbb{H})$, the space of abstract distributions with values in $\mathbb{H}$ and supports in $[0, \infty)$. Then, by taking convolutions with certain functions from $\mathcal{D}$, we transform $\mathbb{W} \in \mathcal{D}'_0(\mathbb{H})$ into $\mathbb{H}$-valued infinitely differentiable wrt t functions and obtain $B\mathbb{W}$ belonging to $\mathcal{G}(\Omega, H_a)$. Doing so we combine the multiplication theory in Colombeau algebras, which found applications in solving non-linear differential equations, mainly hyperbolic ones (see, e.g., [91, 92, 94]) with the theory of regularized semi-groups and the theory of stochastic processes in spaces of abstract distributions. This makes it possible to solve semi-linear abstract stochastic equations with different types of white noise.

As examples of A satisfying the conditions and generating different integrated semi-groups one can take many differential operators $A = A(i\partial/\partial x)$ generating the Petrovsky correct systems given in Section 3.2 as well as these operators disturbed by bounded operators of any nature.

Following the presentation in [18, 19, 91], we extend Colombeau algebras to the case of random variables that take values in spaces of abstract distributions (generalized functions). As usual in the theory of generalized functions, we begin with definition of test functions. These functions, in particular the ones generating certain types of δ-shaped sequences, will be used in the construction of the Colombeau algebras.

Let $\mathcal{A}_0$ be the set of functions $\varphi \in \mathcal{D}$ such that $\int_{\mathbb{R}} \varphi(t) \, dt = 1$ and for each $q \in \mathbb{N}$ let

$$\mathcal{A}_q := \left\{ \varphi \in \mathcal{A}_0 : \int_{\mathbb{R}} t^k \varphi(t) \, dt = 0, \quad k = 1, \ldots, q \right\}.^3$$

Show that the sets $\mathcal{A}_q$ are not void: consider continuous linear functionals J_0, J_k on $\mathcal{D}$:

$$J_0(\varphi) := \langle \varphi, 1 \rangle, \qquad J_k(\varphi) := \langle \varphi, x^k \rangle, \qquad k = 1, 2, \ldots q, \quad \varphi \in \mathcal{D}.$$

It is easy to check that these functionals are linear and independent: if $\sum_{k=0}^{q} a_k J_k = 0$, then $\sum_{k=0}^{q} a_k x^k = 0$, $x \in \mathbb{R}$, hence $a_k = 0$. That means J_0 is not a linear combination of J_k, $k = 1, 2, \ldots q$, and by the Hahn–Banach theorem, there exist $\psi \in \mathcal{D}$ such that $J_0(\psi) = 1$ and $\psi(J_k) = J_k(\psi) = 0$, that is, $\psi \in \mathcal{A}_q$.

In addition to the sets $\mathcal{A}_0, \mathcal{A}_q$, let us introduce the important subset of the functions $\varphi_\varepsilon \in \mathcal{A}_0$:

$$\varphi_\varepsilon(t) := \frac{1}{\varepsilon} \varphi\left(\frac{t}{\varepsilon}\right), \qquad t \in \mathbb{R}, \quad \varepsilon > 0.$$

For the algebra $H_a \subset H$ we define the space of H_a-valued infinitely differentiable transformations $u(\varphi) := u(\varphi(\cdot), t)$ depending on functions $\varphi \in \mathcal{A}_0$ and

---

[3] For simplicity we will consider $\mathcal{D} = \mathcal{D}(\mathbb{R}^n)$ for the case of $n = 1$. Following [91], the sets can be introduced for $n \in \mathbb{N}$.

220   5. Stochastic Cauchy problems in spaces of distributions

$t \in \mathbb{R}$ as follows:

$$\mathcal{E}(H_a) := (C^\infty(\mathbb{R}; H_a))^{\mathcal{A}_0} = \{u : \ \mathcal{A}_0 \to C^\infty(\mathbb{R}; H_a)\}.$$

By the definition, for each $\varphi \in \mathcal{A}_0$, $u(\varphi)$ is an infinitely differentiable H_a-valued function of the argument $t \in \mathbb{R}$. These functions u can be considered as functions of two variables $\varphi \in \mathcal{A}_0$ and $t \in \mathbb{R}$, that is,

$$u : \ \mathcal{A}_0 \times \mathbb{R} \ \to \ H_a : \quad u = u(\varphi, t), \quad \varphi \in \mathcal{A}_0, \ t \in \mathbb{R},$$

and each $u \in C^\infty(\mathbb{R}, H_a)$ depending on φ as on a parameter is infinitely differentiable wrt the second variable.

Differentiation and multiplication of the functions introduced are defined as follows:

$$(u\,v)(\varphi) := u(\varphi)v(\varphi), \qquad u^{(n)}(\varphi) := \frac{d^n}{dt^n}u(\varphi), \qquad \varphi \in \mathcal{A}_0. \qquad (5.2.2)$$

The space of the H_a-valued distributions $\mathcal{D}'(H_a)$, being a subset of the abstract distributions space $\mathcal{D}'(H)$, is embedded into $\mathcal{E}(H_a)$ by the convolution mapping

$$\mathbf{i} : \ \mathcal{D}'(H_a) \to \mathcal{E}(H_a), \qquad (\mathbf{i}\,w)(\varphi) := w * \varphi, \qquad w \in \mathcal{D}'(H_a), \ \varphi \in \mathcal{A}_0,$$

in particular, if φ_{ε_n} is a δ-shaped sequence, then $w * \varphi_{\varepsilon_n} \to w$ as $\varepsilon_n \to 0$.

Thus we have imbedded $\mathcal{D}'(H_a)$ into the differentiable algebra $\mathcal{E}(H_a)$, where the product of elements is defined by (5.2.2) and this product is consistent with the differentiation introduced. The space $\mathcal{E}(H_a)$ has certain necessary properties to be the desirable algebra. Nevertheless, there are obstacles on the way to taking $\mathcal{E}(H_a)$ as the Colombeau algebra of abstract distributions. They are the following:

- *the way to assign the element in $\mathcal{E}(H_a)$ to an element from $\mathcal{D}'(H_a)$ by the embedding is not unique,*

- *if we consider infinitely differentiable (in $t \in \mathbb{R}$) functions w_1, w_2 as elements of $\mathcal{D}'(H_a)$ and embed them into $\mathcal{E}(H_a)$, then the multiplication (5.2.2) would not agree with the usual multiplication of infinitely differentiable functions since we generally have*

$$(w_1 \cdot w_2) * \varphi \neq w_1 * \varphi \cdot w_2 * \varphi, \qquad \varphi \in \mathcal{A}_0.$$

To overcome the difficulties we take into account that $(w_1 \cdot w_2) * \varphi_\varepsilon$ and $w_1 * \varphi_\varepsilon \cdot w_2 * \varphi_\varepsilon$ converge to $w_1 \cdot w_2$ as $\varepsilon \to 0$ and define (the crucial point in the Colombeau theory!) the *linear manifold of moderate elements* $\mathcal{E}_M(H_a)$ from $\mathcal{E}(H_a)$ consisting of $u \in \mathcal{E}(H_a)$ satisfying the condition

(M) for each compact $K \subset \mathbb{R}$ and each $n \in \mathbb{N}_0$, there exists $q \in \mathbb{N}$ such that

$$\sup_{t \in K} \left\| \frac{d^n}{dt^n} u(\varphi_\varepsilon, t) \right\|_H = \mathcal{O}_{\varepsilon \to 0}(\varepsilon^{-q}) \quad \text{for each } \varphi \in \mathcal{A}_q,$$

5.2. Quasi-linear stochastic Cauchy problem in abstract Colombeau spaces 221

and define a "null subset" $N(H_a)$ consisting of elements $u \in \mathcal{E}(H_a)$ that satisfy the condition

(N) for each compact $K \subset \mathbb{R}$ and each $n \in \mathbb{N}_0$, there exists $p \in \mathbb{N}$ such that

$$\sup_{t \in K} \left\| \frac{d^n}{dt^n} u(\varphi_\varepsilon, t) \right\|_H = \mathcal{O}_{\varepsilon \to 0}(\varepsilon^{q-p}) \quad \text{for each } \varphi \in \mathcal{A}_q \text{ and } q \geq p.$$

It is not difficult to verify that all images $\mathbf{i}w$ of elements from $\mathcal{D}'(H_a)$ belong to $\mathcal{E}_M(H_a)$. This is due to the structure theorem for abstract distributions, which states that each $w \in \mathcal{D}'(H_a)$ is locally equal to a derivative of some order k of a continuous function f (see, e.g., [30] and Section 3.4). We have then

$$\begin{aligned}
(w * \varphi_\varepsilon)(t) &= \langle \varphi_\varepsilon(t - \cdot), w(\cdot) \rangle = \langle \varphi_\varepsilon(\cdot), f^{(k)}(t - \cdot) \rangle \\
&= \langle \varphi_\varepsilon^{(k)}(y), f(t - y) \rangle = \frac{1}{\varepsilon^k} \langle \varphi^{(k)}(y), f(t - \varepsilon y) \rangle.
\end{aligned}$$

Hence on a bounded set K we have

$$\sup_{t \in K} \|(w * \varphi_\varepsilon)(t)\| \leq C(K)\varepsilon^{-k}.$$

To complete the definition of the Colombeau algebra of abstract distributions (abstract generalized functions) we introduce

$$\mathcal{G}(H_a) := \mathcal{E}_M(H_a)/N(H_a).$$

Similar to $\mathcal{G}(\mathbb{R})$ (see, e.g., [91]), the algebra $\mathcal{G}(H_a)$ is an associative and commutative H_a-valued differential algebra. Moreover, the elements of the space $\mathcal{E}_M(H_a)$ form a differential algebra. This is indeed the case since, as we have shown, $\mathbf{i}$ maps the elements of $\mathcal{D}'(H_a)$ into $\mathcal{E}_M(H_a)$ and $N(H_a)$ is the differential ideal in $\mathcal{E}_M(H_a)$.

The set $\mathbf{i}^{-1}(N(H_a))$ consists of the null elements of $\mathcal{D}'(H_a)$. To show this let $\mathbf{i}(w) \in N(H_a)$, then for $\varphi_\varepsilon \in \mathcal{A}_q$ with q large enough, $w \leftarrow w * \varphi_\varepsilon \to 0$ as $\varepsilon \to 0$ in $\mathcal{D}'(H_a)$.

At last, the definition of $\mathcal{G}(H_a)$ agrees with the multiplication of infinitely differentiable functions since

$$(w_1 \cdot w_2) - (w_1 * \varphi_\varepsilon) \cdot (w_2 * \varphi_\varepsilon) \in N(H_a)$$

for all infinitely differentiable functions w_1, w_2.

The elements of the algebra $\mathcal{G}(H_a)$ introduced are classes of mappings. We denote them by capitals $Y, V, \ldots$. We will denote a representative of a class $Y \in \mathcal{G}(H_a)$ by the corresponding small letter y and the class Y containing y we will denote by $\{y\}$. Each element of $\mathcal{D}'(H_a)$ is embedded in the corresponding class of $\mathcal{G}(H_a)$ by the mapping $\mathbf{i}$.

The support of an element $V \in \mathcal{G}(H_a)$ is defined as the complement of

222   5. Stochastic Cauchy problems in spaces of distributions

the widest open set, where $V = 0$. We say that V is equal to zero on an open set $\Lambda \subset \mathbb{R}$ if its restriction to $\mathcal{G}_\Lambda(H_a)$ is equal to zero in $\mathcal{G}_\Lambda(H_a)$. The algebra $\mathcal{G}_\Lambda(H_a)$ is defined in the same way as $\mathcal{G}_\Lambda(\mathbb{R}) = \mathcal{G}(\Lambda)$. The imbedding of any $w \in \mathcal{D}'(H_a)$ into $\mathcal{G}_\Lambda(H_a)$ can be done with help of a specially chosen function $\vartheta(\varphi) \in \mathcal{D}(\Lambda)$ such that for any compact set $K \subset \Lambda$, the equality $(\vartheta(\varphi_\varepsilon)w * \varphi_\varepsilon)(x) = \langle w(\cdot), \varphi_\varepsilon(x - \cdot)\rangle$, $x \in K$, holds for ε small enough [91].

Then, if $w \in \mathcal{D}'(H_a)$, like the case of $\mathcal{D}'(\mathbb{R})$, the support of $\{iw\} \in \mathcal{G}(H_a)$ coincides with that of $w \in \mathcal{D}'(H_a)$.

Now we define $\mathcal{G}(\Omega, H_a)$, the algebra of $\mathcal{G}(H_a)$-valued random variables $\{V = V(\omega), \omega \in (\Omega, \mathcal{F}, P)\}$ as a mapping from $(\Omega, \mathcal{F}, P)$ to $\mathcal{G}(H_a)$ measurable in the following sense: there exists a representative $v \in V$ such that for any $\varphi \in \mathcal{A}_0$, $v^{-1}(\varphi, \cdot)$ maps any Borel subset of $\mathcal{B}(C^\infty(\mathbb{R}; H_a))$ onto an element of $\mathcal{F}$, in particular an Borel subset of $\mathcal{B}(\Omega)$, where the Borel σ-algebra $\mathcal{B}(C^\infty(\mathbb{R}; H_a))$ is generated by the system of semi-norms $p_{n,k}(f) = \sup_{t \in [-n,n]} \sup_{i \leq k} \|f^{(i)}(t)\|_H$.

To complete the setting of the problem we define the generalized white noise process $\mathbb{W}_+ = \mathbb{W}_+(\cdot, \omega)$, $\omega \in \Omega$, as an element of the space $\mathcal{D}'_0(\mathbb{H})$ of abstract distributions with values in $\mathbb{H}$ and supports in $[0, \infty)$ and then transform it into an element of $\mathcal{G}(\Omega, \mathbb{H}_a)$.

One way to define a generalized $\mathbb{H}$-valued white noise on $(\Omega, \mathcal{F}, P)$, more precisely Q-white noise, is via a derivative of $\mathbb{H}$-valued Q-Wiener process $\{W_Q(t), t \geq 0\}$ continued by zero on $(-\infty, 0)$:

$$\langle \theta, \mathbb{W}_+(\cdot, \omega)\rangle := -\langle \theta', W_Q(\cdot, \omega)\rangle, \quad \theta \in \mathcal{D}.$$

Another way to do this is based on the ideas of the theory of abstract stochastic distributions (see, e.g., [7, 80]), which we will consider in detail in the next chapter. Let $\mathcal{S} = \mathcal{S}(\mathbb{R})$ be the space of rapidly decreasing test functions. Denote by $\mathcal{S}'(\mathbb{H})$ the space of $\mathbb{H}$-valued distributions over $\mathcal{S}$ and consider a Borel σ-algebra $\mathcal{B}(\Omega)$ generated by the weak topology of $\Omega := \mathcal{S}'(\mathbb{H})$. Then by the generalization of the Bochner–Minlos theorem to the case of Hilbert space valued generalized functions [5], there exist a unique probability measure μ on $\mathcal{B}(\Omega)$ and a trace class operator Q satisfying the condition[4]

$$\int_\Omega e^{i(\langle \theta, \omega\rangle, h)_\mathbb{H}} \, d\mu(\omega) = e^{-\frac{1}{2}\|\theta\|^2 (Qh, h)}, \quad \theta \in \mathcal{S}, \ h \in \mathbb{H}.$$

This makes it possible to define the white noise process on $(\Omega, \mathcal{B}(\Omega), \mu)$ with values in $\mathcal{S}'(\mathbb{H}) \subset \mathcal{D}'(\mathbb{H})$ by the identical mapping:

$$\langle \theta(\cdot), \mathbb{W}(\cdot, \omega)\rangle := \langle \theta, \omega\rangle, \quad \theta \in \mathcal{S}.$$

The process defined above is the generalization of the corresponding real-valued Gaussian process [59]; it has zero mean and $\mathbf{Cov}\,\langle \theta, \mathbb{W}\rangle = \|\theta\|^2 Q$.

---

[4]Here and below, if it is not pointed out especially, $\|\cdot\|$ denotes the norm of $L^2(\mathbb{R})$.

5.2. Quasi-linear stochastic Cauchy problem in abstract Colombeau spaces 223

Define $\mathbb{W}_+$ with support in $[0, \infty)$ via $\hat{\mathbb{W}}_+$ defined as follows:

$$\left\langle \theta(\cdot), \hat{\mathbb{W}}_+(\cdot, \omega) \right\rangle := (-1)^k \left\langle \theta^{(k)}(\cdot), \chi f(\cdot) \right\rangle, \qquad \theta \in \mathcal{S},$$

where, according to the structure theorem for slowly decreasing distributions, f is the (global) continuous primitive of $\omega \in \mathcal{S}'(\mathbb{H})$ of an order k and χ is the Heaviside function. Then, removing from the distribution $\hat{\mathbb{W}}_+$ its component with support at zero, we obtain $\mathbb{W}_+$.

Finally, we map $B\mathbb{W}_+$ into the Colombeau algebra $\mathcal{G}(\Omega, \mathbb{H}_a)$. By convoluting $\mathbb{W}_+$ with a function from $\mathcal{A}_0$ we transform $\mathbb{W}_+(\cdot, \omega) \in \mathcal{D}'_0(\mathbb{H})$ into

$$w(\varphi, t, \omega) := \langle \varphi(t - \cdot), \mathbb{W}_+(\cdot, \omega) \rangle, \qquad \varphi \in \mathcal{A}_0, \quad t \in \mathbb{R}, \quad \omega \in \Omega. \qquad (5.2.3)$$

The function w is infinitely differentiable wrt $t \in \mathbb{R}$ and measurable wrt $\omega \in \Omega$.

Thus we have $w(\varphi, \cdot, \omega) \in C^\infty(\mathbb{R}; \mathbb{H})$ for $\varphi \in \mathcal{A}_0$ and $\omega \in \Omega$ a.s., and $w(\varphi, \cdot, \cdot) \in C^\infty(\mathbb{R}; L^2(\Omega; \mathbb{H}))$.

Let $B \in \mathcal{L}(\mathbb{H}, H_a)$. Then, applying B to w, we obtain that $Bw(\varphi, t, \omega) \in H_a$ and

$$Bw(\varphi, \cdot, \cdot) : \mathcal{A}_0 \to C^\infty(\mathbb{R}; L^2(\Omega; H_a))$$

is a representative of a class in $\mathcal{G}(\Omega, H_a)$. We denote the corresponding class by $B\mathbb{W}$. Since the support of $\mathbb{W}_+ \in \mathcal{D}'(\mathbb{H})$ is $[0, \infty)$, by the definition of support of an element of $\mathcal{G}(\Omega, H_a)$ we have $\operatorname{supp} B\mathbb{W} = [0, \infty)$. That is how we understand the stochastic term considering (5.2.1) in $\mathcal{G}(\Omega, H_a)$.

5.2.2 Solutions to the stochastic Cauchy problem with infinitely differentiable non-linearities

Let $H = \mathbb{H} = L^2(\mathbb{R})$ and the domain of A lies in the set of all continuous functions of $L^2(\mathbb{R})$. Let H_a be the set of all finitely differentiable functions of $L^2(\mathbb{R})$ and $H_a \subseteq \operatorname{dom} A$. Then multiplication of elements of H_a is well defined as point-wise multiplication of continuous functions.

Now for the problem (5.2.1) with the stochastic term $B\mathbb{W}$ constructed above and an infinitely differentiable non-linearity F, which is supposed to be bounded with all its derivatives and satisfy the property $F(0) = 0$, we will search for a solution as an element of the abstract stochastic Colombeau algebra $\mathcal{G}(\Omega, H_a)$. Since H_a is chosen as the set of finitely many times differentiable functions in $L^2(\mathbb{R})$, then $B \in \mathcal{L}(\mathbb{H}, H_a)$ can be taken, for example, as convolution with a finitely-differentiable function from $L^2(\mathbb{R})$ satisfying the condition of convolution existence.

Suppose first that A generates a C_0–semi-group $\{U(t), t \geq 0\}$ in $L^2(\mathbb{R})$. Consider the question of existence of a solution to the problem

$$Y' = AY + F(Y) + B\mathbb{W} + \zeta\{\delta\}, \qquad \operatorname{supp} Y \subseteq [0, \infty), \qquad (5.2.4)$$

as an element of algebra $\mathcal{G}(\Omega, H_a)$. To do this, for an arbitrary $\eta > 0$ we

224   5. Stochastic Cauchy problems in spaces of distributions

consider $Bw(t) := Bw(\varphi, t, w)$, $\varphi \in \mathcal{A}_0$, $w \in \Omega$, with support in $[-\eta, \infty)$ as a representative of the white noise term $BW \in \mathcal{G}(\Omega, H_a)$ constructed above. By the definition of elements of $\mathcal{G}(\Omega, H_a)$, for each fixed $\varphi \in \mathcal{A}_0$, $Bw(t)$ is an infinitely differentiable function of $t \in \mathbb{R}$ with values in H_a, measurable wrt $w \in \Omega$. Let us take an arbitrary $\varphi \in \mathcal{A}_0$ and consider the problem

$$y'(t) = Ay(t) + F(y(t)) + Bw(t) + \varsigma \mathbf{i}\, \delta(t), \ t \geq -\eta, \quad y(t) = 0, \ t \leq -\eta. \quad (5.2.5)$$

We are looking for $y(t) = y(\varphi, t, w)$, $\varphi \in \mathcal{A}_0$, $w \in \Omega$, a solution to this problem belonging to $C^\infty([-\eta, \infty), dom\, A)$ for w a.s.

Consider the equation

$$y(t) = \int_{-\eta}^{t} U(t - s) F(y(s))\, ds + \int_{-\eta}^{t} U(t - s)[Bw(s) + \varsigma \mathbf{i}\, \delta(s)]\, ds =: \mathcal{Q}y(t),$$

$$(5.2.6)$$

$t \geq -\eta$. The operator $\mathcal{Q}$ is Volterra type. Using the differentiability of F and the boundedness of its derivative, let us show that $\mathcal{Q}^k$ (where $k = k(T)$) is a contraction on the segment $[-\eta, T]$.

Since F is differentiable, we have $F(a) - F(b) = F'(c)(a - b)$, $c \in (a, b)$, for any $a, b \in \mathbb{R}$. Then for any $y(\cdot)$ and $z(\cdot)$ we get the point-wise equality

$$F(y(s)) - F(z(s)) = F'(c(s))(y(s) - z(s)), \qquad s \in [-\eta, \infty),$$

where $c(s)$ is an appropriate point from the interval $(y(s), z(s))$ and the following estimate holds:

$$\|F(y(s)) - F(z(s))\| \leq L\|y(s) - z(s)\|, \qquad L = \max_{c \in \mathbb{R}} |F'(c)|.$$

This and the exponential boundedness of C_0-semi-groups $\|U(t)\|_{\mathcal{L}(L^2(\mathbb{R}))} \leq Ce^{at}$ for each $t \in [-\eta, \infty)$ imply

$$\|\mathcal{Q}y(t) - \mathcal{Q}z(t)\| \leq CLe^{a(t+\eta)}(t + \eta) \max_{s \in [-\eta, t]} \|y(s) - z(s)\|.$$

For the squares we have

$$\mathcal{Q}^2 y(t) - \mathcal{Q}^2 z(t) = \int_{-\eta}^{t} U(t - s) F(\mathcal{Q}(y(s)))\, ds - \int_{-\eta}^{t} U(t - s) F(\mathcal{Q}(z(s)))\, ds$$

$$= \int_{-\eta}^{t} U(t - s) \left[F(\mathcal{Q}(y(s))) - F(\mathcal{Q}(z(s))) \right] ds.$$

Then

$$\|\mathcal{Q}^2 y(t) - \mathcal{Q}^2 z(t)\| \leq C^2 L^2 e^{2a(t+\eta)} \frac{(t + \eta)^2}{2} \max_{s \in [-\eta, t]} \|y(s) - z(s)\|,$$

and for each $k \in \mathbb{N}$,

$$\|\mathcal{Q}^k y(t) - \mathcal{Q}^k z(t)\| \leq C^k L^k e^{ka(t+\eta)} \frac{(t + \eta)^k}{k!} \max_{s \in [-\eta, t]} \|y(s) - z(s)\|.$$

5.2. Quasi-linear stochastic Cauchy problem in abstract Colombeau spaces 225

Hence

$$\max_{t\in[-\eta,T]} \|\mathcal{Q}^k y(t) - \mathcal{Q}^k z(t)\| \le C^k L^k e^{ka(T+\eta)} \frac{(T+\eta)^k}{k!} \max_{t\in[-\eta,T]} \|y(t) - z(t)\|.$$

The constant in this estimate can be made less then unity by choosing $k = k(T)$. Hence, for the k chosen, the operator $\mathcal{Q}^k$ is a contraction and the sequence of approximations $y_n(t) = \mathcal{Q}^{nk} y_0(t)$ has the limit in H

$$y(t) = \lim_{n\to\infty} \mathcal{Q}^{nk} y_0(t).$$

Note that if one takes a function $y_0(\cdot)$ infinitely differentiable wrt t as the first point of the approximating sequence, then the function

$$\mathcal{Q} y_0(t) = \int_{-\eta}^t U(t-s) F(y_0(s)) \, ds + \int_{-\eta}^t U(t-s)[Bw(s) + \zeta \mathrm{i}\, \delta(s)] \, ds, \quad t \ge -\eta,$$

is also infinitely differentiable wrt t. Consequently, $y_1(\cdot) = \mathcal{Q}^k y_0(\cdot)$ as well as all subsequent iterations $y_n(\cdot)$ have this property.

It can be shown by the same argument that the sequence $y_n'(\cdot)$ converges to its limit in H uniformly wrt $t \in [-\eta, T]$, hence $y(\cdot)$ is differentiable and $y'(t) = \lim_{n\to\infty} y_n'(t)$. Similarly it can be shown that $y(\cdot)$ is an infinitely differentiable function with values in H.

Now we show that $y_n(t) \in H_a$ if $y_0(t) \in H_a$, $t \ge 0$. Let $t \ge 0$ be fixed. First note that $F(\alpha) = \mathcal{O}(\alpha)$ as $\alpha \to 0$. Due to the infinite differentiability of F and since $F(0) = 0$, $F(\alpha)$ can be decomposed into the Taylor series with the first term proportional to α. Then, since $y_0(t) = y_0(t, \cdot) \in H_a$, for any t the function $y_0(t, \cdot) \in L^2(\mathbb{R})$ is differentiable wrt the second argument and $y_0(t, x) \to 0$ as $x \to \infty$. Thus $F(y_0(t)) = \mathcal{O}(y_0(t))$ as x (the argument of functions from $L^2(\mathbb{R})$) tends to infinity.

Further, the semi-group of operators $U(t)$ maps $L^2(\mathbb{R})$ into $L^2(\mathbb{R})$ and, moreover, $dom\, A \subset L^2(\mathbb{R})$ and some subsets $H_a \subset dom\, A$ are invariant wrt this mapping. It follows that if $\zeta \in dom\, A$ and $H_a = dom\, A$, then

$$U(t-s) F(y_0(s)) : H_a \to H_a \quad \text{and} \quad U(t-s)[Bw(s) + \zeta \mathrm{i}\, \delta(s)] \in H_a;$$

hence $\mathcal{Q}$ acts in H_a and $y_1(t) = \mathcal{Q}^k y_0(t) \in H_a$ as well as $y_n(t)$ for any $n \in \mathbb{N}$.

Thus we obtain $y_n(t) \in H_a$, but in the general case $\lim_{n\to\infty} y_n(t) = y(t)$ does not belong to H_a, since algebra H_a is not closed in the sense of $L^2(\mathbb{R})$-convergence.

If $y(t) \in H_a$, then we show that it is a representative of a class from $\mathcal{G}(\Omega, H_a)$. As is known (see, e.g., [79]), if $\mathbb{F}(t), t \ge 0$, is a differentiable function or $\mathbb{F}(t) \in dom\, A$ for any $t \ge 0$, then the solution of the inhomogeneous abstract Cauchy problem

$$u'(t) = Au(t) + \mathbb{F}(t), \quad t \ge 0, \qquad u(0) = 0,$$

with A generating a C_0–semi-group $\{U(t), t \ge 0\}$ exists and is defined by

$u(t) = \int_0^t U(t-s)\mathbb{F}(s)\,ds$. Since in the case that we consider, $F(y(t))$ as well as any representative $Bw(t)$ of the white noise process and $\zeta i\,\delta$ are $L^2(\mathbb{R})$-valued infinitely differentiable wrt t functions, the solution of (5.2.6) is a solution to the problem (5.2.5).

Let $U(t) = 0$ as $t < 0$. It follows from (5.2.6) that $y(t) = 0$ as $t \leq -\eta$, i.e., the support of the solution obtained lies in $[-\eta, \infty)$.

Now we show that y is a representative of a class $Y \in \mathcal{G}(\Omega, H_a)$, i.e., it satisfies the condition (M) almost everywhere. It follows from the differentiability of F and the condition $F(0) = 0$ that for each $s \in [-\eta, t]$ the equality

$$F(y(\varphi, s, \omega)) = F(y(\varphi, s, \omega)) - F(0) = F'(c)y(\varphi, s, \omega),$$

where $c \in (0, y(\varphi, s, \omega))$, $\varphi \in \mathcal{A}_0$, $\omega \in \Omega$, holds. Now for an arbitrary compact $K \subset \mathbb{R}$ from (5.2.6) we obtain

$$\max_{t \in K} \|y(\varphi_\varepsilon, t, \omega)\| \leq C \max_{t \in K} e^{a(t+\eta)} \|Bw(\varphi_\varepsilon, t, \omega) + \zeta\varphi_\varepsilon(t)\|$$
$$+ C \max_{c \in \mathbb{R}} \|F'(c)\| \max_{t \in K} e^{a(t+\eta)} \int_{-\eta}^t \|y(\varphi_\varepsilon, s, \omega)\|\,ds,$$

and due to the boundedness of F', we have

$$\max_{t \in K} \|y(\varphi_\varepsilon, t, \omega)\| \leq C_1 \max_{t \in K} \|Bw(\varphi_\varepsilon, t, \omega) + \zeta\varphi_\varepsilon(t)\| + C_2 \max_{t \in K} \int_{-\eta}^t \|y(\varphi_\varepsilon, s, \omega)\|\,ds.$$
$$(5.2.7)$$

Since Bw is a representative of the class $B\mathbb{W}$ from $\mathcal{G}(\Omega, H_a)$, the first term in the right-hand side of (5.2.7) for each $\varphi \in \mathcal{A}_q$ grows as $\varepsilon \to 0$ not faster than ε^{-q} for some $q \in \mathbb{N}$. Then, due to the Gronwall–Bellmann inequality,[5] the left-hand side behaves in the same way. This proves the condition (M) with $n = 0$. The behavior of derivatives of $y(\cdot)$ can be checked in the same manner using the fact that F is infinitely differentiable and its derivatives are bounded.

Now let us consider two solutions of (5.2.5) $y_{\eta_1}(\cdot)$ and $y_{\eta_2}(\cdot)$ corresponding $\eta_1 \neq \eta_2$ and verify that the difference $y_{\eta_1}(\cdot) - y_{\eta_1}(\cdot)$ belongs to $N(H_a)$. Denote $\eta = \max\{\eta_1, \eta_2\}$. Then

$$y'_{\eta_1}(t) - y'_{\eta_2}(t) = A(y_{\eta_1}(t) - y_{\eta_2}(t)) + F(y_{\eta_1}(t)) - F(y_{\eta_2}(t)) + g(t), \quad t \geq -\eta,$$
$$y_{\eta_1}(t) - y_{\eta_2}(t) = 0, \quad t \leq -\eta,$$

where $g \in N(H_a)$ is the difference of two representatives of the stochastic

---

[5]The Gronwall–Bellmann inequality states that if

$$y(t) \leq c + \int_{t_0}^t f(s)y(s)\,ds, \quad c > 0, \; t > t_0,$$

for positive continuous functions y and f, then $y(t) \leq c \cdot \exp(\int_{t_0}^t f(s)\,ds)$.

5.2. Quasi-linear stochastic Cauchy problem in abstract Colombeau spaces 227

term BW and the term $\zeta\{\delta\}$ with supports in $[0, \infty)$. Then, similar to (5.2.7), we obtain the estimate

$$\max_{t \in K} \|y_{\eta_1}(\varphi_\varepsilon, t, \omega) - y_{\eta_2}(\varphi_\varepsilon, t, \omega)\| \le C_1 \max_{t \in K} \|g(\varphi_\varepsilon, t, \omega)\|$$

$$+ C_2 \max_{t \in K} \int_{-\eta}^t \|y_{\eta_1}(\varphi_\varepsilon, s, \omega) - y_{\eta_2}(\varphi_\varepsilon, t, \omega)\| \, ds. \quad (5.2.8)$$

Since the first term in this inequality satisfies the condition (N), the Gronwall–Bellman inequality implies $y_{\eta_1}(t) - y_{\eta_2}(t) \in N(H_a)$. Moreover, it follows that supp $Y \subseteq [0, \infty)$ for Y, which representatives y_{η_1} and y_{η_2} are, and that the class Y is uniquely determined. Let $Y_1, Y_2 \in \mathcal{G}(\Omega, H_a)$ be two solutions of (5.2.4) with supports in $[0, \infty)$. Then, for any representatives y_1, y_2 of these classes and each $\eta > 0$, we have estimates similar to (5.2.8). Hence $y_1(t) - y_2(t) \in N(H_a)$, i.e., $Y_1 - Y_2 = 0$ in $\mathcal{G}(\Omega, H_a)$. That is the solution of (5.2.4) is unique in the algebra $\mathcal{G}(\Omega, H_a)$.

In the general case, since the limit of $y_n(t) \notin H_a$, we obtain only approximated solutions of (5.2.6) given by the fundamental sequence $\{y_n\}$ defined by the equalities

$$y_n(t) = \mathcal{Q}^k y_{n-1}(t), \qquad t \le -\eta.$$

Thus we arrive at the following result.

Theorem 5.2.1 *Let A be the generator of a C_0-semi-group $\{U(t), t \ge 0\}$ in $L^2(\mathbb{R})$. Let F be an infinitely differentiable function in $\mathbb{R}$, bounded with all its derivatives and $F(0) = 0$. Let $B \in \mathcal{L}(L^2(\mathbb{R}), H_a)$ and BW be an element of $\mathcal{G}(\Omega, H_a)$ with representative Bw defined by (5.2.3). Then for any $\eta > 0$ and $\varphi \in \mathcal{A}_0$ there exists a unique solution of (5.2.5), $y \in C^\infty([-\eta, \infty), H)$. If $y \in C^\infty([-\eta, \infty), H_a)$, then Equation (5.2.4) has the unique solution $Y = \{y\}$ in the algebra $\mathcal{G}(\Omega, H_a)$ for any $\zeta \in H_a = \text{dom } A$.*

Now consider the case of A generating an integrated semi-group. If operator A generates an exponentially bounded n-times integrated semi-group $\{S(t), t \ge 0\}$, then the solution operators $U(t)$ of the corresponding homogeneous Cauchy problem are defined by $\langle \varphi, U \rangle = (-1)^n \langle \varphi^{(n)}, S \rangle$, $\varphi \in \mathcal{D}$, and instead of (5.2.6) we have the following equation:

$$y(\varphi, t, \omega) = (-1)^n \int_{-\eta}^t S(t-s) F^{(n)}(y(\varphi, s, \omega)) \, ds$$

$$+ (-1)^n \int_{-\eta}^t S(t-s)[(Bw)^{(n)}(\varphi, s, \omega) + \zeta\varphi_\varepsilon^{(n)}(s)] \, ds,$$

$$\varphi \in \mathcal{A}_0, \quad \omega \in \Omega, \quad t \ge -\eta.$$

Here all derivatives exist due to the infinite differentiability of F, Bw, and φ_ε. Using the equality, similarly to the case of C_0-semi-groups, we obtain the

228   5. Stochastic Cauchy problems in spaces of distributions

corresponding approximations y_n and obtain a solution in $\mathcal{G}(\Omega, H_a)$ if the limit of $y_n(t)$ belongs to H_a.

In conclusion, we note that the present section contains just the beginning of the study of abstract stochastic equations with non-linearities in Colombeau algebras. Many questions still remain open. Among them are properties of $\lim y_n$, convergence of solutions $y(\varphi_\varepsilon)$ as $\varepsilon \to 0$, equations with generators of wider classes than we have considered, and equations in arbitrary Hilbert spaces.

Chapter 6

Infinite-dimensional extension of white noise calculus with application to stochastic problems

As we noted in the previous sections, numerous behavior patterns of systems under random perturbations suggest considering differential equations in infinite-dimensional spaces with white noise type inhomogeneities. Among problems for these equations, called abstract stochastic equations, the basic one is the Cauchy problem (P.1) in Hilbert spaces $H, \mathbb{H}$:

$$X'(t) = AX(t) + F(t, X) + B(t, X)\mathbb{W}(t), \quad t \geq 0, \quad X(0) = \zeta, \qquad (6.0.1)$$

with an $\mathbb{H}$-valued white noise $\mathbb{W}$ and A generating a regularized semigroup in H, in particular a C_0-semigroup; $F(t, X) \in H$ and $B(t, X) \in \mathcal{L}(\mathbb{H}, H)$.

Chapter 4 was devoted to solving the problem (6.0.1) by reducing it to the integral problem (I.1) with an Itô integral wrt a Wiener process. That was done on the basis of generalization of the Itô integral wrt Brownian motion to Hilbert spaces. In addition to weak and mild solutions obtained for (I.1), in Chapter 5 we constructed generalized solutions for (6.0.1) in spaces of abstract distributions and Gelfand–Shilov spaces. All the solutions constructed are supposed to be predictable due to the properties of Wiener and white noise processes.

In order to solve equations from a wider class, in particular, equations with singular white noise and with no predictability restrictions on solutions (the latter is connected with the adaptedness property of the processes to the filtration generated by a Wiener process), we need a new stochastic technique. Application of the ideas of the white noise calculus to analysis of stochastic processes with values in Hilbert spaces leads to progress in this direction.

In the present chapter, after some comments on the finite-dimensional white noise calculus, we present in detail its infinite-dimensional extension. That is the theory of abstract stochastic distributions, or generalized random variables spaces $(\mathcal{S})_{-\rho}(\mathbb{H})$, $0 \leq \rho \leq 1$. In these spaces $\mathbb{H}$-valued Q-Wiener and cylindrical Wiener processes, as well as t-derivatives of these processes, will be defined; the latter are called a Q-white noise and a singular white noise, respectively. These spaces and analysis in these spaces presented in Sections 6.1–6.3 will be the key techniques for obtaining solutions to the stochastic problems with singular white noise in Section 6.4.

229

230   6. Infinite-dimensional extension of white noise calculus

6.1 Spaces of Hilbert space-valued generalized random variables: $(\mathcal{S})_{-\rho}(\mathbb{H})$. Basic examples

The white noise probability space plays a fundamental role in our construction of spaces of Hilbert space valued random variables. We begin this section with its definition and main properties, then give some explanation of the finite-dimensional white noise calculus. Then we present in detail spaces of abstract (Hilbert space valued) stochastic distributions or abstract generalized random variables $(\mathcal{S})_{-\rho}(\mathbb{H})$, $0 \le \rho \le 1$. We show that these spaces contain the $\mathbb{H}$-valued Q-Wiener and cylindrical Wiener processes, as well as Q-white noise and singular white noise.

6.1.1 Spaces of $\mathbb{R}$-valued generalized random variables

Let $\mathcal{S}'$ be the space of tempered distributions over the space of rapidly decreasing test functions $\mathcal{S}$. The space $\mathcal{S}$ is countably Hilbert. This means

$$\mathcal{S} = \bigcap_{p \in \mathbb{N}} \mathcal{S}_p, \quad \text{where} \quad \mathcal{S}_p = \left\{ \theta \in L^2(\mathbb{R}) : \ (\theta, \eta)_p < \infty \right\}, \tag{6.1.1}$$

and the scalar product $(\cdot, \cdot)_p$ is defined by

$$(\theta, \eta)_p := (\hat{D}^p \theta, \hat{D}^p \eta)_{L^2(\mathbb{R})}, \quad \text{where} \quad \hat{D} := -\frac{d^2}{dx^2} + x^2 + 1. \tag{6.1.2}$$

Denote by $|\cdot|_p$ the norm generated by this scalar product. Under the definition of $\mathcal{S}_p$ we have that for any p the injection $\mathcal{S}_{p+1} \hookrightarrow \mathcal{S}_p$ is a nuclear operator, hence the space $\mathcal{S}$ is nuclear. Due to this fact, by the Bochner–Minlos–Sazonov theorem (see, e.g., [45], Theorem 4.7) there exists a unique probability measure μ defined on the Borel σ-algebra $\mathcal{B}(\mathcal{S}')$ of subsets of $\mathcal{S}'$, satisfying the condition

$$\int_{\mathcal{S}'} e^{i\langle \theta, \omega \rangle} d\mu(\omega) = e^{-\frac{1}{2}|\theta|_0^2}, \qquad \theta \in \mathcal{S}, \tag{6.1.3}$$

where $|\cdot|_0$ is the norm of the space $L^2(\mathbb{R})$.

The measure μ is called a normalized Gaussian measure on $\mathcal{S}'$ since for any $\theta_1, \theta_2, \ldots, \theta_n \in \mathcal{S}$ orthogonal in $L^2(\mathbb{R})$, the random variable $\omega \mapsto (\langle \theta_1, \omega \rangle, \langle \theta_2, \omega, \rangle, \ldots, \langle \theta_n, \omega \rangle)$ is Gaussian with the probability density

$$\frac{1}{(2\pi)^{\frac{n}{2}} \prod_{i=1}^n |\theta_i|_0} \exp\left(-\frac{1}{2} \sum_{i=1}^n \frac{x_i^2}{|\theta_i|_0^2} \right).$$

6.1. Spaces of Hilbert space valued generalized random variables 231

Equivalently,

$$\mathbf{E}\Big(f\big(\langle\theta_1,\cdot\rangle,\ldots,\langle\theta_n,\cdot\rangle\big)\Big)$$

$$= \frac{1}{(2\pi)^{\frac{n}{2}}\prod_{i=1}^{n}|\theta_i|_0}\int_{\mathbb{R}^n} f(x_1,\ldots,x_n)e^{-\frac{1}{2}\sum_{i=1}^{n}\frac{x_i^2}{|\theta_i|_0^2}}\,dx_1\ldots dx_n \quad (6.1.4)$$

for any $f : \mathbb{R}^n \to \mathbb{R}$ such that the integral in the right-hand side exists. (See Section 4.1, where normalized Gaussian measures were introduced.)

The probability space $(\Omega, \mathcal{F}, P) := (\mathcal{S}', \mathcal{B}(\mathcal{S}'), \mu)$ is called the *white noise probability space*. Why the space is so called will be clear a little later after we show that the "primitives" of elements $\omega \in \mathcal{S}'$ can be cosidered as the paths of a Brownian motion.

Denote by (L^2) the space $L^2(\mathcal{S}', \mu; \mathbb{R})$ of all square integrable wrt μ functions defined on $\mathcal{S}'$ (random variables) with values in $\mathbb{R}$. Denote by $\|\cdot\|_0$ its norm. Using (6.1.4) we can show that for any $\theta, \eta \in \mathcal{S}$ the following equalities hold true:

$$\big(\langle\theta,\cdot\rangle, \langle\eta,\cdot\rangle\big)_{(L^2)} = \mathbf{E}\Big(\langle\theta,\cdot\rangle\langle\eta,\cdot\rangle\Big) = (\eta,\theta)_{L^2(\mathbb{R})}, \qquad (6.1.5)$$

in particular $\|\langle\theta,\cdot\rangle\|_0^2 = \mathbf{E}\langle\theta,\cdot\rangle^2 = |\theta|_0^2$. It follows from here that the mapping $\theta \mapsto \langle\theta,\cdot\rangle$ can be extended by continuity from $\mathcal{S}$ onto $L^2(\mathbb{R})$; thus $\langle\theta,\cdot\rangle$ is well defined as an element of the space (L^2) for any $\theta \in L^2(\mathbb{R})$. The equalities (6.1.5) remain valid for $\theta \in L^2(\mathbb{R})$ and (6.1.4) remains valid for $\theta_1,\ldots,\theta_n \in L^2(\mathbb{R})$. Thus for any $t \geq 0$ the random variable

$$\beta(t) := \langle 1_{[0,t]}, \cdot\rangle \qquad (6.1.6)$$

is well defined as an element of (L^2).

It follows from (6.1.3) that $\beta(t)$ is a Gaussian random variable with mean zero for any $t \geq 0$ and the following properties of the process follow from (6.1.5):

$$\mathbf{E}\big[\beta(t)\beta(s)\big] = (\beta(t), \beta(s))_{L^2(\Omega)} = (1_{[0,t]}, 1_{[0,s]})_{L^2(\mathbb{R})} = \min\{t,s\}, \quad t,s \geq 0,$$

$$\mathbf{E}\big[\beta^2(t)\big] = (\beta(t), \beta(t))_{L^2(\Omega)} = \|1_{[0,t]}\|^2 = t,$$

where $L^2(\Omega) = (L^2)$. For disjoint intervals (t_3, t_4) and (t_1, t_2),

$$\mathbf{E}\big[(\beta(t_4) - \beta(t_3))(\beta(t_2) - \beta(t_1))\big] = (1_{[t_4,t_3]}, 1_{[t_2,t_1]})_{L^2(\mathbb{R})} = 0.$$

Hence the process $\{\beta(t), t \geq 0\}$ satisfies the properties (B1)–(B3) of the definition of Brownian motion (Definition 4.1.5). Moreover, for $0 \leq s < t$ we have

$$\mathbf{E}\big[(\beta(t) - \beta(s))^4\big] = \mathbf{E}\big[\langle 1_{[0,t]}, \cdot\rangle^4\big] = \frac{1}{\sqrt{2\pi(t-s)}}\int_{\mathbb{R}} x^4 e^{-\frac{x^2}{2(t-s)}}\,dx = 3(t-s)^2.$$

$$(6.1.7)$$

232   6. *Infinite-dimensional extension of white noise calculus*

It follows from (6.1.7) by the Kolmogorov continuity theorem (see [109], Section 4.1.1) that $\{\beta(t),\, t \geq 0\}$ has a continuous version, which is denoted by the same symbol and is called a Brownian motion.

Following the common practice in the theory of distributions we can write informally the right-hand side of (6.1.6) as an integral: $\langle \omega, 1_{[0;t]} \rangle = \int_0^t \omega(s)\, ds$. Thus we have

$$\beta(t) = \int_0^t \omega(s)\, ds$$

and the elements $\omega \in \mathcal{S}'$, being the elementary outcomes within the white noise probability space framework, can be thought of as the trajectories of the white noise.

Let $\{\xi_k\}_{k=1}^{\infty}$ be the orthonormal basis of the space $L^2(\mathbb{R})$ consisting of the Hermite functions

$$\xi_k(x) = \pi^{-\frac{1}{4}} \left((k-1)!\right)^{-\frac{1}{2}} e^{-\frac{x^2}{2}} h_{k-1}(x),$$

where $\{h_k(x)\}_{k=0}^{\infty}$ are the Hermite polynomials

$$h_k(x) = (-1)^n e^{\frac{x^2}{2}} \frac{d^k}{dx^k} e^{-\frac{x^2}{2}}.$$

We will use the next well-known estimates for ξ_i later (see e.g., [43]):

$$|\xi_i(x)| = \mathcal{O}\left(i^{-\frac{1}{4}}\right), \quad \left|\int_0^x \xi_i(s)\, ds\right| = \mathcal{O}\left(i^{-\frac{3}{4}}\right), \quad \sup_{x \in \mathbb{R}} |\xi_i(x)| = \mathcal{O}\left(i^{-\frac{1}{12}}\right).$$

$$(6.1.8)$$

Let $\mathcal{T} \subset \left(\mathbb{N} \cup \{0\}\right)^{\mathbb{N}}$ be the set of all finite multiindices. The stochastic Hermite polynomials are defined by the equalities

$$\mathbf{h}_\alpha(\omega) := \prod_k h_{\alpha_k}\left(\langle \xi_k, \omega \rangle\right), \qquad \omega \in \mathcal{S}', \quad \alpha \in \mathcal{T}. \qquad (6.1.9)$$

The product in (6.1.9) is in fact finite since each α is finite and consequently $h_{\alpha_k}(x) = h_0(x) = 1$ for all sufficiently large k.

Let $\alpha, \beta \in \mathcal{T}$ and $n = \max\{k \in \mathbb{N} : \alpha_k \neq 0 \text{ or } \beta_k \neq 0\}$. By equality (6.1.4) and the orthogonality of the Hermite polynomials in the space $L^2\left(\mathbb{R}; \frac{1}{\sqrt{2\pi}} e^{-\frac{x^2}{2}}\, dx\right)$ we have

$$
\begin{aligned}
(\mathbf{h}_\alpha, \mathbf{h}_\beta)_{(L^2)} &= \mathbf{E}\left[\prod_{k=1}^n h_{\alpha_k}\left(\langle \xi_k, \omega \rangle\right) \prod_{k=1}^n h_{\beta_k}\left(\langle \xi_k, \omega \rangle\right)\right] \\
&= \frac{1}{(2\pi)^{\frac{n}{2}} \prod_{i=1}^n |\xi_i|_0} \int_{\mathbb{R}^n} \prod_{k=1}^n h_{\alpha_k}(x_k) \prod_{k=1}^n h_{\beta_k}(x_k) e^{-\frac{1}{2}\sum_{k=1}^n \frac{x_k^2}{|\xi_k|_0^2}}\, dx_1 \ldots dx_n \\
&= \frac{1}{(2\pi)^{\frac{n}{2}}} \prod_{k=1}^n \int_{\mathbb{R}} h_{\alpha_k}(x_k) h_{\beta_k}(x_k) e^{-\frac{1}{2}x_k^2}\, dx_k =
\begin{cases}
0, & \alpha \neq \beta, \\
\alpha!, & \alpha = \beta,
\end{cases}
\end{aligned}
$$

$$(6.1.10)$$

6.1. Spaces of Hilbert space valued generalized random variables   233

where $\alpha! := \prod_k \alpha_k!$. Thus stochastic Hermite polynomials form an orthogonal system in the space $(L^2) = L^2(\mathcal{S}', \mathcal{B}(\mathcal{S}'), \mu)$. Moreover, $\{h_\alpha : \alpha \in \mathcal{T}\}$ is an orthogonal basis in (L^2) (Theorem 2.2.3 in [44]). It follows from this fact and the equality (6.1.10) that the following equality holds for the scalar product and the norm in (L^2):

$$(\Phi, \Psi)_{(L^2)} = \sum_{\alpha \in \mathcal{T}} \alpha! \Phi_\alpha \Psi_\alpha, \qquad \|\Phi\|_{(L^2)}^2 = \sum_{\alpha \in \mathcal{T}} \alpha! \Phi_\alpha^2,$$

where

$$\Phi = \sum_{\alpha \in \mathcal{T}} \Phi_\alpha h_\alpha, \quad \Psi = \sum_{\alpha \in \mathcal{T}} \Psi_\alpha h_\alpha, \quad \Phi_\alpha = \frac{1}{\alpha!} (\Phi, h_\alpha)_{(L^2)}, \quad \Psi_\alpha = \frac{1}{\alpha!} (\Psi, h_\alpha)_{(L^2)}.$$

Due to (6.1.10) one can informally think of the space (L^2) as of $L^2 \left(\mathbb{R}^\infty; \prod_{k=1}^\infty \frac{1}{\sqrt{2\pi}} e^{-\frac{x_k^2}{2}} dx_k \right)$ identifying any element $\omega \in \mathcal{S}'$ with the sequence of its "Fourier coefficients" $\langle \omega, \xi_k \rangle$ wrt the system of Hermite functions. Thus square integrable random variables on the white noise probability space $(\mathcal{S}', \mathcal{B}(\mathcal{S}'), \mu)$ can be considered as functions of infinite real variables. This linear structure of the domain of definition of random variables leads to the following generalization of the Schwartz theory to the infinite variables case.

Consider the construction of the Gelfand triple

$$(\mathcal{S})_\rho \subset (L^2) \subset (\mathcal{S})_{-\rho} \qquad (0 \le \rho \le 1),$$

which is a generalization parameterized by ρ of the triple $\mathcal{S} \subset L^2(\mathbb{R}) \subset \mathcal{S}'$, where $\mathcal{S} = \bigcap \mathcal{S}_p$, $\mathcal{S}' = \bigcup \mathcal{S}'_p$ and spaces $\mathcal{S}_p$ are defined by (6.1.1). It was first introduced in [55] and is used in [59, 44].

Recall that due to the fact that the Hermite functions are the eigenfunctions of the differential operator

$$\hat{D} = -\frac{d^2}{dx^2} + x^2 + 1 \quad \text{with} \quad \hat{D}\xi_i = (2i)\xi_i, \quad i \in \mathbb{N},$$

the spaces $\mathcal{S}_p$ defined by (6.1.1) can be characterized in terms of expansions with respect to Hermite functions $\{\xi_i\}$ in the following way:

$$\mathcal{S}_p = \left\{ \theta = \sum_{i=1}^\infty \theta_i \xi_i \in L^2(\mathbb{R}) : \sum_{i=1}^\infty |\theta_i|^2 (2i)^{2p} < \infty \right\}.$$

The spaces $(\mathcal{S}_p)$ are defined by analogy with $\mathcal{S}_p$:

$$(\mathcal{S}_p) := \left\{ \varphi = \sum_{\alpha \in \mathcal{T}} \varphi_\alpha h_\alpha \in (L^2) : \sum_{\alpha \in \mathcal{T}} (\alpha!) |\varphi_\alpha|^2 (2\mathbb{N})^{2p\alpha} < \infty \right\},$$

with the norms $|\cdot|_p$ generated by the scalar products

$$(\varphi, \psi)_p = \sum_{\alpha \in \mathcal{T}} (\alpha!) \varphi_\alpha \overline{\psi}_\alpha (2\mathbb{N})^{2p\alpha}, \qquad (2\mathbb{N})^{p\alpha} := \prod_{i \in \mathbb{N}} (2i)^{p\alpha_i}.$$

234   *6. Infinite-dimensional extension of white noise calculus*

The spaces $(\mathcal{S}_p)_\rho$ are subspaces of $(\mathcal{S}_p)$ parameterized by an additional parameter $0 \le \rho \le 1$ and defined by analogy with $\mathcal{S}_p$:

$$(\mathcal{S}_p)_\rho = \left\{ \varphi = \sum_{\alpha \in \mathcal{T}} \varphi_\alpha \mathbf{h}_\alpha \in (L^2) : \sum_{\alpha \in \mathcal{T}} (\alpha!)^{1+\rho} |\varphi_\alpha|^2 (2\mathbb{N})^{2p\alpha} < \infty \right\},$$

with the norms $|\cdot|_{p,\rho}$ generated by the scalar products

$$(\varphi, \psi)_{p,\rho} = \sum_{\alpha \in \mathcal{T}} (\alpha!)^{1+\rho} \varphi_\alpha \overline{\psi}_\alpha (2\mathbb{N})^{2p\alpha}.$$

To clarify the analogy with the definition by (6.1.1)–(6.1.2), note that the other way to define the scalar product $(\cdot, \cdot)_{p,\rho}$ when $\rho = 0$ is doing this in terms of the so-called second quantization operator $\Gamma(\hat{D})$, which is usually defined via identifying the space (L^2) with the Fock space $\bigoplus_{n=0}^\infty \hat{L}^2(\mathbb{R}^n)$ with the help of the Wiener–Itô chaos expansion (see, e.g., [59]). To simplify our presentation we define $\Gamma(\hat{D})$ in an equivalent way:

$$\Gamma(\hat{D})\mathbf{h}_\alpha := \prod_{i=1}^\infty (2i)^{\alpha_i} h_{\alpha_i}\big(\langle \xi_i, \cdot \rangle \big).$$

Then, by analogy with (6.1.1), we have

$$(\varphi, \psi)_{p,0} = \left(\Gamma(\hat{D})^p \varphi, \Gamma(\hat{D})^p \psi \right)_{(L^2)}.$$

The space $(\mathcal{S})_\rho$ is defined as $(\mathcal{S})_\rho = \bigcap_{p \in \mathbb{N}} (\mathcal{S}_p)_\rho$ with the projective limit topology and is called *the space of test random variables.*

The space $(\mathcal{S})_{-\rho}$ is defined by

$$(\mathcal{S})_{-\rho} = \bigcup_{p \in \mathbb{N}} (\mathcal{S}_{-p})_{-\rho},$$

with the inductive limit topology, where $(\mathcal{S}_{-p})_{-\rho}$ is the dual to the space $(\mathcal{S}_p)_\rho$. The elements of $(\mathcal{S})_{-\rho}$ are called *generalized random variables.* The space $(\mathcal{S}_{-p})_{-\rho}$ can be identified with the Hilbert space of all formal (countable) expansions $\Phi = \sum_{\alpha \in \mathcal{T}} \Phi_\alpha \mathbf{h}_\alpha$ satisfying the condition

$$\sum_{\alpha \in \mathcal{T}} (\alpha!)^{1-\rho} |\Phi_\alpha|^2 (2\mathbb{N})^{-2p\alpha} < \infty,$$

with the scalar product

$$(\Phi, \Psi)_{-p,-\rho} = \sum_{\alpha \in \mathcal{T}} (\alpha!)^{1-\rho} \Phi_\alpha \overline{\Psi}_\alpha (2\mathbb{N})^{-2p\alpha}.$$

We will denote the norm of the space $(\mathcal{S}_{-p})_{-\rho}$ by $|\cdot|_{-p,-\rho}$.

6.1. Spaces of Hilbert space valued generalized random variables   235

For $\Phi = \sum_{\alpha \in \mathcal{T}} \Phi_\alpha \mathbf{h}_\alpha \in (\mathcal{S})_{-\rho}$ and $\varphi = \sum_{\alpha \in \mathcal{T}} \varphi_\alpha \mathbf{h}_\alpha \in (\mathcal{S})_\rho$, we have

$$\langle \varphi, \Phi \rangle = \sum_{\alpha \in \mathcal{T}} \alpha! \overline{\Phi}_\alpha \varphi_\alpha.$$

The notion of a bounded set plays an important role later. The way it was defined in Section 3.3 is suitable for $(\mathcal{S})_\rho$. The set $M \subseteq (\mathcal{S})_\rho$ is called bounded if for any sequence $\{\varphi_n\} \subseteq M$ and for any $\{\varepsilon_n\} \subset \mathbb{R}$ convergent to zero, $\{\varepsilon_n \varphi_n\}$ is convergent to zero in $(\mathcal{S})_\rho$.

From the definition of spaces $(\mathcal{S})_\rho$ and the definition of bounded sets it is easy to obtain the following characterization of bounded sets in $(\mathcal{S})_\rho$.

Proposition 6.1.1 *A set is bounded in $(\mathcal{S})_\rho$ if and only if it is bounded in any $(\mathcal{S}_p)_\rho$, $p \in \mathbb{N}$.*

Now we define spaces $(\mathcal{S})_{-\rho}(\mathbb{H})$ of Hilbert space valued random variables over the spaces $(\mathcal{S})_\rho$ of test random variables.

6.1.2   Spaces of Hilbert space valued generalized random variables $(\mathcal{S})_{-\rho}(\mathbb{H})$. Examples of important generalized random processes

Let $\mathbb{H}$ be a separable complex Hilbert space with the scalar product $(\cdot, \cdot)$ and the corresponding norm $\|\cdot\|$. By $(L^2)(\mathbb{H})$ we will denote the space of all $\mathbb{H}$-valued functions f on $\mathcal{S}'$, which are square Bochner integrable wrt μ, i.e., $\int_{\mathcal{S}'} \|f(\omega)\| \, d\mu(\omega) < \infty$.

Let $\{e_j\}_{j=1}^\infty$ be an orthonormal basis in $\mathbb{H}$. The family of $\mathbb{H}$-valued functions $\{\mathbf{h}_\alpha e_j\}_{\alpha \in \mathcal{T}, j \in \mathbb{N}}$ forms an orthogonal basis in $(L^2)(\mathbb{H})$. Any $f \in (L^2)(\mathbb{H})$ can be decomposed into the Fourier series with respect to this basis as follows:

$$f = \sum_{\alpha \in \mathcal{T}, j \in \mathbb{N}} f_{\alpha,j} \mathbf{h}_\alpha e_j = \sum_{\alpha \in \mathcal{T}} f_\alpha \mathbf{h}_\alpha = \sum_{j=1}^\infty f_j e_j,$$

$$f_{\alpha,j} \in \mathbb{R}, \qquad f_\alpha = \sum_j f_{\alpha,j} e_j \in \mathbb{H}, \qquad f_j = \sum_{\alpha \in \mathcal{T}} f_{\alpha,j} \mathbf{h}_\alpha \in (L^2),$$

and

$$\|f\|^2_{(L^2)(\mathbb{H})} = \sum_{\alpha \in \mathcal{T}, j \in \mathbb{N}} \alpha! \, |f_{\alpha,j}|^2 = \sum_{\alpha \in \mathcal{T}} \alpha! \, \|f_\alpha\|^2_{\mathbb{H}} = \sum_{j=1}^\infty \|f_j\|^2_{(L^2)}.$$

Denote by $(\mathcal{S})_{-\rho}(\mathbb{H})$ the space of all linear continuous operators $\Phi : (\mathcal{S})_\rho \to \mathbb{H}$ endowed with the topology of uniform convergence on bounded subsets of the space $(\mathcal{S})_\rho$. We will refer to this convergence as strong convergence in $(\mathcal{S})_{-\rho}(\mathbb{H})$ and call elements of this space the $\mathbb{H}$-valued generalized random variables over the space of test functions $(\mathcal{S})_\rho$. The action of $\Phi \in (\mathcal{S})_{-\rho}(\mathbb{H})$ on a test function $\varphi \in (\mathcal{S})_\rho$ will be denoted by $\Phi[\varphi]$.

236   6. Infinite-dimensional extension of white noise calculus

To develop analysis of $(\mathcal{S})_{-\rho}(\mathbb{H})$-valued functions of $t \in \mathbb{R}$ we first describe the structure of this space. The following two propositions can be considered as an extension of the famous structure theorems for the Shwartz spaces $\mathcal{D}'$ and $\mathcal{S}'$, which state that any element in these spaces has finite order (see Section 3.3).

Proposition 6.1.2 *Any* $\Phi \in (\mathcal{S})_{-\rho}(\mathbb{H})$ *is a bounded operator from* $(\mathcal{S}_p)_\rho$ *to* $\mathbb{H}$ *for some* $p \in \mathbb{N}$.

Proof. Suppose the opposite is true. Let $\Phi \in (\mathcal{S})_{-\rho}(\mathbb{H})$. For any $p \in \mathbb{N}$ choose $\varphi_p \in (\mathcal{S}_p)_\rho$ so that $|\varphi_p|_{p,\rho} = 1$ and $\|\Phi[\varphi_p]\| \geq p$. By the inequalities $|\varphi_k|_{p,\rho} \leq |\varphi_k|_{k,\rho}$, which hold true for all $k > p$, the sequence $\left\{ \frac{\varphi_k}{k} \right\}$ converges to zero in the space $(\mathcal{S})_\rho$. At the same time we have $\left\| \Phi\left[\frac{\varphi_k}{k} \right] \right\| \geq 1$, which contradicts the continuity of Φ. $\qquad\square$

Note some properties of the space of test functions $(\mathcal{S})_\rho$. It is a countably Hilbert space since all spaces $(\mathcal{S}_p)_\rho$ are Hilbert and convergence $\varphi_n \to \varphi$ in $(\mathcal{S})_\rho$ is equivalent to convergence $\|\varphi_n - \varphi\|_p \to 0$ for any p. In addition, it is a nuclear space since for any $p \in \mathbb{N}$ the embedding operator

$$I_{p,p+1} : (\mathcal{S}_{p+1})_\rho \hookrightarrow (\mathcal{S}_p)_\rho$$

is Hilbert–Schmidt. To check this, recall that an operator $L : H_1 \to H_2$ is Hilbert–Schmidt if $\|L\|_{\mathrm{HS}} = \sum_i \|Le_i\|_{H_2} < \infty$, where $\{e_i\}$ is a basis in H_1 (see Section 4.1.1), and take the following orthonormal basis $\{e_\alpha^{p,\rho}\}$ of the space $(\mathcal{S}_{p+1})_\rho$:

$$e_\alpha^{p,\rho} = \left\{ \frac{\mathbf{h}_\alpha}{(\alpha!)^{\frac{1+\rho}{2}} (2\mathbb{N})^{(p+1)\alpha}} \right\}.$$

Then we have

$$\|I_{p,p+1}\|_{\mathrm{HS}}^2 = \sum_{\alpha \in \mathcal{T}} \left| \frac{\mathbf{h}_\alpha}{(\alpha!)^{\frac{1+\rho}{2}} (2\mathbb{N})^{(p+1)\alpha}} \right|_{p,\rho}^2 = \sum_{\alpha \in \mathcal{T}} \frac{1}{(2\mathbb{N})^{p\alpha}} =: \mathcal{A}(p). \quad (6.1.11)$$

It is proved in [44] that $\mathcal{A}(p) < \infty$ for any $p > 1$. Thus the series (6.1.11) is convergent.

Due to the nuclearity of $(\mathcal{S})_\rho$, we have the following characterization of the elements of generalized $\mathbb{H}$-valued random variables.

Proposition 6.1.3 *Any* $\Phi \in (\mathcal{S})_{-\rho}(\mathbb{H})$ *is a Hilbert–Schmidt operator from* $(\mathcal{S}_p)_\rho$ *to* $\mathbb{H}$ *for some* $p \in \mathbb{N}$.

Proof. Let $\Phi \in (\mathcal{S})_{-\rho}(\mathbb{H})$. By Proposition 6.1.2 Φ is bounded as an operator acting from $(\mathcal{S}_p)_\rho$ to $\mathbb{H}$ for some $p \in \mathbb{N}$. Denote by $\tilde{\Phi}$ its extension to $(\mathcal{S}_p)_\rho$ by continuity. Then, as an operator acting from $(\mathcal{S}_{p+1})_\rho$ to $\mathbb{H}$, the operator Φ can be written as $\tilde{\Phi} I_{p,p+1}$, which is Hilbert–Schmidt as a composition of Hilbert–Schmidt and bounded operators. $\qquad\square$

In order to investigate the topology of uniform convergence on bounded

6.1. Spaces of Hilbert space valued generalized random variables   237

subsets of $(\mathcal{S})_\rho$ that we introduced in $(\mathcal{S})_{-\rho}(\mathbb{H})$, we will also need the decomposition of this space into countable unions of separable Hilbert spaces.

For any $\Phi \in (\mathcal{S})_{-\rho}(\mathbb{H})$ denote by Φ_j the linear functional, defined on the space $(\mathcal{S})_\rho$ by the equality

$$\langle \varphi, \Phi_j \rangle := (\Phi[\varphi], e_j), \quad \varphi \in (\mathcal{S})_\rho, \quad e_j \in \mathbb{H}.$$

Let p be such that Φ is a Hilbert–Schmidt operator from $(\mathcal{S}_p)_\rho$ to $\mathbb{H}$. Then all Φ_j, $j \in \mathbb{N}$, belong to the dual space $(\mathcal{S}_{-p})_{-\rho}$ with the same parameters p, ρ and therefore can be decomposed into the series

$$\Phi_j = \sum_{\alpha \in \mathcal{T}} \Phi_{\alpha,j} h_\alpha, \quad \text{where} \quad \sum_{\alpha \in \mathcal{T}} (\alpha!)^{1-\rho} |\Phi_{\alpha,j}|^2 (2\mathbb{N})^{-2p\alpha} < \infty.$$

Denote by $\|\Phi\|_{-p,-\rho}$ the Hilbert–Schmidt norm of an operator $\Phi : (\mathcal{S}_p)_\rho \to \mathbb{H}$ acting from $(\mathcal{S}_p)_\rho$ to $\mathbb{H}$. We have

$$\|\Phi\|_{-p,-\rho}^2 = \sum_{\alpha \in \mathcal{T}} \left\| \Phi \left[\frac{h_\alpha}{(\alpha!)^{\frac{1+\rho}{2}} (2\mathbb{N})^{p\alpha}} \right] \right\|_{\mathbb{H}}^2 = \sum_{\alpha \in \mathcal{T}} \sum_{j=1}^{\infty} \left| \left\langle \frac{h_\alpha}{(\alpha!)^{\frac{1+\rho}{2}} (2\mathbb{N})^{p\alpha}}, \Phi_j \right\rangle \right|^2$$

$$= \sum_{\alpha \in \mathcal{T}, j \in \mathbb{N}} (\alpha!)^{1-\rho} |\Phi_{\alpha,j}|^2 (2\mathbb{N})^{-2p\alpha}. \tag{6.1.12}$$

Denote by $(\mathcal{S}_{-p})_{-\rho}(\mathbb{H})$ the space of all Hilbert–Schmidt operators acting from $(\mathcal{S}_p)_\rho$ to $\mathbb{H}$. It is a separable Hilbert space. Operators $h_\alpha \otimes e_j$, $\alpha \in \mathcal{T}$, $j \in \mathbb{N}$, defined by the equality

$$(h_\alpha \otimes e_j)\varphi := (\varphi, h_\alpha)_{(L^2)} e_j, \quad \varphi \in (\mathcal{S}_p)_\rho,$$

form an orthogonal basis in the space. It follows from Proposition 6.1.3 that we have the following decomposition:

$$(\mathcal{S})_{-\rho}(\mathbb{H}) = \bigcup_{p \in \mathbb{N}} (\mathcal{S}_{-p})_{-\rho}(\mathbb{H})$$

and any $\Phi \in (\mathcal{S})_{-\rho}(\mathbb{H})$ has the following decomposition:

$$\Phi[\cdot] = \sum_{j \in \mathbb{N}} \langle \cdot, \Phi_j \rangle e_j = \sum_{\alpha \in \mathcal{T}, j \in \mathbb{N}} \Phi_{\alpha,j}(h_\alpha \otimes e_j) = \sum_{\alpha \in \mathcal{T}} \Phi_\alpha(\cdot, h_\alpha)_{(L^2)},$$

where $\Phi_j = (\Phi[\cdot], e_j) \in (\mathcal{S}_{-p})_{-\rho}$ for some $p \in \mathbb{N}$ and $\Phi_\alpha = \sum_{j \in \mathbb{N}} \Phi_{\alpha,j} e_j \in \mathbb{H}$ with

$$\|\Phi\|_{-p,-\rho}^2 = \sum_{j \in \mathbb{N}} |\Phi_j|_{-p,-\rho}^2 = \sum_{\alpha \in \mathcal{T}, j \in \mathbb{N}} (\alpha!)^{1-\rho} |\Phi_{\alpha,j}|^2 (2\mathbb{N})^{-2p\alpha}$$

$$= \sum_{\alpha \in \mathcal{T}} (\alpha!)^{1-\rho} \|\Phi_\alpha\|^2 (2\mathbb{N})^{-2p\alpha} < \infty.$$

238   *6. Infinite-dimensional extension of white noise calculus*

It is easy to see that

$$(\mathcal{S}_{-p_1})_{-\rho}(\mathbb{H}) \subseteq (\mathcal{S}_{-p_2})_{-\rho}(\mathbb{H}), \qquad p_1 < p_2, \tag{6.1.13}$$

and

$$\|\Phi\|_{-p_1,-\rho} \geq \|\Phi\|_{-p_2,-\rho}, \qquad \Phi \in (\mathcal{S}_{-p_1})_{-\rho}(\mathbb{H}). \tag{6.1.14}$$

In Section 4.1 we introduced an $\mathbb{H}$-valued Q-Wiener process $W_Q = \{W_Q(t),\, t \geq 0\}$ via the Fourier series wrt an orthonormal basis $\{e_j\}$ in $\mathbb{H}$:

$$W_Q(t) = \sum_{j \in \mathbb{N}} \sigma_j \beta_j(t) e_j \qquad (Q e_j = \sigma_j^2 e_j), \tag{6.1.15}$$

where β_j are independent Brownian motions. We also introduced a cylindrical (or weak) Wiener process $W = \{W(t),\, t \geq 0\}$ via the Fourier series

$$W(t) = \sum_{j \in \mathbb{N}} \beta_j(t) e_j, \tag{6.1.16}$$

which is generally divergent in $\mathbb{H}$, but weakly convergent in the sense that for any $h \in \mathbb{H}$ the series $(h, W(t)) = \sum_{j \in \mathbb{N}} \beta_j(t)(h, e_j)$ is convergent.

Now we give a construction (or a constructive proof of existence) of such a sequence of independent Brownian motions. Using this sequence we define Q-Wiener and cylindrical Wiener processes (6.1.15)–(6.1.16) and later the series defining Q-white noise and singular white noise processes in spaces $(\mathcal{S}_{-p})_{-\rho}$.

First, we introduce a sequence of independent identically distributed Brownian motions on the white noise probability space. To do this, take a bijection $n(\cdot, \cdot) : \mathbb{N} \times \mathbb{N} \to \mathbb{N}$ satisfying the condition

$$n(i, j) \geq ij, \quad i, j \in \mathbb{N}. \tag{6.1.17}$$

It can be defined in different ways, for example, by the following table:

i \ j	1	2	3	4	5	6	7	$\cdots$
1	1	3	6	10	15	21	28	$\cdots$
2	2	5	9	14	20	27		
3	4	8	13	19	26			
4	7	12	18	25				
5	11	17	24				$n(i,j)$	
6	16	23						
7	22							
$\cdots$								

Define a sequence of linear operators $\mathfrak{I}_j,\, j \in \mathbb{N}$, in the space $L^2(\mathbb{R})$ by

$$\mathfrak{I}_j f = \sum_{i=1}^{\infty} (f, \xi_i) \xi_{n(i,j)}. \tag{6.1.18}$$

6.1. Spaces of Hilbert space valued generalized random variables 239

Let $L^2(\mathbb{R})_j$ be the closure of the linear span of the set $\{\xi_{n(i,j)}, \, i \in \mathbb{N}\}$. For any $j \in \mathbb{N}$, the operator $\mathfrak{I}_j$ is an isometric isomorphism of the spaces $L^2(\mathbb{R})$ and $L^2(\mathbb{R})_j$ since for any $f, g \in L^2(\mathbb{R})$ we have

$$(\mathfrak{I}_j f, \mathfrak{I}_j g)_{L^2(\mathbb{R})_j} = \sum_{i=1}^{\infty} (f, \xi_i)(g, \xi_i) = (f, g)_{L^2(\mathbb{R})}.$$

Since the spaces $L^2(\mathbb{R})_j$ with different j are spanned by disjoint families of $\{\xi_i\}$, they are pairwise orthogonal subspaces of $L^2(\mathbb{R})$. Moreover, since

$$\{\xi_i\}_{i=1}^{\infty} = \bigcup_{j=1}^{\infty} \{\xi_{n(i,j)}\}_{i=1}^{\infty}, \quad \text{we have} \quad L^2(\mathbb{R}) = \bigoplus_{j=1}^{\infty} L^2(\mathbb{R})_j.$$

Define $1_{[a,b]}^j := \mathfrak{I}_j 1_{[a,b]}$, where $1_{[a,b]}$ is the indicator of $[a,b]$. For any $a, b, c, d \in \mathbb{R}$ the functions $1_{[a,b]}^{j_1}$ and $1_{[c,d]}^{j_2}$ with $j_1 \neq j_2$ are orthogonal in $L^2(\mathbb{R})$.

Now we consider random processes defined by

$$\beta_j(t) := \langle 1_{[0,t]}^j, \cdot \rangle, \qquad j = 1, 2, \ldots, \quad t \in \mathbb{R}.$$

By (6.1.5) and (6.1.2) we have

$$\mathbf{E}\left[\beta_j(t)\beta_j(s)\right] = (1_{[0,t]}^j, 1_{[0,s]}^j)_{L^2(\mathbb{R})} = (1_{[0,t]}^j, 1_{[0,s]}^j)_{L^2(\mathbb{R})_j}$$
$$= (1_{[0,t]}, 1_{[0,s]})_{L^2(\mathbb{R})} = \min\{t; s\}.$$

We also have
$$\mathbf{E}\left[\beta_{j_1}(t)\beta_{j_2}(s)\right] = (1_{[0,t]}^{j_1}, 1_{[0,s]}^{j_2})_{L^2(\mathbb{R})} = 0$$

for $j_1 \neq j_2$. It follows that $\{\beta_j(t), \, t \geq 0\}_{j=1}^{\infty}$ is a sequence of independent Brownian motions. We have the following decompositions for them:

$$\beta_j(t) = \langle 1_{[0,t]}^j, \cdot \rangle = \left\langle \sum_{i=1}^{\infty} \int_0^t \xi_i(s) \, ds \, \xi_{n(i,j)}, \cdot \right\rangle$$
$$= \sum_{i=1}^{\infty} \int_0^t \xi_i(s) \, ds \, \langle \xi_{n(i,j)}, \cdot \rangle = \sum_{i=1}^{\infty} \int_0^t \xi_i(s) \, ds \, \mathbf{h}_{\epsilon_{n(i,j)}},$$

where $\epsilon_n := (0, 0, \ldots, \underset{n}{1}, 0, \ldots)$.

Now using these equalities we define Q-Wiener and cylindrical Wiener processes in spaces $(\mathcal{S})_{-\rho}(\mathbb{H})$, $0 \leq \rho \leq 1$.

For a cylindrical Wiener process W defined by the informal (generally divergent) series (6.1.16) we obtain the following Fourier series with respect to stochastic Hermit polynomials $\{\mathbf{h}_{\epsilon_n}\}$:

$$W(t) = \sum_{j \in \mathbb{N}} \beta_j(t)e_j = \sum_{n \in \mathbb{N}} W_{\epsilon_n}(t) \, \mathbf{h}_{\epsilon_n}, \qquad t \in \mathbb{R}, \tag{6.1.19}$$

240   6. *Infinite-dimensional extension of white noise calculus*

where

$$W_{\epsilon_n}(t) := \int_0^t \xi_{i(n)}(s)\, ds\, e_{j(n)} \in \mathbb{H}$$

and $i(n), j(n) \in \mathbb{N}$ are such that $n(i(n), j(n)) = n$.

Let Q be a positive trace class operator in a Hilbert space $\mathbb{H}$ defined by the following decomposition:

$$Qh = \sum_{j=1}^{\infty} \sigma_j^2 (h, e_j) e_j, \quad h \in \mathbb{H}, \quad \text{or} \quad Q = \sum_{j=1}^{\infty} \sigma_j^2 (e_j \otimes e_j).^{[1]} \qquad (6.1.20)$$

The fact that $Q \in \mathcal{L}_{\mathrm{Tr}}(\mathbb{H})$ (the space of all trace class operators acting in $\mathbb{H}$) implies $Tr\, Q = \sum_{j=1}^{\infty} \sigma_j^2 < \infty$. For the Q-Wiener process W_Q defined by (6.1.15), the following equalities hold:

$$W_Q(t) = \sum_{j \in \mathbb{N}} \sigma_j \beta_j(t) e_j = \sum_{n \in \mathbb{N}} W_{\epsilon_n}^Q(t)\, \mathbf{h}_{\epsilon_n}, \qquad (6.1.21)$$

where

$$W_{\epsilon_n}^Q(t) := \sigma_j \int_0^t \xi_{i(n)}(s)\, ds\, e_{j(n)} \in \mathbb{H}, \qquad t \in \mathbb{R}.$$

As we know, $W_Q(t) \in (L^2)(\mathbb{H})$ for all $t \in \mathbb{R}$, but $W(t) \notin (L^2)(\mathbb{H})$. At the same time, for any $h \in \mathbb{H}$, we have

$$\mathbf{E}\big(W(t), h\big)^2 = \sum_{j \in \mathbb{N}} (e_j, h)^2 \mathbf{E}\big[\beta_j^2(t)\big] = t\|h\|^2.$$

That means $\big(W(t), h\big) \in (L^2) = L^2(\mathcal{S}', \mathcal{B}(\mathcal{S}'), \mu)$. Moreover, using the estimate $\int_0^t \xi_i(s)\, ds = \mathcal{O}\big(i^{-\frac{3}{4}}\big)$ from (6.1.8) and the condition (6.1.17), we obtain

$$\|W(t)\|_{-1,-\rho}^2 = \sum_{i,j \in \mathbb{N}} \left|\int_0^t \xi_i(s)\, ds\right|^2 \big(2n(i,j)\big)^{-2} \le \sum_{i,j \in \mathbb{N}} O\left(i^{-\frac{3}{2}-2} j^{-2}\right) < \infty.$$

$$(6.1.22)$$

Therefore, $W(t) \in (\mathcal{S}_{-1})_{-\rho}(\mathbb{H}) \subset (\mathcal{S})_{-\rho}(\mathbb{H})$ for any $0 \le \rho \le 1$.

Now define the $\mathbb{H}$-valued Q-white noise by the equality obtained by informal differentiation of (6.1.21):

$$\mathbb{W}_Q(t) := \sum_{i,j \in \mathbb{N}} \sigma_j \xi_i(t)\, \mathbf{h}_{\epsilon_{n(i,j)}} e_j = \sum_{n \in \mathbb{N}} \mathbb{W}_{\epsilon_n}^Q(t)\, \mathbf{h}_{\epsilon_n}, \qquad (6.1.23)$$

where

$$\mathbb{W}_{\epsilon_n}^Q(t) = \sigma_j \xi_{i(n)}(t)\, e_{j(n)} \in \mathbb{H}, \qquad t \in \mathbb{R},$$

---

[1] Recall that for $v \in V$, $u \in U$, where V and U are Hilbert spaces, we denote by $v \otimes u$ the operator acting from U to V, defined by the equality $(v \otimes u)h := v(u, h)_U$.

6.2. *Analysis of* $(\mathcal{S})_{-\rho}(\mathbb{H})$*-valued processes* 241

and the cylindrical (or singular) white noise by the equality

$$\mathbb{W}(t) := \sum_{i,j \in \mathbb{N}} \xi_i(t) \, \mathbf{h}_{\epsilon_{n(i,j)}} e_j = \sum_{n \in \mathbb{N}} \mathbb{W}_{\epsilon_n}(t) \, \mathbf{h}_{\epsilon_n}, \qquad (6.1.24)$$

$$\mathbb{W}_{\epsilon_n}(t) = \xi_{i(n)}(t) \, e_{j(n)} \in \mathbb{H}, \qquad t \in \mathbb{R},$$

obtained by informal differentiation of (6.1.19).

Using the estimate $\xi_i(t) = \mathcal{O}\left(i^{-\frac{1}{4}}\right)$ from (6.1.8), similarly to (6.1.22), we have $\|\mathbb{W}_Q(t)\|_{-1,-\rho}^2 < \infty$ and $\|\mathbb{W}(t)\|_{-1,-\rho}^2 < \infty$; thus both Q-white noise and cylindrical white noise are in $(\mathcal{S}_{-1})_{-\rho}(\mathbb{H}) \subset (\mathcal{S})_{-\rho}(\mathbb{H})$, $\rho \in [0,1]$.

In the next section we introduce differentiation and integration for $(\mathcal{S})_{-\rho}(\mathbb{H})$-valued functions wrt parameter t and show that for all $t \in \mathbb{R}$ we have

$$\frac{d}{dt} W_Q(t) = \mathbb{W}_Q(t) \qquad \text{and} \qquad \frac{d}{dt} W(t) = \mathbb{W}(t).$$

6.2 Analysis of $(\mathcal{S})_{-\rho}(\mathbb{H})$-valued processes

To introduce differentiation and integration of $(\mathcal{S})_{-\rho}(\mathbb{H})$-valued functions of $t \in \mathbb{R}$, we first describe in more detail the topology in $(\mathcal{S})_{-\rho}(\mathbb{H})$, which is defined as the topology of uniform convergence on bounded subsets of $(\mathcal{S})_{\rho}$. For this we need the notion of boundedness in the space $(\mathcal{S})_{-\rho}(\mathbb{H})$, which is defined in a similar way as in $(\mathcal{S})_{\rho}$.

Definition 6.2.1 *A set* $\mathcal{M} \subseteq (\mathcal{S})_{-\rho}(\mathbb{H})$ *is called bounded if for any sequence* $\{\Phi_n\} \subseteq \mathcal{M}$ *and for any* $\{\varepsilon_n\} \subset \mathbb{R}$ *convergence* $\varepsilon_n \to 0$ *implies that* $\{\varepsilon_n \Phi_n\}$ *is convergent to zero in* $(\mathcal{S})_{-\rho}(\mathbb{H})$.

The following propositions give characterizations of bounded sets in $(\mathcal{S})_{-\rho}(\mathbb{H})$.

Proposition 6.2.1 *A set* $\mathcal{M}$ *is bounded in the space* $(\mathcal{S})_{-\rho}(\mathbb{H})$ *if and only if for any bounded* $M \subset (\mathcal{S})_{\rho}$ *the set*

$$\{\Phi[\varphi] : \Phi \in \mathcal{M}, \ \varphi \in M\}$$

is bounded in $\mathbb{H}$.

Proof. To prove the "only if" part, let $\mathcal{M}$ be a bounded subset of $(\mathcal{S})_{-\rho}(\mathbb{H})$. Suppose there exists a bounded $M \subset (\mathcal{S})_{\rho}$ such that for any $n \in \mathbb{N}$ there exist $\varphi_n \in M$ and $\Phi_n \in \mathcal{M}$ such that $\|\Phi_n[\varphi_n]\| > n$. Then we have $\sup_{k \in \mathbb{N}} \left\|\frac{1}{n}\Phi_n[\varphi_k]\right\| \geq \left\|\frac{1}{n}\Phi_n[\varphi_n]\right\| > 1$ and consequently $\{\frac{1}{n}\Phi_n\}$ is not uniformly convergent to zero on the bounded set $\{\varphi_k, \ k \in \mathbb{N}\} \subseteq M$. Thus $\{\frac{1}{n}\Phi_n\}$ is not convergent to zero in $(\mathcal{S})_{-\rho}(\mathbb{H})$.

The "if" part is evident. $\square$

242   6. Infinite-dimensional extension of white noise calculus

Proposition 6.2.2 *A set* $\mathcal{M} \subset (\mathcal{S})_{-\rho}(\mathbb{H})$ *is bounded if and only if there exist such* $p \in \mathbb{N}$ *and* $C > 0$ *that for any* $\Phi \in \mathcal{M}$ *the inequality* $\|\Phi[\varphi]\| \leq C|\varphi|_{p,\rho}$ *holds true for all* $\varphi \in (\mathcal{S})_{\rho}$.

Proof. First we prove the necessity of this condition. Suppose for any $p \in \mathbb{N}$ there exist $\Phi_p \in \mathcal{M}$ and $\varphi_p \in M$ such that $\|\Phi_p[\varphi_p]\| > p|\varphi_p|_{p,\rho}$. Denote

$$\psi_n := \frac{\varphi_n}{|\varphi_n|_{n,\rho}}.$$

The set $M = \{\psi_n : n \in \mathbb{N}\}$ is bounded in $(\mathcal{S})_{\rho}$ since for any $p \in \mathbb{N}$ we have $|\psi_n|_{p,\rho} = \dfrac{|\varphi_n|_{p,\rho}}{|\varphi_n|_{n,\rho}} \leq 1$ when $n > p$. By Proposition 6.2.1, the set $\{\Phi[\varphi] : \Phi \in \mathcal{M}, \varphi \in M\}$ is bounded in $\mathbb{H}$, which contradicts the inequality $\|\Phi[\psi_n]\| > n$.

To prove the sufficiency, take p and $C > 0$ so that for any $\Phi \in \mathcal{M}$ and $\varphi \in (\mathcal{S})_{\rho}$ it holds that

$$\|\Phi[\varphi]\| \leq C|\varphi|_{p,\rho}. \tag{6.2.1}$$

Take a bounded $M \subset (\mathcal{S})_{\rho}$. Since by Proposition 6.1.1 it is bounded in any $(\mathcal{S}_p)_{\rho}$, it follows from (6.2.1) that the set $\{\Phi[\varphi] : \Phi \in \mathcal{M}, \varphi \in M\}$ is bounded in $\mathbb{H}$. By Proposition 6.2.1 the assertion follows. $\qquad\square$

Proposition 6.2.3 *If a set* $\mathcal{M}$ *is bounded in* $(\mathcal{S})_{-\rho}(\mathbb{H})$, *then* $\mathcal{M} \subset (\mathcal{S}_{-p})_{-\rho}(\mathbb{H})$ *for some* $p \in \mathbb{N}$ *and* $\mathcal{M}$ *is bounded in* $(\mathcal{S}_{-p})_{-\rho}(\mathbb{H})$.

Proof. Let $\mathcal{M}$ be bounded in $(\mathcal{S})_{-\rho}(\mathbb{H})$. It follows from Proposition (6.2.2) that any $\Phi \in \mathcal{M}$ is bounded as operator from $(\mathcal{S}_p)_{\rho}$ to $\mathbb{H}$ for some $p \in \mathbb{N}$ with

$$\|\Phi\|_{\mathcal{L}((\mathcal{S}_p)_{\rho};\mathbb{H})} \leq C$$

for some $C > 0$. Denoting by $\tilde{\Phi}$ the extension of Φ by continuity to $(\mathcal{S}_p)_{\rho}$ and taking an arbitrary orthonormal basis $\{\zeta_i\}_{i=1}^{\infty}$ in $(\mathcal{S}_{p+1})_{\rho}$, we obtain

$$\|\Phi\|_{\mathcal{L}_{\text{HS}}\left((\mathcal{S}_{p+1})_{\rho};\mathbb{H}\right)}^2 = \|\tilde{\Phi}I_{p,p+1}\|_{\mathcal{L}_{\text{HS}}\left((\mathcal{S}_{p+1})_{\rho};\mathbb{H}\right)}^2 = \sum_{i=1}^{\infty}\left\|\tilde{\Phi}I_{p,p+1}\zeta_i\right\|_{\mathbb{H}}^2 \leq$$

$$\leq C^2 \sum_{i=1}^{\infty}\|I_{p,p+1}\zeta_i\|_{\mathbb{H}}^2 = C^2\|I_{p,p+1}\|_{\mathcal{L}_{\text{HS}}\left((\mathcal{S}_{p+1})_{\rho};(\mathcal{S}_p)_{\rho}\right)}. \quad\square$$

The next proposition gives a characterization of strong convergence in $(\mathcal{S})_{-\rho}(\mathbb{H})$.

Proposition 6.2.4 *Let* $\Phi_n = \sum_{\alpha}\Phi_{\alpha,n}\mathbf{h}_{\alpha}$, $\Psi = \sum_{\alpha}\Psi_{\alpha}\mathbf{h}_{\alpha} \in (\mathcal{S})_{-\rho}(\mathbb{H})$. *The following assertions are equivalent:*

(i) $\{\Phi_n\}$ *converges to* Ψ *in the space* $(\mathcal{S})_{-\rho}(\mathbb{H})$;

6.2. Analysis of $(\mathcal{S})_{-\rho}(\mathbb{H})$-valued processes 243

(ii) for any $\alpha \in \mathcal{T}$ $\lim_{n\to\infty} \|\Phi_{\alpha,n} - \Psi_\alpha\| = 0$, each Φ_n and Ψ belong to $(\mathcal{S}_{-p})_{-\rho}(\mathbb{H})$ for some $p \in \mathbb{N}$ and $\{\Phi_n\}$ is bounded in this space;

(iii) all elements of the sequence $\{\Phi_n\}$ and Ψ belong to $(\mathcal{S}_{-p})_{-\rho}(\mathbb{H})$ for some $p \in \mathbb{N}$ and $\lim_{n\to\infty} \|\Phi_n - \Psi\|_{-p,-\rho} = 0$.

Proof. $(i) \implies (ii)$. Let $\{\Phi_n\}$ converge to Ψ in the space $(\mathcal{S})_{-\rho}(\mathbb{H})$. Then for any $\alpha \in \mathcal{T}$ we have

$$\|\Phi_{\alpha,n} - \Psi_\alpha\| = \frac{1}{\alpha!}\|\Phi^{(n)}[\mathbf{h}_\alpha] - \Psi[\mathbf{h}_\alpha]\| \to 0 \quad \text{as} \quad n \to \infty.$$

By Proposition 6.1.3, $\Psi \in (\mathcal{S}_{-p})_{-\rho}$ for some $p \in \mathbb{N}$. For any bounded $M \subset (\mathcal{S}_\rho)$, for sufficiently large n, and for all $\varphi \in M$, it holds that $\|\Phi_n[\varphi] - \Phi[\varphi]\| < 1$ and consequently

$$\|\Phi_n[\varphi]\| \le 1 + \|\Psi\|_{-p,-\rho}|\varphi|_{p,\rho} \le 1 + \|\Psi\|_{-p,-\rho}C_p,$$

where $C_p = \sup_{\varphi \in M} |\varphi|_{p,\rho}$. By Proposition 6.2.1, the sequence $\{\Phi_n\}$ is bounded in $(\mathcal{S})_{-\rho}(\mathbb{H})$. It follows from Proposition 6.2.3 that the sequence belongs to some $(\mathcal{S}_{-q})_{-\rho}(\mathbb{H})$ and is bounded in it.

$(ii) \implies (iii)$. Let $\{\Phi_n\}$ and Ψ satisfy (ii). By (6.1.13) and (6.1.14), one can assume that there exists such q that for all $p > q$ the sequence $\{\Phi_n\}$ and Ψ belong to $(\mathcal{S}_{-p})_{-\rho}(\mathbb{H})$ and $\{\Phi_n\}$ are bounded by the norm of each of these spaces by some $C > 0$.

Let Index $\alpha := \max\{n \in \mathbb{N} : \alpha_n \neq 0\}$. The following estimate holds true:

$$\|\Phi_n - \Psi\|^2_{-(p+1),-\rho}$$

$$= \sum_{\text{Index}\,\alpha \le k} (\alpha!)^{1-\rho}\|\Phi_{\alpha,n} - \Psi_\alpha\|^2 (2\mathbb{N})^{-2(p+1)\alpha}$$

$$+ \sum_{\text{Index}\,\alpha > k} (\alpha!)^{1-\rho}\|\Phi_{\alpha,n} - \Psi_\alpha\|^2 (2\mathbb{N})^{-2(p+1)\alpha}$$

$$\le \max_{\text{Index}\,\alpha \le k} \left[(\alpha!)^{1-\rho}\|\Phi_{\alpha,n} - \Psi_\alpha\|^2\right] \cdot \sum_{\text{Index}\,\alpha \le k} (2\mathbb{N})^{-2(p+1)\alpha}$$

$$+ \sum_{\text{Index}\,\alpha > k} \left[(\alpha!)^{1-\rho}\left(2\|\Phi_{\alpha,n}\|^2 + 2\|\Psi_\alpha\|^2\right)(2\mathbb{N})^{-2p\alpha}\right](2\mathbb{N})^{-2\alpha}$$

$$\le \max_{\text{Index}\,\alpha \le k} \left[(\alpha!)^{1-\rho}\|\Phi_{\alpha,n} - \Psi_\alpha\|^2\right] \cdot \mathcal{A}(2p+1) + 4C^2 \cdot \sum_{\text{Index}\,\alpha > k} (2\mathbb{N})^{-2\alpha},$$

where $\mathcal{A}(2p+1)$ is defined in (6.1.11) and $\mathcal{A}(p) < \infty$ for $p > 1$.

Now for any $\varepsilon > 0$ first choose k so that

$$\sum_{\text{Index}\,\alpha > k} (2\mathbb{N})^{-2\alpha} < \frac{\varepsilon}{8C^2},$$

244   6. Infinite-dimensional extension of white noise calculus

then choose N so that for all $n > N$ it holds that

$$\max_{\text{Index } \alpha \le k} \left[(\alpha!)^{1-\rho} \|\Phi_{\alpha,n} - \Psi_\alpha\|^2 \right] < \frac{\varepsilon}{2\mathcal{A}(2p+2)}.$$

Then $\|\Phi_n - \Psi\|^2_{-(p+1),-\rho} < \varepsilon$ for all $n > N$.

$(iii) \implies (i)$ is evident. $\square$

We will understand the limit of a function $\Phi(\cdot) : \mathbb{R} \to (\mathcal{S})_{-\rho}(\mathbb{H})$ at a point $t_0 \in \mathbb{R}$ in the sense of strong convergence in the space $(\mathcal{S})_{-\rho}(\mathbb{H})$. The derivative will be defined as usual with the limit being understood in the above sense.

The next corollary follows from Proposition 6.2.4.

Corollary 6.2.1 *Let $t_0 \in (a,b)$ and $\Phi(t) = \sum_\alpha \Phi_\alpha(t)\mathbf{h}_\alpha \in (\mathcal{S})_{-\rho}(\mathbb{H})$ for all $t \in (a,b) \setminus \{t_0\}$. Let $\Psi = \sum_\alpha \Psi_\alpha \mathbf{h}_\alpha \in (\mathcal{S})_{-\rho}(\mathbb{H})$, then the next assertions are equivalent:*

(i) $\lim_{t \to t_0} \Phi(t) = \Psi$ *in the space* $(\mathcal{S})_{-\rho}(\mathbb{H})$;

(ii) $\lim_{t \to t_0} \|\Phi_\alpha(t) - \Psi_\alpha\| = 0$ *for any $\alpha \in \mathcal{T}$ and there exist $\delta > 0$, $p \in \mathbb{N}$ and $M > 0$ such that $\|\Phi(t)\|_{-p,-\rho} \le M$ for any $t \in (a,b)$ with $0 < |t-t_0| < \delta$, $\Psi \in (\mathcal{S}_{-p})_{-\rho}(\mathbb{H})$;*

(iii) *there exist $\delta > 0$, $p \in \mathbb{N}$ such that $\Phi(t) \in (\mathcal{S}_{-p})_{-\rho}(\mathbb{H})$ for any $t \in (a,b)$ with $0 < |t - t_0| < \delta$, $\Psi \in (\mathcal{S}_{-p})_{-\rho}(\mathbb{H})$ and $\lim_{t \to t_0} \|\Phi(t) - \Psi\|_{-p,-\rho} = 0$.*

The proof entirely repeats the steps of the proof of Proposition 6.2.4 and thus is omitted. Applying Corollary 6.2.1, we obtain the following statement.

Corollary 6.2.2 *Let $t_0 \in (a,b)$ and $\Phi(t) = \sum_\alpha \Phi_\alpha(t)\mathbf{h}_\alpha \in (\mathcal{S})_{-\rho}(\mathbb{H})$ for all $t \in (a,b) \setminus \{t_0\}$, then the next assertions are equivalent:*

(i) $\Phi(t)$ *is differentiable at t_0 with* $\dfrac{d}{dt}\Phi(t_0) = \Psi$;

(ii) *for any $\alpha \in \mathcal{T}$ the function $\Phi_\alpha : (a,b) \to \mathbb{H}$ is differentiable at the point t_0,*

$$\Psi := \sum_\alpha \Phi'_\alpha(t_0)\mathbf{h}_\alpha \in (\mathcal{S}_{-p})_{-\rho}(\mathbb{H})$$

and there exist $\delta > 0$, $p \in \mathbb{N}$, $C > 0$ such that $\left\| \dfrac{\Phi(t) - \Phi(t_0)}{t - t_0} \right\|_{-p,-\rho} \le C$ *for any $t \in (a,b)$ with $0 < |t - t_0| < \delta$;*

(iii) $\dfrac{d\Phi}{dt} := \lim_{t \to t_0} \dfrac{\Phi(t) - \Phi(t_0)}{t - t_0}$ *exists in the space $(\mathcal{S}_{-p})_{-\rho}(\mathbb{H})$ for some p.*

Making use of this corollary one can prove that the cylindrical Wiener process $W(t)$ defined by (6.1.19) (and, of course, Q-Wiener process $W_Q(t)$ defined by (6.1.21)) is differentiable everywhere in $\mathbb{R}$ and its derivative coincides

$$6.2. \ Analysis \ of \ (\mathcal{S})_{-\rho}(\mathbb{H})\text{-valued processes} \qquad 245$$

with the white noise $\mathbb{W}(t)$, defined by (6.1.24). This is indeed the case since for any $t_0 \in \mathbb{R}$ and any $n \in \mathbb{N}$ we have $\dfrac{d W_{\epsilon_n}}{dt}(t_0) = \mathbb{W}_{\epsilon_n}(t_0)$. Moreover, using the estimate $\sup_{t \in \mathbb{R}} |\xi_i(t)| = \mathcal{O}(i^{-\frac{1}{12}})$ from (6.1.8) we obtain

$$\left\| \frac{\mathbb{W}(t) - \mathbb{W}(t_0)}{t - t_0} \right\|_{-p,-\rho}^2 = \sum_{i,j \in \mathbb{N}} \left\| \frac{1}{t - t_0} \int_{t_0}^{t} \xi_i(\tau) \, d\tau e_j \right\|^2 (2\mathbb{N})^{-2p \epsilon_{n(i,j)}}$$

$$\leq \sum_{i,j \in \mathbb{N}} \left(\sup_{t \in \mathbb{R}} |\xi_i(t)| \right)^2 (2n(i,j))^{-2p} \leq C \sum_{i,j \in \mathbb{N}} i^{-2p - \frac{1}{6}} j^{-2p} < \infty$$

for any $p \geq 1$, which shows that condition (ii) of Corollary 6.2.2 is fulfilled.

Similarly, using the estimate $\sup_{t \in \mathbb{R}} |\xi_i(t)| = \mathcal{O}(i^{-\frac{1}{12}})$ and the well-known property of Hermite functions

$$\xi_1'(t) = \xi_2(t), \qquad \xi_i'(t) = \sqrt{\frac{i}{2}} \xi_{i-1}(t) + \sqrt{\frac{i+1}{2}} \xi_{i+1}(t), \quad i = 2, 3, \ldots,$$

which implies the estimate

$$\sup_{t \in \mathbb{R}} |\xi_i^{(n)}(t)| = \mathcal{O}\left(i^{-\frac{1}{12} + \frac{n}{2}}\right), \tag{6.2.2}$$

one can show that $\mathbb{W}(t)$ is infinitely differentiable as a $(\mathcal{S})_{-\rho}(\mathbb{H})$-valued function.

We will call a function $\Phi(\cdot) : \mathbb{R} \to (\mathcal{S})_{-\rho}(\mathbb{H})$ *integrable on a measurable set* $\mathcal{G} \subset \mathbb{R}$ if there exists $p \in \mathbb{N}$ such that for any $t \in \mathcal{G}$, $\Phi(t) \in (\mathcal{S}_{-p})_{-\rho}(\mathbb{H})$ and Φ is Bochner integrable on $\mathcal{G}$ as a function with values in the Hilbert space $(\mathcal{S}_{-p})_{-\rho}(\mathbb{H})$.

It follows from the equality (6.1.12) expressing the norm $\| \cdot \|_{-p,-\rho}$ that for any $\alpha \in \mathcal{T}$ we have the estimate

$$\|\Phi_\alpha\|_{\mathbb{H}}^2 \leq \frac{(2\mathbb{N})^{2p\alpha}}{(\alpha!)^{1-\rho}} \|\Phi\|_{-p,-\rho}^2$$

which implies that if $\Phi(t) = \sum_\alpha \Phi_\alpha(t) \mathbf{h}_\alpha$ is integrable on $\mathcal{G}$, then for any $\alpha \in \mathcal{T}$ the function $\Phi_\alpha(t)$ is Bochner integrable on $\mathcal{G}$ as an $\mathbb{H}$-valued function. Moreover, we have the following sufficient condition of integrability.

Proposition 6.2.5 *Let* $\Phi(\cdot) : \mathbb{R} \to (\mathcal{S})_{-\rho}(\mathbb{H})$ *be defined by decomposition*

$$\Phi(t) := \sum_{\alpha \in \mathcal{T}} \Phi_\alpha(t) \mathbf{h}_\alpha.$$

If for any $\alpha \in \mathcal{T}$, $\Phi_\alpha : \mathbb{R} \to \mathbb{H}$ *is square Bochner integrable on* $\mathcal{G} \subset \mathbb{R}$ *with the Lebesque measure* $P(\mathcal{G}) < \infty$, $\Phi(t) \in (\mathcal{S}_{-q})_{-\rho}(\mathbb{H})$ *for all* $t \in \mathcal{G}$, *and*

$$\sum_\alpha (\alpha!)^{1-\rho} \int_{\mathcal{G}} \|\Phi_\alpha(t)\|_{\mathbb{H}}^2 dt \, (2\mathbb{N})^{-2q\alpha} < \infty \tag{6.2.3}$$

246   *6. Infinite-dimensional extension of white noise calculus*

for some $q \in \mathbb{N}$, then $\Phi(t)$ is integrable on $\mathcal{G}$ and

$$\int_{\mathcal{G}} \Phi(t)\, dt = \sum_{\alpha} \int_{\mathcal{G}} \Phi_{\alpha}(t)\, dt\, \mathbf{h}_{\alpha}. \tag{6.2.4}$$

Proof. Let $\{\alpha^{(k)}\}_{n=1}^{\infty}$ be a fixed ordering of the set $\mathcal{T}$ of multiindices. Let it be such that

$$\lim_{k\to\infty} |\alpha^{(k)}| = \infty \quad \text{and} \quad \lim_{k\to\infty} \operatorname{Index} \alpha^{(k)} = \lim_{n\to\infty} \max\{n \in \mathbb{N}: \alpha_n^{(k)} \neq 0\} = \infty.$$

Since $\Phi(t) \in (\mathcal{S}_{-q})_{-\rho}(\mathbb{H})$, the sequence defined by

$$F_n(t) := \sum_{k=1}^{n} \Phi_{\alpha^{(k)}}(t)\mathbf{h}_{\alpha^{(k)}},$$

converges to $\Phi(t)$ in this space for any $t \in \mathcal{G}$. It follows from the equality

$$\|\Phi_{\alpha^{(k)}}(t)\mathbf{h}_{\alpha^{(k)}}\|_{-p,-\rho} = (\alpha!)^{\frac{1-\rho}{2}} \|\Phi_{\alpha^{(k)}}(t)\|_{\mathbb{H}} (2\mathbb{N})^{-p\alpha}$$

that any $\Phi_{\alpha^{(k)}}(t)\mathbf{h}_{\alpha^{(k)}}$, $k \in \mathbb{N}$, and consequently all $F_n(t)$ are Bochner integrable as $(\mathcal{S}_{-p})_{-\rho}(\mathbb{H})$-valued functions for any $p \in \mathbb{N} \cup \{0\}$; thus we have $\int_{\mathcal{G}} \|F_n(t)\|_{-p,-\rho}\, dt < \infty$ for any $p \in \mathbb{N} \cup \{0\}$. It is also easy to see that $\int_{\mathcal{G}} \Phi_{\alpha^{(k)}}(t)\mathbf{h}_{\alpha^{(k)}}\, dt = \int_{\mathcal{G}} \Phi_{\alpha^{(k)}}(t)\, dt\, \mathbf{h}_{\alpha^{(k)}}$ (note that the left-hand side is a Bochner integral of an $(\mathcal{S}_{-p})_{-\rho}(\mathbb{H})$-valued function and the integral in the right-hand side is a Bochner integral of an $\mathbb{H}$-valued function); thus

$$\int_{\mathcal{G}} F_n(t)\, dt = \sum_{k=1}^{n} \int_{\mathcal{G}} \Phi_{\alpha^{(k)}}(t)\, dt\, \mathbf{h}_{\alpha^{(k)}}. \tag{6.2.5}$$

Using the condition (6.2.3) we obtain

$$\int_{\mathcal{G}} \|F_n(t)\|_{-q,-\rho}\, dt \leq \sqrt{P(\mathcal{G})} \left(\int_{\mathcal{G}} \|F_n(t)\|_{-q,-\rho}^2\, dt \right)^{\frac{1}{2}}$$

$$= \sqrt{P(\mathcal{G})} \left(\sum_{k=1}^{n} ((\alpha^{(k)})!)^{1-\rho} \int_{\mathcal{G}} \|\Phi_{\alpha^{(k)}}(t)\|_{\mathbb{H}}^2\, dt\, (2\mathbb{N})^{-2q\alpha^{(k)}} \right)^{\frac{1}{2}}$$

$$\leq \sqrt{P(\mathcal{G})} \left(\sum_{k=1}^{\infty} ((\alpha^{(k)})!)^{1-\rho} \int_{\mathcal{G}} \|\Phi_{\alpha^{(k)}}(t)\|_{\mathbb{H}}^2\, dt\, (2\mathbb{N})^{-2q\alpha^{(k)}} \right)^{\frac{1}{2}} =: C,$$

where $P(\mathcal{G})$ is the Lebesque measure of $\mathcal{G}$. It follows from here that since $\|F_n(t)\|_{-q,-\rho} \to \|\Phi(t)\|_{-q,-\rho}$ for $t \in \mathcal{G}$, by the Fatou lemma, $\int_{\mathcal{G}} \|\Phi(t)\|_{-q,-\rho}\, dt < \infty$ and

$$\lim_{n\to\infty} \int_{\mathcal{G}} \|F_n(t)\|_{-q,-\rho}\, dt = \int_{\mathcal{G}} \|\Phi(t)\|_{-q,-\rho}\, dt.$$

6.3. S-transform and Wick product. Hitsuda–Skorohod integral   247

Therefore, $\Phi(t)$ is Bochner integrable on $\mathcal{G}$ as a $(\mathcal{S}_{-q})_{-\rho}(\mathbb{H})$-valued function. We also obtain $\int_{\mathcal{G}} \|\Phi(t)\|_{-q,-\rho}\, dt < C$ and we have

$$\int_{\mathcal{G}} \|F_n(t) - \Phi(t)\|_{-q,-\rho}\, dt \le \int_{\mathcal{G}} \|F_n(t)\|_{-q,-\rho}\, dt + \int_{\mathcal{G}} \|\Phi(t)\|_{-q,-\rho}\, dt \le 2M.$$

Since $\|F_n(t) - \Phi(t)\|_{-q,-\rho} \to 0$, by the Fatou lemma we deduce that

$$\int_{\mathcal{G}} \|F_n(t) - \Phi(t)\|_{-q,-\rho}\, dt \to 0$$

and by

$$\left\| \int_{\mathcal{G}} F_n(t)\, dt - \int_{\mathcal{G}} \Phi(t)\, dt \right\|_{-q,-\rho} \le \int_{\mathcal{G}} \|F_n(t) - \Phi(t)\|_{-q,-\rho}\, dt,$$

we finally obtain $\lim\limits_{n\to\infty} \int_{\mathcal{G}} F_n(t)\, dt = \int_{\mathcal{G}} \Phi(t)\, dt$ in $(\mathcal{S}_{-p})_{-\rho}(\mathbb{H})$. Therefore, (6.2.4) follows from (6.2.5). $\qquad\square$

6.3   S-transform and Wick product. Hitsuda–Skorohod integral. Main properties. Connection with Itô integral

Consider the function defined on $\mathcal{S}'$ by $\mathcal{E}_\theta(\cdot) := e^{\langle \theta, \cdot \rangle - \frac{1}{2}|\theta|_0^2}$. It is called the exponential function associated to θ, or renormalized exponential function. It plays an important role in white noise analysis, in particular in the definition of the S-transform. We have the following decomposition of $\mathcal{E}_\theta$ into the series of the stochastic Hermite polynomials:

$$\mathcal{E}_\theta = \sum_{\alpha \in \mathcal{T}} \mathcal{E}_{\alpha,\theta}\, \mathbf{h}_\alpha.$$

Let us show that for the Fourier coefficients the following equalities hold:

$$\mathcal{E}_{\alpha,\theta} = \frac{1}{\alpha!} \prod_{i=1}^{\infty} (\theta, \xi_i)_{L^2(\mathbb{R})}^{\alpha_i}. \tag{6.3.1}$$

We use the formula (6.1.4) and the following calculation. Take $\theta \in \mathcal{S}$, a stochastic Hermite polynomial $\mathbf{h}_\alpha(\cdot) = \prod_{i=1}^{n} h_{\alpha_i}(\langle \cdot, \xi_i \rangle)$, and consider the decomposition

$$\theta = \sum_{i=1}^{n} (\theta, \xi_i)_{L^2(\mathbb{R})} \xi_i + \theta^{\perp},$$

248   6. *Infinite-dimensional extension of white noise calculus*

where $\theta^\perp$ is orthogonal to $\xi_1, \ldots, \xi_n$ in $L^2(\mathbb{R})$. Then

$$
\mathcal{E}_{\alpha,\theta} = \frac{1}{\alpha!}\left(\mathcal{E}_\theta, \mathbf{h}_\alpha\right)_{(L^2)} = \frac{1}{\alpha!}\mathbf{E}\left[\mathcal{E}_\theta \mathbf{h}_\alpha\right] = \frac{1}{\alpha!}\int_{\mathcal{S}'} e^{\langle \omega, \theta\rangle - \frac{1}{2}|\theta|_0^2}\mathbf{h}_\alpha(\omega)\, d\mu(\omega)
$$

$$
= \frac{1}{\alpha!}\int_{\mathcal{S}'} \exp\Big(\sum_{i=1}^n \langle \omega, \xi_i\rangle (\theta, \xi_i)_{L^2(\mathbb{R})} + \langle \omega, \frac{\theta^\perp}{|\theta^\perp|_0}\rangle|\theta^\perp|_0 -
$$

$$
- \frac{1}{2}\Big(\sum_{i=1}^n (\theta, \xi_i)_{L^2(\mathbb{R})} + |\theta^\perp|_0^2\Big)\Big) \cdot \prod_{i=1}^n h_{\alpha_i}\big(\langle \omega, \xi_i\rangle\big)d\mu(\omega),
$$

and by the formula (6.1.4), where

$$
x_1 := (\theta, \xi_1), \ldots, \quad x_n := (\theta, \xi_n), \quad x_{n+1} := \theta^\perp / |\theta^\perp|_0,
$$

we obtain

$$
f(x_1, \ldots, x_{n+1}) := \exp\Big(\sum_{i=1}^n x_i(\theta, \xi_i)_{L^2(\mathbb{R})} + x_{n+1}|\theta^\perp|_0 -
$$

$$
- \frac{1}{2}\Big(\sum_{i=1}^n (\theta, \xi_i)_{L^2(\mathbb{R})}^2 + |\theta^\perp|_0^2\Big)\Big) \cdot \prod_{i=1}^n h_{\alpha_i}(x_i)e^{-\frac{1}{2}\sum_{i=1}^n x_i^2}.
$$

It follows that

$$
\mathcal{E}_{\alpha,\theta} = \frac{1}{\alpha!(2\pi)^{\frac{n+1}{2}}}\int_{\mathbb{R}^{n+1}} f(x_1, \ldots, x_{n+1})dx_1...dx_{n+1}
$$

$$
= \frac{1}{\alpha!}\prod_{i=1}^n \frac{1}{\sqrt{2\pi}}\int_{\mathbb{R}} \exp\Big(x_i(\theta, \xi_i)_{L^2(\mathbb{R})} - \frac{1}{2}(\theta, \xi_i)_{L^2(\mathbb{R})}^2 h_{\alpha_i}(x_i)e^{-\frac{1}{2}x_i^2}\Big)\, dx_i
$$

$$
\cdot \frac{1}{\sqrt{2\pi}}\int_{\mathbb{R}} \exp\Big(x|\theta^\perp|_0 - \frac{1}{2}|\theta^\perp|_0^2 - \frac{1}{2}x^2\Big)\, dx.
$$

Recalling that for the generating function of the Hermite polynomials

$$
\psi(x, t) := e^{xt - \frac{t^2}{2}} = \sum_{n=0}^\infty \frac{t^n}{n!}h_n(x),
$$

the following equality holds for $n = 0, 1, 2, \ldots$:

$$
\big(\psi(\cdot, t), h_n\big)_{L^2\left(\mathbb{R};\, \frac{1}{\sqrt{2\pi}}e^{-\frac{x^2}{2}}dx\right)} = \frac{1}{\sqrt{2\pi}}\int_{\mathbb{R}} e^{xt - \frac{t^2}{2}}h_n(x)e^{-\frac{x^2}{2}}\, dx = t^n,
$$

we obtain (6.3.1)

$$
\mathcal{E}_{\alpha,\theta} = \frac{1}{\alpha!}\prod_{i=1}^n (\theta, \xi_i)_{L^2(\mathbb{R})}^{\alpha_i} = \frac{1}{\alpha!}\prod_{i=1}^\infty (\theta, \xi_i)_{L^2(\mathbb{R})}^{\alpha_i}.
$$

6.3. S-transform and Wick product. Hitsuda–Skorohod integral 249

Now taking into account the following estimate for $\mathcal{E}_\theta$:

$$|\mathcal{E}_\theta|_{p,\rho} \leq 2^{\rho/2} \exp\left[(1-\rho)^{\frac{2\rho-1}{1-\rho}} |\theta|_p^{\frac{2}{1-\rho}}\right], \tag{6.3.2}$$

(see, e.g., [59]), we see that for any $\theta \in \mathcal{S}$ the exponential function $\mathcal{E}_\theta$ belongs to $(\mathcal{S})_\rho$ for any $0 \leq \rho < 1$. This allows us to define the S-transform of $\Phi \in (\mathcal{S})_{-\rho}(\mathbb{H})$, $0 \leq \rho < 1$, by the equality

$$(S\Phi)(\theta) := \Phi[\mathcal{E}_\theta], \qquad \theta \in \mathcal{S}.$$

We see that the S-transform of $\Phi \in (\mathcal{S})_{-\rho}(\mathbb{H})$ is an $\mathbb{H}$-valued function of $\theta \in \mathcal{S}$. Note that if $\Phi \in (L^2)(\mathbb{H})$, we have

$$(S\Phi)(\theta) = \int_{\mathcal{S}'} \Phi(\omega)\mathcal{E}_\theta(\omega)\, d\mu(\omega) = \mathbf{E}(\Phi\mathcal{E}_\theta) \tag{6.3.3}$$

and the equality (6.3.3) holds for all $\theta \in L^2(\mathbb{R})$.

A very important property of the exponential functions $\mathcal{E}_\theta$ is that they form a linearly dense subset in $(\mathcal{S})_\rho$ $(0 \leq \rho < 1)$ and hence in (L^2) and in any $(\mathcal{S}_p)_\rho$. It follows from here that the equality $(S\Phi)(\theta) = 0$ for all $\theta \in \mathcal{S}$ implies $\Phi = 0$. Thus any element of $(\mathcal{S})_{-\rho}$ $(0 \leq \rho < 1)$ is uniquely determined by its S-transform.

Since any $\Phi \in (\mathcal{S})_{-\rho}(\mathbb{H})$ belongs to $(\mathcal{S}_{-p})_{-\rho}(\mathbb{H})$ for some $p \in \mathbb{N}$, it follows from the estimate (6.3.2) that for any $\Phi \in (\mathcal{S})_{-\rho}(\mathbb{H})$ there exists $p \in \mathbb{N}$ such that

$$\|(S\Phi)(\theta)\| = \|\Phi[\mathcal{E}_\theta]\| \leq \|\Phi\|_{-p,-\rho}\|\mathcal{E}_\theta\|_{p,\rho}$$
$$\leq 2^{\rho/2}\|\Phi\|_{-p,-\rho} \exp\left[(1-\rho)^{\frac{2\rho-1}{1-\rho}} |\theta|_p^{\frac{2}{1-\rho}}\right]. \tag{6.3.4}$$

It happens that an estimate of this type is sufficient for an $\mathbb{H}$-valued function acting from $\mathcal{S}$ to $\mathbb{H}$ to be an S-transform of a generalized $\mathbb{H}$-valued random variable. Namely, the following characteristic theorem is true.

Theorem 6.3.1 *Let* $\Phi \in (\mathcal{S})_{-\rho}(\mathbb{H})$, $0 \leq \rho < 1$. *Then a function* $F = S\Phi$ *satisfies the conditions:*

1) *for all* $\theta, \eta \in \mathcal{S}$ *the function* $F(\theta + z\eta)$ *is an entire analytic function of* $z \in \mathbb{C}$;

2) *there exist* $C > 0, a > 0, p \in \mathbb{N}$ *such that*

$$\|F(\theta)\| \leq C \exp\left[a|\theta|_p^{\frac{2}{1-\rho}}\right], \qquad \theta \in \mathcal{S}.$$

If a function $F : \mathcal{S} \to \mathbb{H}$ *satisfies conditions 1 and 2, there exists a* unique function $\Phi \in (\mathcal{S})_{-\rho}(\mathbb{H})$ such that $F = S\Phi$ and for any q such that $e^2 \left(\frac{2a}{1-\rho} \right)^{1-\rho} \sum_{i=1}^{\infty} (2i)^{-2(q-p)} < 1$ it holds that

$$\|\Phi\|_{-q,-\rho} \leq C \left(1 - e^2 \left(\frac{2a}{1-\rho} \right)^{1-\rho} \sum_{i=1}^{\infty} (2i)^{-2(q-p)} \right)^{-1/2}.$$

We omit the proof as it almost completely repeats the one in the $\mathbb{R}$-valued case (see, e.g., [59]).

Example 6.3.1 *Consider the S-transforms of the Q-white noise and the cylindrical white noise. We have*

$$[S\mathbb{W}_Q(t)](\theta) = \mathbb{W}_Q(t)[\mathcal{E}_\theta] = \sum_{i,j\in\mathbb{N}} \xi_i(t)\sigma_j e_j (\xi_{n(i,j)}, \theta)_{L^2(\mathbb{R})}, \qquad (6.3.5)$$

$$[S\mathbb{W}(t)](\theta) = \mathbb{W}(t)[\mathcal{E}_\theta] = \sum_{i,j\in\mathbb{N}} \xi_i(t)e_j (\xi_{n(i,j)}, \theta)_{L^2(\mathbb{R})}, \qquad t \in \mathbb{R}.$$

We also have

$$\left\| [S\mathbb{W}_Q(\cdot)](\theta) \right\|^2_{L^2(\mathbb{R};\mathbb{H})} = \sum_{i,j\in\mathbb{N}} \sigma_j^2 \left| (\xi_{n(i,j)}, \theta)_{L^2(\mathbb{R})} \right|^2$$

and since the functions $\xi_i(t)e_j$, $i,j \in \mathbb{N}$, form an orthonormal basis in the space $L^2(\mathbb{R};\mathbb{H})$, we have

$$\left\| [S\mathbb{W}(\cdot)](\theta) \right\|^2_{L^2(\mathbb{R};\mathbb{H})} = \sum_{i,j\in\mathbb{N}} \left| (\xi_{n(i,j)}, \theta)_{L^2(\mathbb{R})} \right|^2 = |\theta|^2_{L^2(\mathbb{R})}.$$

Let H be another separable Hilbert space. The space $\mathcal{L}_{\mathrm{HS}}(\mathbb{H}; H)$ of all Hilbert–Schmidt operators acting from $\mathbb{H}$ to H is a separable Hilbert space; therefore, we can introduce the space $(\mathcal{S})_{-\rho}\big(\mathcal{L}_{\mathrm{HS}}(\mathbb{H}; H)\big)$ of $\mathcal{L}_{\mathrm{HS}}(\mathbb{H}; H)$-valued generalized random variables over the space $(\mathcal{S})_\rho$ of test functions in the same manner as was done in Section 4.1.

Consider $\Psi \in (\mathcal{S})_{-\rho}\big(\mathcal{L}_{\mathrm{HS}}(\mathbb{H}; H)\big)$ and $\Phi \in (\mathcal{S})_{-\rho}(\mathbb{H})$. Their S-transforms satisfy conditions 1 and 2 of the Theorem 6.3.1. For any $\theta \in \mathcal{S}$ we have $S\Psi(\theta) \in \mathcal{L}_{\mathrm{HS}}(\mathbb{H}; H)$, $S\Phi(\theta) \in \mathbb{H}$; therefore, the values of the function $F(\theta) := S\Psi(\theta)S\Phi(\theta)$ belong to H and for any $\theta, \eta \in \mathcal{S}$ the function $F(\theta + z\eta)$ of $z \in \mathbb{C}$ is analytic. We have

$$\|S\Psi(\theta)S\Phi(\theta)\|_H \leq \|S\Psi(\theta)\|_{\mathcal{L}_{\mathrm{HS}}(\mathbb{H};H)}\|S\Phi(\theta)\|_{\mathbb{H}} \leq C_1 C_2 \exp\left[(a_1 + a_2)|\theta|_p^{\frac{2}{1-\rho}}\right],$$

where C_1, C_2, a_1, a_2 are the constants from condition 2 of Theorem 6.3.1, which holds true for Ψ and Φ correspondingly (we evidently can presume these conditions to be fulfilled with the same p). Thus F is an S-transform of some generalized random variable $\Theta \in (\mathcal{S})_{-\rho}(H)$.

This justifies the following definition.

6.3. *S-transform and Wick product. Hitsuda–Skorohod integral* 251

Definition 6.3.1 *Let* $\Psi \in (\mathcal{S})_{-\rho}\big(\mathcal{L}_{HS}(\mathbb{H};H)\big)$, $\Phi \in (\mathcal{S})_{-\rho}(\mathbb{H})$ $(0 \leq \rho < 1)$. *A generalized random variable* $\Theta \in (\mathcal{S})_{-\rho}(H)$ *such that*

$$S\Theta = S\Psi S\Phi$$

is called the Wick product of Ψ *and* Φ *and is denoted by* $\Psi \diamond \Phi$.

The following equalities follow from decomposition (6.3.1):

$$S\Psi(\theta) = \sum_{\alpha \in \mathcal{T}} \Psi_\alpha \prod_{i=1}^{\infty} (\theta, \xi_i)^{\alpha_i}_{L^2(\mathbb{R})}, \qquad S\Phi(\theta) = \sum_{\alpha \in \mathcal{T}} \Phi_\alpha \prod_{i=1}^{\infty} (\theta, \xi_i)^{\alpha_i}_{L^2(\mathbb{R})},$$

where $\Psi_\alpha \in \mathcal{L}_{HS}(\mathbb{H};H)$, $\Phi_\alpha \in \mathbb{H}$. It follows that

$$S\Psi(\theta)S\Phi(\theta) = \sum_{\gamma \in \mathcal{T}} \left(\sum_{\alpha+\beta=\gamma} \Psi_\alpha \Phi_\beta \right) \prod_{i=1}^{\infty} (\theta, \xi_i)^{\gamma_i}_{L^2(\mathbb{R})}.$$

By the uniqueness of the S-transform we obtain

$$\Psi \diamond \Phi = \sum_{\gamma \in \mathcal{T}} \left(\sum_{\alpha+\beta=\gamma} \Psi_\alpha \Phi_\beta \right) \mathbf{h}_\gamma.$$

Let Q be a positive trace class operator in $\mathbb{H}$ defined by (6.1.20), where $\{e_j\}$ is the fixed orthonormal basis in $\mathbb{H}$ consisting of eigenvectors of Q with $Qe_j = \sigma_j^2 e_j$. Recall the space $\mathbb{H}_Q$ defined in Section 4.1 as the space $Q^{\frac{1}{2}}(\mathbb{H})$ endowed with the scalar product $(u, v)_{\mathbb{H}_Q} = (Q^{-\frac{1}{2}}u, Q^{-\frac{1}{2}}v)_{\mathbb{H}}$.

The following proposition shows that under a condition on decreasing of the sequence $\{\sigma_j^2\}$, along with $\mathbb{W}(t) \in (\mathcal{S})_{-\rho}(\mathbb{H})$, we have $\mathbb{W}(t) \in (\mathcal{S})_{-\rho}(\mathbb{H}_Q)$.

Proposition 6.3.1 *For any* $t \in \mathbb{R}$ *and any positive* $Q \in \mathcal{L}_{Tr}(\mathbb{H})$ *it holds that* $\mathbb{W}_Q(t) \in (\mathcal{S})_{-\rho}(\mathbb{H}_Q)$ *for all* $\rho \in [0,1)$. *If, moreover, the condition*

$$\sum_{j=1}^{\infty} \sigma_j^{-2} j^{-2p} < \infty \quad \text{for some} \ \ p \in \mathbb{N} \tag{6.3.6}$$

holds, then $\mathbb{W}(t) \in (\mathcal{S})_{-\rho}(\mathbb{H}_Q)$ *for all* $t \in \mathbb{R}$.

Proof. The first assertion follows from the estimate

$$\|\mathbb{W}^Q_{\epsilon_{n(i,j)}}\|^2_{\mathbb{H}_Q}(2\mathbb{N})^{-2p\epsilon_{n(i,j)}} = |\xi_i(t)|^2 (2n(i,j))^{-2p} \leq \frac{|\xi_i(t)|^2}{(2ij)^{2p}} = \mathcal{O}\big(i^{-2p-\frac{1}{2}} j^{-2p}\big).$$

The second assertion follows from the estimate

$$\|\mathbb{W}_{\epsilon_{n(i,j)}}\|^2_{\mathbb{H}_Q}(2\mathbb{N})^{-2p\epsilon_{n(i,j)}} = |\xi_i(t)|^2 \sigma_j^{-2} (2n(i,j))^{-2p}$$

$$\leq \frac{|\xi_i(t)|^2}{\sigma_j^2 (2ij)^{2p}} = \mathcal{O}\big(\sigma_j^{-2} i^{-2p-\frac{1}{2}} j^{-2p}\big). \qquad \square$$

252   6. Infinite-dimensional extension of white noise calculus

We will use this property of $\mathbb{W}(t)$ in the construction of solutions to stochastic Cauchy problems with singular white noise.

Let again H be another separable Hilbert space. Consider $\mathcal{L}(\mathbb{H}; H)$, the space of all linear bounded operators from $\mathbb{H}$ to H. Since it is not a separable Hilbert space, we cannot introduce the space of $\mathcal{L}(\mathbb{H}; H)$-valued generalized random variables as the space $(\mathcal{S})_{-\rho}\big(\mathcal{L}_{\mathrm{HS}}(\mathbb{H}; H)\big)$ above. Nevertheless, we will introduce the notion of a generalized operator-valued random variable due to the following proposition.

Proposition 6.3.2 *Any generalized $\mathcal{L}(\mathbb{H}; H)$-valued random variable Φ belongs to $(\mathcal{S})_{-\rho}\big(\mathcal{L}_{\mathrm{HS}}(\mathbb{H}_Q; H)\big)$.*

Proof. First note that by the same argument as in the proof of Proposition 6.1.2 one can show that any generalized $\mathcal{L}(\mathbb{H}; H)$-valued random variable Φ belongs to $\mathcal{L}\big((\mathcal{S}_p)_\rho; \mathcal{L}(\mathbb{H}; H)\big)$ for some $p \in \mathbb{N}$ and thus we have

$$\|\Phi[\varphi]\|_{\mathcal{L}_{\mathrm{HS}}(\mathbb{H}_Q; H)} \leq \|\Phi\|_{\mathcal{L}\big((\mathcal{S}_p)_\rho; \mathcal{L}(\mathbb{H}; H)\big)} \sqrt{\sum_{j=1}^{\infty} \sigma_j^2 \cdot \|\varphi\|_{p,\rho}}, \qquad \varphi \in (\mathcal{S})_\rho.$$

It follows that Φ is a continuous operator from $(\mathcal{S})_\rho$ to $\mathcal{L}_{\mathrm{HS}}(\mathbb{H}_Q; H)$. $\square$

This justifies the following definition of a generalized operator-valued random variable.

Definition 6.3.2 *A linear continuous operator $\Phi : (\mathcal{S})_\rho \to \mathcal{L}(\mathbb{H}; H)$ is called a generalized $\mathcal{L}(\mathbb{H}; H)$-valued random variable.*

It follows from Propositions 6.3.1 and 6.3.2 that for any generalized $\mathcal{L}(\mathbb{H}; H)$-valued random process $\Phi(t)$, the Wick product $\Phi(t) \diamond \mathbb{W}_Q(t)$ is well defined for all t and belongs to the space $(\mathcal{S})_{-\rho}(\mathbb{H})$ since we can consider $\Phi(t)$ as an $(\mathcal{S})_{-\rho}\big(\mathcal{L}_{\mathrm{HS}}(\mathbb{H}_Q; \mathbb{H})\big)$ valued process and $W(t)$ as an $(\mathcal{S})_{-\rho}\big(\mathbb{H}_Q\big)$ valued one.

Taking the operator Q under the condition (6.3.6) and again considering $\Phi(t)$ as an $(\mathcal{S})_{-\rho}\big(\mathcal{L}_{\mathrm{HS}}(\mathbb{H}_Q; \mathbb{H})\big)$-valued process, we obtain that the Wick product $\Phi(t) \diamond \mathbb{W}(t)$ is also well defined and belongs to the space $(\mathcal{S})_{-\rho}(\mathbb{H})$ for all $t \in \mathbb{R}$. This justifies the following definition.

Definition 6.3.3 *We will call a generalized $\mathcal{L}(\mathbb{H}; H)$-valued random process $\Phi(t)$ Hitsuda–Skorohod integrable with respect to the Q-white noise $\mathbb{W}_Q(t)$ or singular white noise $\mathbb{W}(t)$ on $[0, T]$ if $\Phi(t) \diamond \mathbb{W}_Q(t)$ or $\Phi(t) \diamond \mathbb{W}(t)$, respectively, is integrable on $[0, T]$ as an $(\mathcal{S})_{-\rho}(H)$-valued function. In such a case we will refer to the integrals*

$$\int_0^T \Phi(t) \diamond \mathbb{W}_Q(t)\, dt \quad and \quad \int_0^T \Phi(t) \diamond \mathbb{W}(t)\, dt$$

as Hitsuda–Skorohod integrals of $\Phi(t)$.

6.4. Generalized solutions to stochastic Cauchy problems   253

The following result establishes the relationship between abstract Itô integrals and Hitsuda–Skorohod integrals.

Theorem 6.3.2 *For any predictable $\mathcal{L}_{HS}(\mathbb{H}_Q; H)$-valued process Φ satisfying the condition*

$$\mathbf{E}\left[\int_0^T \|\Phi(t)\|^2_{\mathcal{L}_{HS}(\mathbb{H}_Q;H)}\, dt\right] < \infty, \tag{6.3.7}$$

any Q-Wiener process W, and the corresponding Q-white noise $\mathbb{W}$, it holds that

$$\int_0^T \Phi(t)\, dW(t) = \int_0^T \Phi(t) \diamond \mathbb{W}(t)\, dt. \tag{6.3.8}$$

For any predictable $\mathcal{L}_{HS}(\mathbb{H}; H)$-valued process Φ satisfying the condition

$$\mathbf{E}\left[\int_0^T \|\Phi(t)\|^2_{\mathcal{L}_{HS}(\mathbb{H};H)}\, dt\right] < \infty,$$

the equality (6.3.8) holds for any cylindrical Wiener process W and the corresponding singular white noise $\mathbb{W}$.

We will prove this important result for the special case of a deterministic integrand Φ in the next section while studying solutions to stochastic linear equations in spaces of generalized Hilbert space valued random processes and the relationship between these solutions and weak solutions constructed in Chapter 4. The proof of Theorem 6.3.2 in the general case, which is due to [6], will be presented in the conclusion of the section.

6.4 Generalized solutions to stochastic Cauchy problems in spaces of abstract stochastic distributions

In the present section we construct generalized wrt ω solutions to the problem

$$X'(t) = AX(t) + B\mathbb{W}(t), \quad t \geq 0, \quad X(0) = \zeta, \tag{6.4.1}$$

with additive and multiplicative $(B = B(t, X))$ white noise defined in spaces of generalized Hilbert space valued random variables (or abstract stochastic distributions) introduced in Section 6.2.

To obtain generalized solutions for the problem with a white noise (Q-white noise or singular white noise) in the spaces of generalized Hilbert space valued random variables, we use the technique of Fourier series wrt stochastic polynomials, Wick products, and Hitsuda–Skorohod integrals introduced in these spaces. In the construction of generalized wrt ω solutions, a stochastic convolution, being the main part of the solutions, is again needed. This is

254   *6. Infinite-dimensional extension of white noise calculus*

similar to the case of stochastic problems in the integral form and the case of generalized wrt t stochastic problems in differential form. The stochastic convolution here is defined with the help of Hitsuda–Skorohod integrals.

In the case of the linear Cauchy problem with additive noise we also study the relationship between generalized wrt ω solutions and weak solutions of the corresponding integral problem.

6.4.1 Equations with additive noise. Example: stochastic heat equation

In order to consider stochastic differential equations in Hilbert spaces as differential equations in spaces of generalized Hilbert space valued random variables, we first extend the action of linear operators acting in separable Hilbert spaces to the corresponding spaces of generalized random variables.

Let first $A \in \mathcal{L}(H_1, H_2)$, where H_1 and H_2 are separable Hilbert spaces. Define its action as an operator from $(\mathcal{S})_{-\rho}(H_1)$ to $(\mathcal{S})_{-\rho}(H_2)$ by the equality

$$A\Phi := \sum_{\alpha \in \mathcal{T}} A\Phi_\alpha \mathbf{h}_\alpha \quad \text{for} \quad \Phi = \sum_{\alpha \in \mathcal{T}} \Phi_\alpha \mathbf{h}_\alpha \in (\mathcal{S})_{-\rho}(H_1). \tag{6.4.2}$$

Defined in such a way, A becomes a linear continuous operator acting from $(\mathcal{S})_{-\rho}(H_1)$ to $(\mathcal{S})_{-\rho}(H_2)$.

If A is unbounded, define $(dom\, A) \subseteq (\mathcal{S})_{-\rho}(H_1)$, $\rho \in [0, 1)$, as the set of all $\sum_{\alpha \in \mathcal{T}} \Phi_\alpha \mathbf{h}_\alpha \in (\mathcal{S})_{-\rho}(H_1)$ such that $\Phi_\alpha \in dom\, A$ for all $\alpha \in \mathcal{T}$ and the condition

$$\sum_{\alpha \in \mathcal{T}} (\alpha!)^{1-\rho} \|A\Phi_\alpha\|_{H_2}^2 (2\mathbb{N})^{-2p\alpha} < \infty$$

holds true for some $p \in \mathbb{N}$.

Then the equality (6.4.2) defines on $(dom\, A)$ a linear operator acting from $(\mathcal{S})_{-\rho}(H_1)$ to $(\mathcal{S})_{-\rho}(H_2)$. It is easy to check its closedness for A being closed as an operator from H_1 to H_2. As a consequence, the following is true.

Proposition 6.4.1 *Let A be a linear closed operator from H_1 to H_2. Then for any $\Phi \in (dom\, A) \subseteq (\mathcal{S})_{-\rho}(H_1)$, we have $\big[S\Phi\big](\theta) \in dom\, A \subseteq H_1$ and*

$$\big[SA\Phi\big](\theta) = A\big[S\Phi\big](\theta), \qquad \theta \in \mathcal{S}.$$

Now let $\mathbb{H}$ and H be separable Hilbert spaces, A be a closed linear operator acting in H, and $B \in \mathcal{L}(\mathbb{H}, H)$. Consider the stochastic Cauchy problem studied in Chapter 4, which is written in the integral form with the Itô integral as follows:

$$X(t) = \int_0^t AX(t)\, dt + \int_0^t B\, dW(t), \quad t \geq 0, \qquad X(0) = \zeta,$$

where $W(t)$ is an $\mathbb{H}$-valued Q-Wiener or cylindrical Wiener process. It is written via differentials:

$$dX(t) = AX(t)dt + BdW(t), \quad t \geq 0, \qquad X(0) = \zeta. \tag{6.4.3}$$

6.4. Generalized solutions to stochastic Cauchy problems

It follows from the relationship between the Itô and the Hitsuda–Skorohod integrals formulated in Theorem 6.3.2 that in the spaces of generalized Hilbert space valued random variables the problem (6.4.3) can be written as

$$X(t) = \int_0^t AX(t)\,dt + \int_0^t B \diamond \mathbb{W}(t)\,dt, \qquad t \geq 0. \tag{6.4.4}$$

We begin our study of stochastic Cauchy problems in the spaces of generalized Hilbert space valued random variables with the linear equation with additive noise. Taking into account that in this case $B \diamond \mathbb{W}(t) = B\mathbb{W}(t)$ since B is deterministic, we differentiate Equation (6.4.4) and obtain that the Cauchy problem (6.4.3) in the spaces introduced above of generalized random variables takes the form

$$X'(t) = AX(t) + B\mathbb{W}(t), \quad t \geq 0, \qquad X(0) = \zeta, \tag{6.4.5}$$

where $\mathbb{W}(t)$, $t \geq 0$, is the $\mathbb{H}$-valued Q-white noise or singular white noise defined by (6.1.23)–(6.1.24). The Cauchy problem (6.4.5) formally looks like the generalized wrt t Cauchy problem (6.4.1); nevertheless, the white noise and hence a solution to the problem have different senses here. In particular, the white noise $\mathbb{W}(t)$ is defined for any $t \geq 0$, in contrast to the case of $\mathbb{W}(\cdot)$ in the space of distributions $\mathcal{D}_0'(\mathbb{H})$.

Now we establish the existence and uniqueness of a solution to this problem in the space $(\mathcal{S})_{-\rho}(H)$, i.e., the existence and uniqueness of an $(\mathcal{S})_{-\rho}(H)$-valued differentiable function $X(t)$, $t \geq 0$, satisfying (6.4.5).

Theorem 6.4.1 *Let A be the generator of a C_0-semigroup $\{U(t), t \geq 0\}$ in a separable Hilbert space H, $\mathbb{W}$ be a singular white noise in $(\mathcal{S})_{-\rho}(\mathbb{H})$, where $\mathbb{H}$ is another separable Hilbert space, and $B \in \mathcal{L}(\mathbb{H}, H)$. Then*

$$X(t) = U(t)\zeta + \int_0^t U(t-s)B\mathbb{W}(s)\,ds \tag{6.4.6}$$

is the unique solution to the Cauchy problem (6.4.5) in the space $(\mathcal{S})_{-\rho}(H)$ for any $\zeta \in (\mathrm{dom}\,A)$.

Proof. Let

$$X(t) = \sum_\alpha X_\alpha(t)\mathbf{h}_\alpha \in (\mathcal{S})_{-\rho}(H), \qquad \zeta = \sum_\alpha \zeta_\alpha \mathbf{h}_\alpha \in (\mathrm{dom}\,A).$$

The process $\{X(t), t \geq 0\}$ is a solution of the problem (6.4.5) only if the functions $X_\alpha(t)$ are the solutions of the following Cauchy problems in the space H:

$$X'_{\epsilon_n}(t) = AX_{\epsilon_n}(t) + B\mathbb{W}_{\epsilon_n}(t), \quad t \geq 0, \qquad X_{\epsilon_n}(0) = \zeta_{\epsilon_n}, \tag{6.4.7}$$

for any $\alpha = \epsilon_n$, where $n \in \mathbb{N}$, and

$$X'_\alpha(t) = AX_\alpha(t), \quad t \geq 0, \qquad X_\alpha(0) = \zeta_\alpha, \tag{6.4.8}$$

256   6. *Infinite-dimensional extension of white noise calculus*

for any $\alpha \neq \epsilon_n$.

Since A is the generator of a C_0-semigroup, the corresponding homogeneous Cauchy problem is well-posed (see Section 1.1) and $X_\alpha(t) := U(t)\zeta_\alpha$, $t \geq 0$, is the unique solution to the problem (6.4.8) for each $\alpha \neq \epsilon_n$, $n \in \mathbb{N}$, since $\zeta_\alpha \in dom\, A$ for all $\alpha \in \mathcal{T}$.

For any $n \in \mathbb{N}$, by (6.1.24), we have $B\mathbb{W}_{\epsilon_n}(t) = Be_j\xi_i(t)$, where $i, j \in \mathbb{N}$ are such that $n = n(i, j)$. Hence it is infinitely differentiable for all $t \in \mathbb{R}$. Therefore, the function

$$v_n(t) := \int_0^t U(t - s)B\mathbb{W}_{\epsilon_n}(s)\, ds = \int_0^t U(s)B\mathbb{W}_{\epsilon_n}(t - s)\, ds$$

is differentiable and by the well-known properties of C_0-semigroups is continuous and belongs to $dom\, A$ for $t \geq 0$. Thus the Cauchy problem (6.4.7) has the unique solution given by

$$X_{\epsilon_n}(t) = U(t)\zeta_{\epsilon_n} + \int_0^t U(t - s)B\mathbb{W}_{\epsilon_n}(s)\, ds, \qquad n \in \mathbb{N}. \tag{6.4.9}$$

Consider $X(t) = \sum_\alpha X_\alpha(t)\mathbf{h}_\alpha$ with $X_\alpha(t) = U(t)\zeta_\alpha$, $t \geq 0$, for $\alpha \neq \epsilon_n$, $n \in \mathbb{N}$, and $X_\alpha(t)$ defined by (6.4.9) for all the other $\alpha \in \mathcal{T}$. Let us show that $X(t) \in (\mathcal{S})_{-0}(H)$ and the equality (6.4.6) is true.

Since U is a C_0-semigroup, there exist $M > 0$ and $a \in \mathbb{R}$ such that

$$\|U(t)\| \leq Me^{at}, \qquad t \geq 0. \tag{6.4.10}$$

It follows from the estimate

$$\int_0^t \left\|U(t - s)B\mathbb{W}_{\epsilon_n}(s)\right\|_H^2 ds \leq M^2\|B\|^2 \int_0^t e^{2a(t-s)}|\xi_{i(n)}(s)|^2\, ds \leq M^2\|B\|^2 e^{2at}$$

that for $p \geq 1$ we have

$$\sum_{n \in \mathbb{N}} \int_0^t \left\|U(t - s)B\mathbb{W}_{\epsilon_n}(s)\right\|_H^2 ds\, (2\mathbb{N})^{-2p\epsilon_n} < \infty.$$

By Proposition 6.2.5, it follows from here that the integral $\int_0^t U(t-s)B\mathbb{W}(s)ds$ exists as an element of $(\mathcal{S})_{-0}(H)$ for all $t \geq 0$ and

$$\int_0^t U(t - s)B\mathbb{W}(s)ds = \sum_{n=1}^\infty \int_0^t U(t - s)B\mathbb{W}_{\epsilon_n}(s)\, ds.$$

We evidently have $U(t)\zeta = \sum_{\alpha \in \mathcal{T}} U(t)\zeta_\alpha\mathbf{h}_\alpha \in (\mathcal{S})_{-0}(H)$. Thus, for $X_\alpha(t)$ defined by (6.4.7)–(6.4.8), $X(t) = \sum_\alpha X_\alpha(t)\mathbf{h}_\alpha \in (\mathcal{S})_{-0}(H)$, $t \geq 0$, and the equality (6.4.6) holds true.

To complete the proof, it is sufficient to show that $X(t)$ is differentiable for

6.4. Generalized solutions to stochastic Cauchy problems 257

$t \geq 0$. Then the equality in (6.4.5) follows from (6.4.7), (6.4.8), and closedness of A.

Let $t \in [0, T)$; then, since $\zeta_\alpha \in dom\, A$ for any $\alpha \in \mathcal{T}$, we have

$$\left\| \frac{U(t+h)\zeta_\alpha - U(t)\zeta_\alpha}{h} \right\| = \frac{1}{|h|} \left\| \int_t^{t+h} U(s)A\zeta_\alpha\, ds \right\| \leq Me^{aT} \|A\zeta_\alpha\|.$$

Since $\zeta \in dom\, A \subset (\mathcal{S})_{-0}(H)$, we have

$$\|A\zeta\|_{-p,-0}^2 = \sum_{\alpha \in \mathcal{T}} (\alpha!)\|A\zeta_\alpha\|^2 (2\mathbb{N})^{-2p\alpha} < \infty$$

for some $p \in \mathbb{N}$; thus, for all $h \in \mathbb{R}$ such that $t + h \in [0, T]$, we have

$$\left\| \frac{U(t+h)\zeta - U(t)\zeta}{h} \right\| \leq Me^{aT} \|A\zeta\|_{-p,-0}. \qquad (6.4.11)$$

We also have

$$\left\| \frac{1}{h} \left(\int_0^{t+h} U(t+h-s)\mathbb{W}_{\epsilon_{n(i,j)}}(s)\, ds - \int_0^t U(t-s)\mathbb{W}_{\epsilon_n}(s)\, ds \right) \right\|$$

$$= \frac{1}{|h|} \left\| \int_t^{t+h} U(s)\xi_i(t+h-s)Be_j\, ds \right.$$

$$\left. + \int_0^t U(s)\big(\xi_i(t+h-s) - \xi_i(t-s)\big)Be_j \right\| \leq$$

$$\leq Me^{aT} \|B\| \Big(\sup_{[0,T]} |\xi_i(t)| + T \sup_{[0,T]} |\xi_i'(t)| \Big) = \mathcal{O}(i^{\frac{5}{12}})$$

by the estimate (6.2.2), uniformly with respect to h such that $t + h \in [0, T]$. It follows from here that

$$\left\| \frac{1}{h} \left(\int_0^{t+h} U(t+h-s)B\mathbb{W}(s)\, ds - \int_0^t U(t-s)B\mathbb{W}(s)\, ds \right) \right\|_{-p,\rho} \leq C,$$

$$\qquad (6.4.12)$$

for $p \geq 2$, some $C > 0$, and h such that $t + h \in [0, T]$. It follows from (6.4.11) and (6.4.12) that $\left\| \dfrac{X_\alpha(t+h) - X_\alpha(t)}{h} \right\|_{-p,-0}$ is bounded for arbitrary $t + h \in [0, T]$ and for some $p \in \mathbb{N}$. By Corollary 6.2.2, it follows from here and the differentiability of all $X_\alpha,\ \alpha \in \mathcal{T}$, that $X'(t)$ exists. $\qquad \square$

Example 6.4.1 *Consider the following Cauchy problem for the heat equation:*

$$\frac{\partial u(t,x)}{\partial t} = \triangle u(t,x) =: Au(t,x), \qquad t \geq 0, \quad x = (x_1, \dots, x_m) \in \mathcal{G} \subset \mathbb{R}^m,$$

$$u(t,x) = 0, \qquad t \geq 0, \quad x \in \partial\mathcal{G},$$

$$u(0,x) = \zeta(x), \qquad x \in \mathcal{G},$$

258   *6. Infinite-dimensional extension of white noise calculus*

and the following stochastic perturbation of the problem:

$$\frac{dX(t)}{dt} = AX(t) + \mathbb{W}(t), \qquad u(0) = \zeta. \tag{6.4.13}$$

By $\partial\mathcal{G}$ we denote the boundary of $\mathcal{G} \subset \mathbb{R}^m$. Then $A = \Delta$ is the operator in the Hilbert space $\mathbb{H} = L^2(\mathcal{G})$ with

$$dom\, A = \left\{ u \in L^2(\mathcal{G}) : u \in \mathcal{H}^2(\mathcal{G}) \cap \mathcal{H}_0^1(\mathcal{G}) \right\},$$

where $\mathcal{H}^2$ and $\mathcal{H}_0^1$ are Sobolev spaces. Suppose $\mathcal{G} = [0,1]^m$. In this case the set of functions

$$\left\{ \phi_{n_1,\dots,n_m}(x_1,\dots,x_m) := 2^{\frac{m}{2}} \prod_{k=1}^m \sin\left(\pi n_k x_k\right), \quad n_1,\dots,n_m \in \mathbb{N} \cup \{0\} \right\}$$

consists of eigenfunctions of A and forms an orthonormal basis in $\mathbb{H}$. The corresponding eigenvalues

$$\left\{ -\sum_{k=1}^m \pi^2 n_k^2, \quad n_1,\dots,n_m \in \mathbb{N} \cup \{0\} \right\}$$

form its spectrum. (Compare with Example 4.2.2, where we considered the backward Cauchy problem.)

Let us fix an ordering of the sets of the eigenfunctions and eigenvalues, and denote them by $\{e_j\}_{j=1}^\infty$ and $\{\lambda_j\}_{j=1}^\infty$, respectively. Operator A generates a C_0-semigroup, given by the formula

$$U(t)u = \sum_{j=1}^\infty e^{\lambda_j t}(e_j, u)_{\mathbb{H}} e_j.$$

By Theorem 6.4.1, the problem (6.4.13) has a unique solution in the space $(\mathcal{S})_{-\rho}(\mathbb{H})$ and we have the explicit formula (6.4.6) for it. Thus we obtain

$$X(t) = \sum_{j=1}^\infty e^{\lambda_j t}(e_j, \zeta)_{\mathbb{H}} e_j + \sum_{i,j=1}^\infty \int_0^t e^{\lambda_j(t-s)} \xi_i(s)\, ds\, \mathbf{h}_{\epsilon_{n(i,j)}} e_j.$$

Consider the norm of $X(t)$ in $(\mathcal{S}_{-p})_{-\rho}(H)$. We have

$$\|X(t)\|_{-p,-\rho}^2 = \|U(t)\zeta\|_{\mathbb{H}}^2 + \sum_{i,j \in \mathbb{N}} \left| \int_0^t e^{\lambda_j(t-s)} \xi_i(s)\, ds \right|^2 \left(2n(i,j)\right)^{-2p}. \tag{6.4.14}$$

It is easy to see that it is finite for any $p \geq 1$; thus the solution lies in $(\mathcal{S}_{-1})_{-0}$.

Note that since we have

$$\sum_{i \in \mathbb{N}} \left| \int_0^t e^{\lambda_j(t-s)} \xi_i(s)\, ds \right|^2 = \left\| e^{\lambda_j(t-\cdot)} 1_{[0,t]} \right\|_{L^2(\mathbb{R})}^2$$

$$= \int_0^t e^{2\lambda_j(t-s)}\, ds = \frac{1 - e^{2\lambda_j t}}{2|\lambda_j|} \leq \frac{1}{2|\lambda_j|},$$

6.4. Generalized solutions to stochastic Cauchy problems 259

the series in the right-hand side of equality (6.4.14) converges for $p = 0$ and $\rho = 0$ only if $m = 1$. Hence this is the only case when the solution belongs to the space $(L^2)(H) = (\mathcal{S}_{-0})_{-0}(H)$.

6.4.2 Equations with multiplicative noise. Example: equation of age structured population

Let H and $\mathbb{H}$ be separable Hilbert spaces, A be a linear closed operator acting in H, $B(\cdot) \in \mathcal{L}(H, \mathcal{L}(\mathbb{H}; H))$, $\zeta \in (dom\, A) \subseteq (\mathcal{S})_{-\rho}(H)$. Consider the Cauchy problem

$$dX(t) = AX(t)\, dt + B(t, X(t))\, dW(t), \quad t \geq 0, \qquad X(0) = \zeta,$$

where $W(t)$ is the $\mathbb{H}$-valued cylindrical Wiener process. It corresponds to the following Itô integral equation:

$$X(t) = \zeta + \int_0^t AX(s)\, ds + \int_0^t B(s, X(s))\, dW(s), \quad t \geq 0.$$

Replacing the Itô integral with the Hitsuda–Skorohod integral and differentiating with respect to t, we come to the Cauchy problem

$$\frac{dX(t)}{dt} = AX(t) + B(t, X(t)) \diamond \mathbb{W}(t), \quad t \geq 0, \qquad X(0) = \zeta. \qquad (6.4.15)$$

We will study the existence and uniqueness of its solution in the space $(\mathcal{S})_{-\rho}(H)$, where $\rho \in [0, 1)$, i.e., the existence and uniqueness of an $(\mathcal{S})_{-\rho}(H)$-valued differentiable function satisfying (6.4.15). Note that if Q is a nuclear operator acting in $\mathbb{H}$ and satisfying the condition of Proposition 6.3.1 for some $p \in \mathbb{N}$, it follows from the fact that, for any $X(t) \in (\mathcal{S})_{-\rho}(H)$ we have $B(X(t)) \in (\mathcal{S})_{-\rho}(\mathcal{L}_{\mathrm{HS}}(\mathbb{H}_Q; \mathbb{H}))$, that the Wick product in Equation (6.4.15) is well defined.

Applying the S-transform to the problem (6.4.15), we obtain the following problem:

$$\frac{d}{dt}\hat{X}(t, \theta) = A\hat{X}(t, \theta) + B(\hat{X}(t, \theta))\hat{\mathbb{W}}(t, \theta), \quad t \geq 0, \quad \theta \in \mathcal{S}, \quad \hat{X}(0, \theta) = \hat{\zeta}(\theta),$$
$$(6.4.16)$$

where $\hat{X}(t, \theta) = S[X(t)](\theta)$, $\hat{\mathbb{W}}(t, \theta) = S[\mathbb{W}(t)](\theta)$, and $\hat{\Phi}(\theta) = S\Phi(\theta)$.

We will suppose later that the operator B in the equation satisfies the following condition:

(B) operator $C(\cdot) : dom\, A \to \mathcal{L}(H)$, defined by the equality

$$C(x)y := AB(x)y - B(Ax)y, \quad x \in dom\, A, \quad y \in \mathbb{H},$$

is bounded.

260   6. Infinite-dimensional extension of white noise calculus

Note that, by the uniform boundedness principle, it follows from condition (B) that there exists $M_{AB} > 0$ such that the following estimate holds true:

$$\|C(x)y\| \leq M_{AB}\|x\| \cdot \|y\|, \quad x \in dom\, A, \quad y \in \mathbb{H}. \tag{6.4.17}$$

Let A be the generator of a C_0-semigroup $\{U(t),\, t \geq 0\}$ and $M > 0$ and $a \in \mathbb{R}$ be such that (6.4.10) holds. To prove the existence of a solution of the problem (6.4.16) we introduce a sequence of linear operators $\{T_k(t,\theta)\}$, $t \geq 0$, $\theta \in \mathcal{S}$, as follows:

$$T_0(t,\theta) = U(t),$$

$$T_k(t,\theta)x = \int_0^t U(t-s)B\big(T_{k-1}(s,\theta)x\big)\hat{W}(s,\theta)\,ds, \quad x \in H, \quad k = 1,2,\dots.$$

To obtain the main result on existence we need some lemmas describing the properties of T_k.

Lemma 6.4.1 *For any $t \geq 0$, $\theta \in \mathcal{S}$, and $k \in \mathbb{N} \cup \{0\}$ the following estimate holds true:*

$$\|T_k(t,\theta)\|_{\mathcal{L}(H)} \leq M^{k+1}\|B\|^k e^{at}|\theta|_{L^2(\mathbb{R})}^k \sqrt{\frac{t^k}{k!}}, \tag{6.4.18}$$

where $M > 0$ and $a \in \mathbb{R}$ are the constants from the estimate (6.4.10) and $\|B\| = \|B\|_{\mathcal{L}(H,\mathcal{L}(\mathbb{H};H))}$.

Proof. Suppose (6.4.18) holds true for some $k \in \mathbb{N}$; then for any $x \in H$ we have

$$\|T_{k+1}(t,\theta)x\| = \left\|\int_0^t U(t-s)B\big(T_k(s,\theta)x\big)\hat{W}(s,\theta)\,ds\right\|$$

$$\leq \int_0^t \left\|U(t-s)B\big(T_k(s,\theta)x\big)\hat{W}(s,\theta)\right\|\,ds$$

$$\leq M\|B\|\int_0^t e^{a(t-s)}\|T_k(s,\theta)x\|\|\hat{W}(s,\theta)\|\,ds$$

$$\leq M^{k+2}\|B\|^{k+1}e^{at}|\theta|_{L^2(\mathbb{R})}^k \int_0^t \sqrt{\frac{s^k}{k!}}\,\|\hat{W}(s,\theta)\|\,ds\,\|x\|$$

$$\leq M^{k+2}\|B\|^{k+1}e^{at}|\theta|_{L^2(\mathbb{R})}^k \left(\int_0^t \frac{s^k}{k!}\,ds\right)^{1/2}\left(\int_0^t \|\hat{W}(s,\theta)\|^2\,ds\right)^{1/2}\|x\|$$

$$\leq M^{k+2}\|B\|^{k+1}e^{at}|\theta|_{L^2(\mathbb{R})}^k \sqrt{\frac{t^{k+1}}{(k+1)!}}\,\|\hat{W}(\cdot,\theta)\|_{L^2(\mathbb{R};H)}\,\|x\|$$

$$\leq M^{k+2}\|B\|^{k+1}e^{at}|\theta|_{L^2(\mathbb{R})}^{k+1} \sqrt{\frac{t^{k+1}}{(k+1)!}}\,\|x\|.$$

Since the estimate (6.4.18) is true for $k = 0$, it follows by induction that it is true for any $k \in \mathbb{N}$. $\qquad\square$

6.4. Generalized solutions to stochastic Cauchy problems 261

Lemma 6.4.2 *For any $t \geq 0$, $\theta \in \mathcal{S}$, $k \in \mathbb{N} \cup \{0\}$, $\zeta \in \text{dom}\, A$, it holds that*

$$\|AT_k(t,\theta)\hat{\zeta}(\theta)\| \leq M^{k+1}\|B\|^{k-1}|\theta|^k_{L^2(\mathbb{R})}e^{at}\sqrt{\frac{t^k}{k!}}$$
$$\cdot \left(\|B\| \cdot \|A\hat{\zeta}(\theta)\| + kM_{AB}\|\hat{\zeta}(\theta)\|\right), \quad (6.4.19)$$

where $M > 0$ and $a \in \mathbb{R}$ are the constants from the estimate (6.4.10), M_{AB} is the constant from the estimate (6.4.17), and $\|B\| = \|B\|_{\mathcal{L}(H,\mathcal{L}(\mathbb{H};H))}$.

Proof. For $k = 0$, using the properties of C_0-semigroups, we obtain

$$\|AT_0(t,\theta)\hat{\zeta}(\theta)\| = \|AU(t)\hat{\zeta}(\theta)\| = \|U(t)A\hat{\zeta}(\theta)\| \leq Me^{at}\|\hat{\zeta}(\theta)\|. \quad (6.4.20)$$

We further have

$$AT_k(t,\theta)\hat{\zeta}(\theta) = \int_0^t AU(t-s)B\big(T_{k-1}(s,\theta)\hat{\zeta}(\theta)\big)\hat{\mathbb{W}}(s,\theta)\,ds$$
$$= \int_0^t U(t-s)AB\big(T_{k-1}(s,\theta)\hat{\zeta}(\theta)\big)\hat{\mathbb{W}}(s,\theta)\,ds$$
$$= \int_0^t U(t-s)\left[B\big(AT_{k-1}(s,\theta)\hat{\zeta}(\theta)\big)\hat{\mathbb{W}}(s,\theta) + C\big(T_{k-1}(s,\theta)\hat{\zeta}(\theta)\big)\hat{\mathbb{W}}(s,\theta)\right]ds.$$

If (6.4.19) is true for some $k \in \mathbb{N}$, by the above representation and the estimate (6.4.18), we obtain

$$\left\|AT_{k+1}(t,\theta)\hat{\zeta}(\theta)\right\|$$
$$\leq \int_0^t Me^{a(t-s)}\left[M^{k+1}\|B\|^k|\theta|^k_{L^2(\mathbb{R})}e^{as}\sqrt{\frac{s^k}{k!}}\left(\|B\|\|A\hat{\zeta}(\theta)\| + kM_{AB}\|\hat{\zeta}(\theta)\|\right)\right.$$
$$\left.\cdot \|\hat{\mathbb{W}}(s,\theta)\| + M_{AB}M^{k+1}\|B\|^k e^{as}|\theta|^k_{L^2(\mathbb{R})}\sqrt{\frac{s^k}{k!}}\|\hat{\zeta}(\theta)\|\|\hat{\mathbb{W}}(s,\theta)\|\right]ds$$
$$= M^{k+2}\|B\|^k|\theta|^k_{L^2(\mathbb{R})}e^{at}\left(\|B\|\|A\hat{\zeta}(\theta)\| + (k+1)M_{AB}\|\hat{\zeta}(\theta)\|\right)$$
$$\cdot \int_0^t \sqrt{\frac{s^k}{k!}}\|\hat{\mathbb{W}}(s,\theta)\|\,ds$$
$$\leq M^{k+2}\|B\|^k|\theta|^k_{L^2(\mathbb{R})}e^{at}\left(\|B\|\|A\hat{\zeta}(\theta)\| + (k+1)M_{AB}\|\hat{\zeta}(\theta)\|\right)$$
$$\cdot \left(\int_0^t \frac{s^k}{k!}\,ds\right)^{1/2}\left(\int_0^t \|\hat{\mathbb{W}}(s,\theta)\|^2\,ds\right)^{1/2}$$
$$\leq M^{k+2}\|B\|^k|\theta|^{k+1}_{L^2(\mathbb{R})}e^{at}\sqrt{\frac{t^{k+1}}{(k+1)!}}\left(\|B\|\|A\hat{\zeta}(\theta)\| + (k+1)M_{AB}\|\hat{\zeta}(\theta)\|\right).$$

262   6. *Infinite-dimensional extension of white noise calculus*

The assertion follows from here and (6.4.20) by induction. $\square$

Consider the series

$$T(t,\theta) = \sum_{k=0}^{\infty} T_k(t,\theta). \tag{6.4.21}$$

It follows from Lemma 6.4.1 that for any $n, m \in \mathbb{N}$ the following estimate is true:

$$\sum_{k=n}^{n+m} \|T_k(t,\theta)\| \le M e^{at} \sum_{k=n}^{n+m} \frac{\left(M\sqrt{2}\|B\|\|\theta\|_{L^2(\mathbb{R})}\sqrt{t}\right)^k}{\sqrt{k!}} \cdot \frac{1}{\sqrt{2^k}}$$

$$\le M e^{at} \left(\sum_{k=n}^{n+m} \frac{\left(2M^2\|B\|^2|\theta|^2_{L^2(\mathbb{R})}t\right)^k}{k!}\right)^{1/2} \left(\sum_{k=n}^{n+m} \frac{1}{2^k}\right)^{1/2} . \tag{6.4.22}$$

It follows that the series (6.4.21) is absolutely convergent in $\mathcal{L}(H)$ for any $t \ge 0$, $\theta \in \mathcal{S}$. Thus $T(t,h) \in \mathcal{L}(H)$.

Proposition 6.4.2 *For any $\zeta \in dom\, A$ and $\theta \in \mathcal{S}$ the function $\hat{X}(t,\theta) = T(t,\theta)\hat{\zeta}(\theta)$ is the unique solution of the problem (6.4.16).*

Proof. It follows from Proposition 6.4.1 and properties of C_0-semigroups that $T_0(t,\theta)\hat{\zeta}(\theta) \in dom\, A$ for any $\zeta \in (dom\, A)$, $t \ge 0$ and $\theta \in \mathcal{S}$. The condition (B) implies $B(dom\, A)\hat{W}(t,\theta) \subseteq dom\, A$ for all $t \ge 0$ and $\theta \in \mathcal{S}$. It follows by induction that $T_k(t,\theta)\hat{\zeta}(\theta) \in dom\, A$ for all $\zeta \in (dom\, A)$, $k \in \mathbb{N}$, $t \ge 0$, and $\theta \in \mathcal{S}$. It also follows from (B) that $B\big(T_k(s,\theta)\hat{\zeta}(\theta)\big)\hat{W}(t,\theta) \in dom\, A$. Moreover, we have

$$\frac{d}{dt}U(t-s)B\big(T_k(s,\theta)\hat{\zeta}(\theta)\big)\hat{W}(t,\theta) = AU(t-s)B\big(T_k(s,\theta)\hat{\zeta}(\theta)\big)\hat{W}(t,\theta),$$

$$t \ge 0, \quad \theta \in \mathcal{S}.$$

Thus for any $\zeta \in (dom\, A)$ we obtain

$$\frac{d}{dt}T_0(t,\theta)\hat{\zeta}(\theta) = AT_0(t,\theta)\hat{\zeta}(\theta), \tag{6.4.23}$$

$$\frac{d}{dt}T_k(t,\theta)\hat{\zeta}(\theta) = \int_0^t AU(t-s)B\big(T_{k-1}(s,\theta)\hat{\zeta}(\theta)\big)\hat{W}(s,\theta)\,ds$$

$$+ B\big(T_{k-1}(t,\theta)\hat{\zeta}(\theta)\big)\hat{W}(t,\theta). \tag{6.4.24}$$

Since A is closed we can rewrite the equality (6.4.24) as

$$\frac{d}{dt}T_k(t,\theta)\hat{\zeta}(\theta) = AT_k(t,\theta)\hat{\zeta}(\theta) + B\big(T_{k-1}(t,\theta)\hat{\zeta}(\theta)\big)\hat{W}(t,\theta). \tag{6.4.25}$$

6.4. Generalized solutions to stochastic Cauchy problems

By Lemma 6.4.2 we obtain the estimate

$$\sum_{k=n+1}^{m} \|AT_k(t,\theta)\hat\zeta(\theta)\|$$

$$\leq Me^{at}\left(\sum_{k=n+1}^{m} \frac{(\sqrt{2}M\|B\|\|\theta|_{L^2(\mathbb{R})}\sqrt{t})^k}{\sqrt{k!}} \cdot \frac{1}{\sqrt{2^k}}\right)\|A\hat\zeta(\theta)\|$$

$$+ \frac{M}{\|B\|}e^{at}\left(\sum_{k=n+1}^{m} \frac{(\sqrt{2}M\|B\|\|\theta|_{L^2(\mathbb{R})}\sqrt{t})^k}{\sqrt{k!}} \cdot \frac{k}{\sqrt{2^k}}\right)M_{AB}\|\hat\zeta(\theta)\|$$

$$\leq Me^{at}\left(\sum_{k=n+1}^{m} \frac{(2M^2\|B\|^2|\theta|_{L^2(\mathbb{R})}^2 t)^k}{k!}\right)^{1/2} \cdot \left(\sum_{k=n+1}^{m} \frac{1}{2^k}\right)^{1/2}\|A\hat\zeta(\theta)\|$$

$$+ \frac{M}{\|B\|}e^{at}\left(\sum_{k=n+1}^{m} \frac{(2M^2\|B\|^2|\theta|_{L^2(\mathbb{R})}^2 t)^k}{k!}\right)^{1/2} \cdot \left(\sum_{k=n+1}^{m} \frac{k^2}{2^k}\right)^{1/2}M_{AB}\|\hat\zeta(\theta)\|.$$

It follows from this estimate that the series $\sum_{k=0}^{\infty} AT_k(t,\theta)\hat\zeta(\theta)$ is convergent in the space H for all $\theta \in \mathcal{S}$, $\zeta \in (dom\, A)$. Summing the equalities (6.4.23) and (6.4.25) with respect to $k \in \mathbb{N}$ we obtain in the right-hand side the series which is convergent in H for all $t \geq 0$, $\theta \in \mathcal{S}$. Thus we have proved that $\hat X(t,\theta) = T(t,\theta)\hat\zeta(\theta)$ is the solution of the problem (6.4.16).

To prove uniqueness, note that if $\hat X(\cdot,\theta)$ is a solution of the problem (6.4.16) for some $\theta \in \mathcal{S}$, then it is a solution of the equation

$$\hat X(t,\theta) = U(t)\hat\zeta(\theta) + \int_0^t U(t-s)B(\hat X(s,\theta))\hat{\mathbb{W}}(s,\theta)\,ds, \qquad t \geq 0.$$

(The inverse is generally speaking not true.) Thus it is sufficient to prove that the equation

$$\hat X(t,\theta) - \int_0^t U(t-s)B(\hat X(s,\theta))\hat{\mathbb{W}}(s,\theta)\,ds = 0, \qquad t \geq 0, \qquad (6.4.26)$$

has only trivial solution $\hat X(\cdot,\theta) \equiv 0$ on $[0,\infty)$ for any $\theta \in \mathcal{S}$.

First note that

$$\|\hat{\mathbb{W}}(t,\theta)\|_{\mathbb{H}}^2 = \sum_{j\in\mathbb{N}}\left(\sum_{i\in\mathbb{N}}(\xi_{n(i,j)},\theta)\xi_i(t)\right)^2$$

is bounded on any $[0,T]$. This follows from the fact that $|(\xi_i,\theta)| = \mathcal{O}(i^{-p})$ for

264   6. Infinite-dimensional extension of white noise calculus

any $p \in \mathbb{N}$, since $\theta \in \mathcal{S}$. Combined with the estimates (6.1.8) and (6.1.17) this implies

$$\left|(\xi_{n(i,j)}, \theta)\xi_i(t)\right| = \mathcal{O}\left(\frac{1}{i^{p+\frac{1}{12}}j^p}\right) \quad \text{for any} \quad p \in \mathbb{N}.$$

It follows that there exists a constant $K > 0$ such that

$$\left\|\int_0^t U(t-s)B(\hat{X}(s,\theta))\hat{\mathbb{W}}(s,\theta)\,ds\right\|_H$$

$$\leq \int_0^t Me^{a(t-s)}\|B\|\|\hat{X}(s,\theta)\|_H\|\hat{\mathbb{W}}(s,\theta)\|_{\mathbb{H}}\,ds$$

$$\leq K\int_0^t \|\hat{X}(s,\theta)\|_H ds, \quad t \in [0,T].$$

It is easy to prove using this estimate and the standard Volterra equations technique that a certain power of the integral operator

$$\int_0^t U(t-s)B(\cdot)\hat{\mathbb{W}}(s,\theta)\,ds$$

is a contraction in the space of all continuous H-valued functions on $[0,T]$ endowed with the norm $\|X\| = \max_{t \in [0,T]} \|X(t)\|_H$. This implies the uniqueness of the solution of (6.4.26). $\qquad\square$

Theorem 6.4.2 *Let A be a linear densely defined in H generator of a C_0-semigroup and $B(\cdot) : H \to \mathcal{L}(\mathbb{H}; H)$ satisfy the conditions* (B). *Then the Cauchy problem* (6.4.15) *has a unique solution in the space* $(\mathcal{S})_{-0}(H)$ *for any $\zeta \in (\text{dom }A) \subseteq (\mathcal{S})_{-0}(H)$.*

Proof. It follows from Proposition 6.4.2 that under the conditions of the theorem the problem (6.4.16) has the unique solution $\hat{X}(t,\theta) = T(t,\theta)\hat{\zeta}(\theta)$ for any $\zeta \in (\text{dom }A)$, $\theta \in \mathcal{S}$. From (6.4.22) the estimate follows:

$$\|T(t,\theta)\| \leq \sum_{k=0}^{\infty}\|T_k(t,\theta)\| \leq Me^{at}\sum_{k=0}^{\infty}\frac{(M\sqrt{2}\|B\|\|\theta\|_0\sqrt{t})^k}{\sqrt{k!}} \cdot \frac{1}{\sqrt{2^k}}$$

$$\leq Me^{at}\left(\sum_{k=0}^{\infty}\frac{(2M^2\|B\|^2|\theta|_0^2 t)^k}{k!}\right)^{1/2}\left(\sum_{k=0}^{\infty}\frac{1}{2^k}\right)^{1/2}$$

$$= M\sqrt{2}\,e^{at}\exp\left(M^2\|B\|^2|\theta|_0^2 t\right).$$

By (6.3.4) we have

$$\|\hat{\zeta}(\theta)\| \leq \|\zeta\|_{-p,-0}\exp\left(|h|_p^2\right), \quad \theta \in \mathcal{S},$$

6.4. Generalized solutions to stochastic Cauchy problems 265

for some $p \in \mathbb{N}$. Consequently, for all $t \geq 0$ we have the following estimate:

$$\|\hat{X}(t,\theta)\| \leq M\sqrt{2}\, e^{at} \exp\left(M^2\|B\|^2|\theta|_0^2 t + |\theta|_p^2\right)\|\zeta\|_{-p,-0}$$
$$\leq M\sqrt{2}\, e^{at} \exp\left((M^2\|B\|^2 t + 1)|\theta|_p^2\right)\|\zeta\|_{-p,-0}, \qquad \theta \in \mathcal{S}.$$

It follows from here that, for any $t \geq 0$, $\hat{X}(t,\theta)$ is the S-transform of the unique generalized random variable $X(t) \in (\mathcal{S})_{-0}(H)$, which is the unique solution of the problem (6.4.16). $\qquad\square$

Consider an example of introducing a multiplicative stochastic perturbation into a partial differential equation.

Example 6.4.2 *Equation of age structured population.*

We consider a simplified example of the equation arising in population dynamics. Let us start with the following deterministic equation:

$$\frac{\partial u(t,s)}{\partial t} = -\frac{\partial u(t,s)}{\partial s} - m(s)u(t,s), \quad t \geq 0, \quad 0 \leq s \leq 1. \tag{6.4.27}$$

This is the McKendrick–von Foerster equation of age structured population. Here t is time, s denotes age, $u(t,s)$ is the density function, so that $u(t,s)\,ds$ represents the amount of species of age within the interval $[s, s+ds]$ in the population at the time t. The structure of the population is changing by means of the processes of aging and death. Aging is modeled by the first term in the right-hand side since the operator $-\frac{\partial}{\partial s}$ is the generator of the right-shift semigroup. The term $m(s)$ represents the rate of death at age s. Suppose $m \in L_\infty[0,1]$. For simplicity consider the boundary condition

$$u(t,0) = 0, \qquad t > 0. \tag{6.4.28}$$

The initial structure of the population is described by condition

$$u(0,s) = \phi(s), \qquad 0 \leq s \leq 1. \tag{6.4.29}$$

The problem (6.4.27)–(6.4.29) can be written as the Cauchy problem

$$u'(t) = Au(t), \quad t \geq 0, \qquad u(0) = \phi, \tag{6.4.30}$$

in the Hilbert space $H = L^2[0,1]$, where A is the operator defined by

$$[A\phi](s) = -\frac{d}{ds}\phi(s) - m(s)\phi(s),$$

with the domain

$$dom\, A = \{\phi \in H : \phi' \in H,\ \phi(0) = 0,\ t > 0\}.$$

Making use of the methods of semigroup perturbation theory, one can show that A is the generator of a C_0-semigroup in H (see, e.g., [51]).

266   6. *Infinite-dimensional extension of white noise calculus*

Suppose now that the process of death is subject to random fluctuations as a consequence of the influence of the external environment. It is natural to suppose the function m to represent the mean value of the death rate in the population. Thus we have to replace this function in the equation by $m + \mu(t)$, where $\mu(t)$ is a "noise term". We face a problem connected with the fact that it is not possible to use the above defined Gaussian white noises (Q-white noise and cylindrical white noise) here in a straightforward manner, as, for any t $\mu(t)$, must be a function of s such that multiplication by it is a bounded operator in $H = L^2[0,1]$. To overcome this problem we take $\mathbb{H} = L^2[0,1]$ and consider the following operator:

$$[B(u)v](s) := \varepsilon(s)u(s) \int_0^1 \psi(s-\tau)v(\tau)\,d\tau, \quad u \in H, \quad v \in \mathbb{H},$$

where $\psi \in C_0^\infty(\mathbb{R})$ and $\varepsilon \in L_\infty[0,1]$ are fixed functions. Taking an appropriate function as the factor ψ in the convolution (it can be, for example, an appropriate member of a sequence converging in a sense to the Dirac δ-function), we obtain that B is the operator of multiplication by a "smooth approximation of v" as a function of u.

For any $u \in H$ and $v \in \mathbb{H}$ we have

$$\|B(u)v\|_H \leq \sup_{t \in \mathbb{R}} |\psi(t)| \cdot \|u\|_H \|v\|_{\mathbb{H}}.$$

Thus $B(\cdot) \in \mathcal{L}\big(H; \mathcal{L}(\mathbb{H}; H)\big)$.

Consider the stochastic perturbation of the Cauchy problem (6.4.30) having the form (6.4.15) with the above defined operator B. Since the values of $\mathbb{W}(t)$ for each t are represented by the series

$$\mathbb{W}(t) := \sum_{i,j \in \mathbb{N}} \xi_i(t)e_j(s)\mathbf{h}_{\epsilon_{n(i,j)}}(\omega),$$

where $\{e_j\}$ is a fixed orthonormal basis in $\mathbb{H}$ $\big(\mathbb{H} = L^2[0,1]\big)$ and since the series is divergent in $\mathbb{H}$ for any $\omega \in \mathcal{S}'$, one can informally think of these values as of irregular functions of s. When in the equation we substitute "$\diamond \mathbb{W}(t)$" as the v variable of the operator B we obtain a sort of smooth approximation of these functions. So the operator $B(\cdot) \diamond \mathbb{W}(t)$ in the equation can be thought of as an operator of specific multiplication by smoothed values of white noise, which seems a natural way of introducing a stochastic perturbation of the operator of multiplication by $m(s)$.

For any $v \in \mathbb{H}$, $u \in dom\ A$ we have

$$[C(u)v](s) := [AB(u)v - B(Au)v](s) = -u(s) \int_0^1 \psi'(s-\tau)v(\tau)\,d\tau.$$

Thus $C(\cdot)v$ is a bounded operator in H and condition (B) is fulfilled. Thus the Cauchy problem (6.4.30) satisfies the conditions of Theorem 6.4.1 and consequently has a unique solution in the space $(\mathcal{S})_{-0}(H)$.

6.4.3 Relationship between weak and generalized wrt ω solutions

In this subsection we establish a connection between the generalized wrt ω solution of (6.4.1) and a weak solution of the Cauchy problem for the corresponding integral Itô equation (6.4.3). Similarly to Section 5.1, where the relationship between a generalized wrt t solution and a weak solution was based on the connection between the stochastic convolution written in the form of the Itô integral and generalized stochastic convolution, here the relationship will be proved on the basis of the connection between the Hitsuda–Skorohod integral and the Itô integral.

We consider the solution (6.4.6) obtained for the stochastic Cauchy problem in a space of stochastic distributions $(\mathcal{S})_{-0}(H)$ with the generator of a C_0-semigroup, white noise $\mathbb{W}$, and initial data $\zeta \in (dom\, A)$. As noted, solutions in the space $(\mathcal{S})_{-0}(H)$ can be constructed for the problem with a Q-white noise and singular white noise.

Now we show that the generalized wrt ω solution coincides with the corresponding weak solution under the conditions of existence for both of the solutions. Since it was shown in Section 4.1 that the case of a cylindrical Wiener process can be reduced to the case of a Q_1-Wiener process with a specially constructed trace class operator Q_1, we restrict ourselves to the case of Q-Wiener processes.

Theorem 6.4.3 *Let A be the generator of a C_0-semigroup U, $\Phi(t) = U(t)B$ satisfies (6.3.7), $\zeta \in (dom\, A)$, and W_Q be a Q-Wiener process. Then the generalized wrt ω solution to the Cauchy problem (6.4.5) and the weak solution to (6.4.3) coincide.*

Proof. The generalized wrt ω solution to (6.4.5) is obtained in Theorem 6.4.1 in the form (6.4.6):

$$X(t) = U\zeta + \int_0^t U(t-s)B\mathbb{W}_Q(s)\,ds, \qquad t \geq 0,$$

where

$$\int_0^t U(t-s)B\mathbb{W}_Q(s)\,ds := \sum_{i,j \in \mathbb{N}} \sigma_j \int_0^t U(t-s)Be_j\xi_i(s)\,ds\mathbf{h}_{\epsilon_{n(i,j)}}.$$

In the case that we consider, to prove the coincidence of the solutions it is enough to show the coincidence of the integrals:

$$\int_0^t U(t-s)B\,dW_Q(s) = \int_0^t U(t-s)B\mathbb{W}_Q(s)\,ds. \qquad (6.4.31)$$

First, for $\mathbb{W}_Q$ defined by (6.1.23) we show that the integral in the right-hand side of (6.4.31) belongs to the space $(L^2)(H) = L^2(\Omega, \mathcal{F}, \mu; H)$, where $\Omega = \mathcal{S}'$,

268   6. *Infinite-dimensional extension of white noise calculus*

$\mathcal{F} = \mathcal{B}(\mathcal{S}')$, and μ is a Bochner–Minlos measure on $\mathcal{B}(\mathcal{S}')$ (see (6.1.3)). This result follows from the equalities

$$\sum_{i,j\in\mathbb{N}} \sigma_j^2 \left\| \int_0^t U(t-s)Be_j\xi_i(s)\,ds \right\|_H^2$$

$$= \sum_{j,k\in\mathbb{N}} \sum_{i\in\mathbb{N}} \left(\int_0^t \xi_i(s)\,(\sigma_j U(t-s)Be_j,\, g_k)_H\,ds \right)^2$$

$$= \sum_{j,k\in\mathbb{N}} \left\| \mathbf{1}_{[0,t]}(\sigma_j U(t-\cdot)Be_j,\, g_k)_{L^2(\mathbb{R})}^2 \right\|$$

$$= \sum_{j,k\in\mathbb{N}} \int_0^t |(U(t-\cdot)BQ^{\frac{1}{2}},\, g_k)_H|^2\,ds$$

$$= \sum_{j\in\mathbb{N}} \int_0^t \sum_{k\in\mathbb{N}} (U(t-\cdot)BQ^{\frac{1}{2}}e_j,\, g_k)_H^2\,ds$$

$$= \int_0^t \sum_{j\in\mathbb{N}} \|U(t-s)BQ^{\frac{1}{2}}e_j\|_H^2\,ds = \int_0^t \|U(t-s)B\|_{\mathcal{L}_{\mathrm{HS}}(\mathbb{H}_Q,H)}^2\,ds.$$

Here $g_k = \sigma_k e_k$. The fact that the integrals in (6.4.31) coincide as elements of $(L^2)(H)$ can be proved by obtaining this equality for elementary functions and passing to the limit. Let $\Phi(s)$ be an elementary function approximating $U(t-s)B$. We have

$$\int_0^t \Phi(s)\mathbb{W}\,ds = \sum_{i,j\in\mathbb{N}} \sigma_j \int_0^t \Phi(s)e_j\xi_i(s)\,ds\mathbf{h}_{\epsilon_{n(i,j)}}$$

$$= \sum_{i,j\in\mathbb{N}} \sigma_j \sum_{k=0}^{N-1} \int_{t_{k-1}}^{t_k} \Phi_k\xi_i(s)\,ds\,e_j\mathbf{h}_{\epsilon_{n(i,j)}}$$

$$= \sum_{k=0}^{N-1} \Phi_k \sum_{i,j\in\mathbb{N}} \sigma_j \int_{t_{k-1}}^{t_k} \xi_i(s)\,ds\,e_j\mathbf{h}_{\epsilon_{n(i,j)}}$$

$$= \sum_{k=0}^{N-1} \Phi_k\left[W(t_k) - W(t_{k-1})\right] = \int_0^t \Phi(t)\,dW(t).$$

Passing to the limit as $N \to \infty$, we obtain (6.4.31). It follows that the generalized wrt ω solution and the weak solution coincide if both exist. $\square$

Thus, while proving the connection between weak and generalized wrt ω solutions we have proved the relationship (6.3.8) between the Itô and Hitsuda–Skorohod integrals in the simplest case, when the integrand $U(t-s)B$ is a deterministic operator-valued function.

Now we present the proof of Theorem 6.3.2 in the general case, i.e., we show that the Hitsuda–Skorohod integral is a generalization of the Itô integral

6.4. Generalized solutions to stochastic Cauchy problems 269

wrt a Wiener process. This relationship can be used for connection of weak and generalized wrt ω solutions for a wider class of equations than that considered above and not only for such considerations. The proof uses the ideas of [26], where this connection is proved in the one-dimensional case. We generalize it to the infinite-dimensional situation. For simplicity, we will consider the case of a Q-Wiener process and the corresponding Q-white noise.

Let $\{\mathcal{B}_t,\, t \geq 0\}$ be the σ-algebra generated by the random variables $(W_Q(s), x)_{\mathbb{H}}$, where $0 \leq s \leq t$, $x \in \mathbb{H}$. Recall that the family $\{\mathcal{B}_t\}$ is called the *filtration* generated by the Q-Wiener process $\{W_Q(t),\, t \geq 0\}$ (see Section 4.1). It is easy to see that $\{\mathcal{B}_t,\, t \geq 0\}$ coincides with the σ-algebra generated by the random variables of the form $(W(s), x)_{\mathbb{H}}$, where $0 \leq s \leq t$, $x \in \mathbb{H}$. Recall also that the Brownian motions $\beta_j(t)$, $t \geq 0$, $j \in \mathbb{N}$, are martingales wrt $\mathcal{B}_t$.

Let $\mathcal{H}$ be a separable Hilbert space. An $\mathcal{H}$-valued random process $\Phi(t)$, $t \geq 0$, is called $\mathcal{B}_t$-adapted if $\Phi(t)$ is $\mathcal{B}_t$-measurable for each $t \geq 0$. We will further consider Itô integrals wrt an $\mathbb{H}$-valued Q-Wiener process for predictable integrands $\Phi(t)$, $t \in [0,T]$, with values in $\mathcal{L}_{\mathrm{HS}}(\mathbb{H}_Q; H)$. Recall that an $\mathcal{H}$-valued process is called predictable if it is measurable as a mapping from $\big([0,T] \times \mathcal{S}', \mathcal{P}_T\big)$ to $\big(\mathcal{H}, \mathcal{B}(\mathcal{H})\big)$, where $\mathcal{P}_T$ is the predictable σ-algebra of subsets of $[0,T] \times \mathcal{S}'$. The latter is defined as the σ-algebra generated by the sets of the form

$$(s,t] \times B, \quad 0 \leq s < t \leq T, \quad B \in \mathcal{B}_s.$$

We will further need a few lemmas which give characterization of $\mathcal{B}_t$-measurable random variables in terms of their S-transforms. They use the operators $\mathfrak{J}_j$, $j \in \mathbb{N}$, defined by (6.1.18), which are isometrical isomorphisms of $L^2(\mathbb{R})$ and the spaces $L^2(\mathbb{R})_j$, and orthogonal projectors π_j, $j \in \mathbb{N}$, of $L^2(\mathbb{R})$ onto the spaces $L^2(\mathbb{R})_j$ defined by

$$\pi_j \xi_n = \begin{cases} \xi_n, & n \in \{n(i,j),\ i \in \mathbb{N}\}, \\ 0, & n \notin \{n(i,j),\ i \in \mathbb{N}\}. \end{cases} \tag{6.4.32}$$

Lemma 6.4.3 *Let $\mathcal{H}$ be a separable Hilbert space. For any $\Theta, \Phi \in (L^2)(\mathcal{H})$ the equality $\Theta = \mathbf{E}\big(\Phi|\mathcal{B}_t\big)$ holds true if and only if*

$$S\Theta(\theta) = S\Phi\Big(\sum_{j=1}^{\infty} \theta_{t,j} \Big), \tag{6.4.33}$$

for any $\theta \in \mathcal{S}$, where $\theta_{t,j} := \mathfrak{J}_j\big(\mathfrak{J}_j^{-1}\pi_j\theta \cdot 1_{[0,t]}\big)$, operators $\mathfrak{J}_j$ are defined by (6.1.18), and π_j, $j \in \mathbb{N}$, are defined by (6.4.32).

Proof. Let $\theta_{t,j}^{\perp} = \mathfrak{J}_j\big(\mathfrak{J}_j^{-1}\pi_j\theta \cdot 1_{[0,t]^c}\big)$ for $\theta \in \mathcal{S}$, $j \in \mathbb{N}$. We have

$$\pi_j\theta = \mathfrak{J}_j\mathfrak{J}_j^{-1}\pi_j\theta = \mathfrak{J}_j\big(\mathfrak{J}_j^{-1}\pi_j\theta \cdot 1_{[0,t]} + \mathfrak{J}_j^{-1}\pi_j\theta \cdot 1_{[0,t]^c}\big) = \theta_{t,j} + \theta_{t,j}^{\perp}.$$

270   6. *Infinite-dimensional extension of white noise calculus*

Moreover, the functions $\theta_{t,j}$ and $\theta_{t,j}^\perp$ are orthogonal in $L^2(\mathbb{R})$:

$$(\theta_{t,j}, \theta_{t,j}^\perp)_{L^2(\mathbb{R})} = \left(\mathfrak{J}_j\big(\mathfrak{J}_j^{-1}\pi_j\theta \cdot 1_{[0,t]}\big), \mathfrak{J}_j\big(\mathfrak{J}_j^{-1}\pi_j\theta \cdot 1_{[0,t]^c}\big)\right)_{L^2(\mathbb{R})}$$
$$= \big(\mathfrak{J}_j^{-1}\pi_j\theta \cdot 1_{[0,t]}, \mathfrak{J}_j^{-1}\pi_j\theta \cdot 1_{[0,t]^c}\big)_{L^2(\mathbb{R})} = 0.$$

Since for any orthogonal in $L^2(\mathbb{R})$ functions θ and η we have

$$\mathcal{E}_{\theta+\eta} = e^{\langle \cdot, \theta+\eta\rangle - \frac{1}{2}\|\theta+\eta\|_{L^2(\mathbb{R})}^2} = e^{\langle \cdot, \theta\rangle - \frac{1}{2}\|\theta\|_{L^2(\mathbb{R})}^2} e^{\langle \cdot, \eta\rangle - \frac{1}{2}\|\eta\|_{L^2(\mathbb{R})}^2} e^{(\theta,\eta)_{L^2(\mathbb{R})}} = \mathcal{E}_\theta\mathcal{E}_\eta,$$
(6.4.34)

it implies that

$$S\Theta\left(\sum_{j=1}^n \pi_j\theta\right) = \mathbf{E}\left(\Theta\mathcal{E}_{\sum_{j=1}^n \pi_j\theta}\right) = \mathbf{E}\left(\Theta\prod_{j=1}^n \mathcal{E}_{\pi_j\theta}\right) = \mathbf{E}\left(\Theta\prod_{j=1}^n \mathcal{E}_{\theta_{t,j}+\theta_{t,j}^\perp}\right).$$

Again, using the property (6.4.34), we obtain

$$S\Theta\left(\sum_{j=1}^n \pi_j\theta\right) = \mathbf{E}\left(\Theta\prod_{j=1}^n \mathcal{E}_{\theta_{t,j}}\prod_{j=1}^n \mathcal{E}_{\theta_{t,j}^\perp}\right).$$

Note that for any $s \in [0;t]$ it holds that

$$\mathbf{E}\left(\beta_j(s)\langle \cdot, \theta_{t,j}^\perp\rangle\right) = \left(\langle \cdot, \mathfrak{J}_j 1_{[0,s]}\rangle, \langle \cdot, \mathfrak{J}_j\big(\mathfrak{J}_j^{-1}\pi_j\theta \cdot 1_{[0,t]^c}\big)\rangle\right)_{(L^2)}$$
$$= \left(\mathfrak{J}_j 1_{[0,s]}, \mathfrak{J}_j\big(\mathfrak{J}_j^{-1}\pi_j\theta \cdot 1_{[0,t]^c}\big)\right)_{L^2(\mathbb{R})} = \big(1_{[0,s]}, \mathfrak{J}_j^{-1}\pi_j\theta \cdot 1_{[0,t]^c}\big)_{L^2(\mathbb{R})} = 0.$$

Thus the random variables $\langle \cdot, \theta_{t,j}^\perp\rangle$, and consequently $\mathcal{E}_{\theta_{t,j}^\perp}$, $j \in \mathbb{N}$, are independent of $\mathcal{B}_t$. Approximating θ in $L^2(\mathbb{R})$ by finite step functions, one can easily prove that the random variables $\langle \cdot, \theta_{t,j}\rangle$, $j \in \mathbb{N}$, and consequently the functions $\mathcal{E}_{\theta_{t,j}}$, are $\mathcal{B}_t$-measurable. Thus, if $\Theta = \mathbf{E}\big(\Phi|\mathcal{B}_t\big)$, by the properties of conditional expectations, we have

$$S\Theta\left(\sum_{j=1}^n \pi_j\theta\right) = \mathbf{E}\left(\mathbf{E}(\Phi|\mathcal{B}_t)\prod_{j=1}^n \mathcal{E}_{\theta_{t,j}}\prod_{j=1}^n \mathcal{E}_{\theta_{t,j}^\perp}\right)$$
$$= \mathbf{E}\left(\mathbf{E}\left(\Phi\prod_{j=1}^n \mathcal{E}_{\theta_{t,j}}\Big|\mathcal{B}_t\right)\right)\mathbf{E}\left(\prod_{j=1}^n \mathcal{E}_{\theta_{t,j}^\perp}\right) = \mathbf{E}\left(\Phi\prod_{j=1}^n \mathcal{E}_{\theta_{t,j}}\right)\mathbf{E}\left(\prod_{j=1}^n \mathcal{E}_{\theta_{t,j}^\perp}\right).$$

Again using the equality (6.4.34), we obtain

$$S\Theta\left(\sum_{j=1}^n \pi_j\theta\right) = \mathbf{E}\left(\Phi\mathcal{E}_{\sum_{j=1}^n \theta_{t,j}}\right)\mathbf{E}\left(\mathcal{E}_{\sum_{j=1}^n \theta_{t,j}^\perp}\right) = S\Phi\left(\sum_{j=1}^n \theta_{t,j}\right). \quad (6.4.35)$$

Since convergence of a sequence θ_n to θ in $L^2(\mathbb{R})$ implies convergence of $\mathbf{E}\big(\Phi\mathcal{E}_{\theta_n}\big)$ to $\mathbf{E}\big(\Phi\mathcal{E}_\theta\big)$ in $\mathcal{H}$ for any $\Phi \in (L^2)(\mathcal{H})$, we obtain the equality (6.4.33) by letting $n \to \infty$ in the equality (6.4.35). $\qquad\square$

6.4. Generalized solutions to stochastic Cauchy problems

Corollary 6.4.1 $\Phi \in (L^2)(\mathcal{H})$ *is* $\mathcal{B}_t$*-measurable if and only if*

$$\mathrm{S}\Phi(\theta) = \mathrm{S}\Phi\left(\sum_{j\in\mathbb{N}}\theta_{t,j}\right), \qquad \theta_{t,j} := \mathfrak{I}_j\big(\mathfrak{I}_j^{-1}\pi_j\theta \cdot 1_{[0,t]}\big), \qquad \theta\in\mathcal{S}.$$

Lemma 6.4.4 *If a random variable* $\Phi \in (L^2)(\mathcal{H})$ *is* $\mathcal{B}_t$*-measurable, then for any* $k\in\mathbb{N}$, $b>t>0$ *it holds that*

$$\mathrm{S}\Big(\Phi\langle\cdot, 1_{(t,b]}^k\rangle\Big)(\theta) = (1_{(t,b]}^k, \theta)_{L^2(\mathbb{R})}\mathrm{S}\Phi(\theta), \qquad \theta\in\mathcal{S}. \tag{6.4.36}$$

Proof. We have

$$\mathrm{S}\Big(\Phi\langle\cdot, 1_{(t,b]}^k\rangle\Big)(\theta) = \mathbf{E}\left(\Phi\langle\cdot, 1_{(t,b]}^k\rangle\mathcal{E}_\theta\right) = e^{-\frac{|\theta|_0}{2}}\mathbf{E}\left(\Phi\frac{d}{d\alpha}e^{\alpha\langle\cdot, 1_{(t,b]}^k\rangle + \langle\cdot, \theta\rangle}\Big|_{\alpha=0}\right)$$

$$= e^{-\frac{|\theta|_0}{2}}\frac{d}{d\alpha}\mathbf{E}\left(\Phi\, e^{\langle\cdot, \alpha 1_{(t,b]}^k + \theta\rangle - \frac{1}{2}|\alpha 1_{(t,b]}^k + \theta|_0^2} \cdot e^{\frac{1}{2}|\alpha 1_{(t,b]}^k + \theta|_0^2}\right)\Big|_{\alpha=0}$$

$$= e^{-\frac{|\theta|_0}{2}}\frac{d}{d\alpha}\left(e^{\frac{1}{2}|\alpha 1_{(t,b]}^k + \theta|_0^2}\mathrm{S}\Phi\big(\alpha 1_{(t,b]}^k + \theta\big)\right)\Big|_{\alpha=0}. \tag{6.4.37}$$

We further have

$$\frac{d}{d\alpha}e^{\frac{1}{2}|\alpha 1_{(t,b]}^k + \theta|_0^2}\Big|_{\alpha=0} = \frac{d}{d\alpha}e^{\frac{1}{2}\left(\alpha^2|1_{(t,b]}^k|_0^2 + 2\alpha(1_{(t,b]}^k, \theta)_{L^2(\mathbb{R})} + |\theta|_0^2\right)}\Big|_{\alpha=0}$$

$$= (1_{(t,b]}^k, \theta)_{L^2(\mathbb{R})}e^{\frac{|\theta|_0}{2}}.$$

Moreover, by $\mathcal{B}_t$-measurability of Φ, Corollary 6.4.1, and the equality

$$(1_{(t,b]}^k)_{t,j} = \mathfrak{I}_j\big(\mathfrak{I}_j^{-1}\pi_j 1_{(t,b]}^k \cdot 1_{[0,t]}\big) = \begin{cases}\mathfrak{I}_j\big(0 \cdot 1_{[0,t]}\big) = 0, & k\neq j, \\ \mathfrak{I}_j\big(1_{(t,b]} \cdot 1_{[0,t]}\big) = 0, & k=j,\end{cases}$$

we obtain the equality

$$\mathrm{S}\Phi\big(\alpha 1_{(t,b]}^k + \theta\big) = \mathrm{S}\Phi\left(\sum_{j\in\mathbb{N}}\big(\alpha(1_{(t,b]}^k)_{t,j} + \theta_{t,j}\big)\right) = \mathrm{S}\Phi\left(\sum_{j\in\mathbb{N}}\theta_{t,j}\right),$$

which implies $\dfrac{d}{d\alpha}\mathrm{S}\Phi\big(\alpha 1_{(t,b]}^k + \theta\big) = 0$. Thus from the equality (6.4.37) the equality (6.4.36) follows. $\qquad\square$

Theorem 6.4.4 *For any predictable* $\mathcal{L}_{HS}(\mathbb{H}_Q; H)$*-valued process satisfying the condition*

$$\mathbf{E}\left[\int_0^T \|\Phi(t)\|_{\mathcal{L}_{HS}(\mathbb{H}_Q;H)}^2 dt\right] < \infty \tag{6.4.38}$$

it holds that

$$\int_0^T \Phi(t)\, dW_Q(t) = \int_0^T \Phi(t) \diamond \mathbb{W}_Q(t)\, dt. \tag{6.4.39}$$

272   6. *Infinite-dimensional extension of white noise calculus*

Proof. To prove the assertion, recall that the Itô integral wrt the Q-Wiener process is first defined for the so-called elementary processes, i.e., for processes having the form

$$\Phi(t) = \sum_{k=0}^{N-1} \Phi_k 1_{(t_k, t_{k+1}]}(t), \qquad (6.4.40)$$

where $0 = t_0 < t_1 < \cdots < t_N = T$ and Φ_k are $\mathcal{L}(\mathbb{H}; H)$-valued $\mathcal{B}_{t_k}$-measurable random variables for all $k = 0, 1, \ldots, N-1$. Then the definition is extended to all predictable $\mathcal{L}_{\mathrm{HS}}(\mathbb{H}_Q; H)$-valued integrands satisfying (6.4.38). Using the equality

$$\mathbf{E}\left\| \int_0^T \Phi(t)\, dW_Q(t) \right\|_H^2 = \|\|\Phi\|\|_T := \mathbf{E}\left[\int_0^T \|\Phi(t)\|_{\mathcal{L}_{\mathrm{HS}}(\mathbb{H}_Q; H)}^2 dt \right],$$

which can be verified for any elementary $\Phi(t)$, and using the fact that any predictable process $\Phi(t)$ with values in $\mathcal{L}_{\mathrm{HS}}(\mathbb{H}_Q; H)$ can be approximated by a sequence of elementary processes converging to Φ with respect to the norm $\|\| \cdot \|\|_T$, one can define the integral $\int_0^T \Phi(t)\, dW_Q(t)$ as the limit in $(L^2)(H)$ of the corresponding sequence of integrals of the elementary processes.

Thus it is sufficient for us to prove the equality (6.4.39) for an elementary process $\Phi(t)$, given by (6.4.40). Since the operators $e_i \otimes e_j$, $i, j \in \mathbb{N}$, form a linearly dense subset in $\mathcal{L}_{\mathrm{HS}}(\mathbb{H}_Q; H)$, we can presume without loss of generality that the Φ_k are of the form

$$\Phi_k = \sum_{i,j=1}^{M} \Phi_{k,i,j}(e_i \otimes e_j), \qquad \Phi_{k,i,j} \in (L^2),$$

where the functions $\Phi_{k,i,j}$ are $\mathcal{B}_{t_k}$-measurable for all $i, j = 1, \ldots, M$ and $k = 0, 1, \ldots N-1$. Consider the S-transform of the left-hand side of the equality (6.4.39). For any $\theta \in \mathcal{S}$ we have

$$S\left[\int_0^T \Phi(t)\, dW_Q(t) \right](\theta) = S\left[\sum_{k=0}^{N-1} \Phi_k \big(W_Q(t_{k+1}) - W_Q(t_k) \big) \right](\theta)$$

$$= \sum_{k=0}^{N-1} \sum_{i,j=1}^{M} \sigma_j S\left[\Phi_{k,i,j} \langle 1_{(t_k, t_{k+1}]}^j, \cdot \rangle \right](\theta) e_i.$$

6.4. Generalized solutions to stochastic Cauchy problems 273

By Lemma 6.4.4 we obtain

$$
\mathrm{S}\left[\int_0^T \Phi(t)\, dW_Q(t)\right](\theta) = \sum_{k=0}^{N-1} \sum_{i,j=1}^{M} \sigma_j\left(1_{(t_k,t_{k+1}]}^j, \theta\right)_{L^2(\mathbb{R})} \mathrm{S}\Phi_{k,i,j}(\theta) e_i
$$

$$
= \sum_{k=0}^{N-1} \sum_{i,j=1}^{M} \sigma_j\left(1_{(t_k,t_{k+1}]}, \mathfrak{J}_j^{-1}\pi_j\theta\right)_{L^2(\mathbb{R})} \mathrm{S}\Phi_{k,i,j}(\theta) e_i
$$

$$
= \sum_{k=0}^{N-1} \sum_{i,j=1}^{M} \sigma_j \int_{t_k}^{t_{k+1}} \left[\mathfrak{J}_j^{-1}\pi_j\theta\right](t)\, dt\, \mathrm{S}\Phi_{k,i,j}(\theta) e_i
$$

$$
= \sum_{k=0}^{N-1} \int_{t_k}^{t_{k+1}} \sum_{i=1}^{M}\sum_{j=1}^{M} \mathrm{S}\Phi_{k,i,j}(\theta)\left[\sigma_j e_i \mathfrak{J}_j^{-1}\pi_j\theta\right](t)\, dt.
$$

Recalling formula (6.3.5) and the definition of Φ_k, we finally obtain

$$
\mathrm{S}\left[\int_0^T \Phi(t)\, dW_Q(t)\right](\theta) = \sum_{k=0}^{N-1} \int_{t_k}^{t_{k+1}} \mathrm{S}\Phi_k(\theta)\mathrm{S}W_Q(t)(\theta)\, dt
$$

$$
= \int_0^T \mathrm{S}\left[\Phi(t)\diamond W_Q(t)\right](\theta)\, dt
$$

$$
= \mathrm{S}\left[\int_0^T \Phi(t)\diamond W_Q(t)\, dt\right](\theta).
$$

By the uniqueness of the S-transform this equality implies (6.4.39). $\square$

The statement establishing a connection between the Itô integral wrt the cylindrical Wiener process and Hitsuda–Skorohod integral wrt the singular white noise process is proved in a similar manner.

Bibliography

[1] S. Albeverio, Z. Haba, and F. Russo. A two-space dimensional semilinear heat equation perturbed by white noise. *Probab. Theory Related Fields*, (3):319–366, 2001.

[2] E. Allen. *Modeling with Ito Stochastic Differential Equations*. Springer, Dordrecht, 2007.

[3] E. Allen, L. J. S. Allen, A. Arciniega, and P. E. Greenwood. Construction of equivalent stochastic differential equation models. *Stochastic Analysis and Applications*, 26(2):274–297, 2008.

[4] D. Alpay and D. Levanony. Linear stochastic systems: A white noise approach. *Acta Appl. Math.*, 110:545–572, 2010.

[5] M. A. Alshanskiy. White noise model with values in a Hilbert space. *Izv. VUZov, Matematika*, (2):10–18, 2004.

[6] M. A. Alshanskiy. Ito and Hitsuda–Skorohod integrals in the infinite-dimensional case. *Siberian Electronic Mathematical Reports*, 11:185–199, 2014.

[7] M. A. Alshanskiy and I. V. Melnikova. Regularized and generalized solutions of infinite dimensional stochastic problems. *Mat. sbornik*, (11):3–30, 2011.

[8] W. Arendt. Vector valued Laplace transforms and Cauchy problems. *Israel J. Math.*, 59(3):327–352, 1987.

[9] W. Arendt, C.J.K. Batty, M. Hieber, and F. Neubrander. *Vector-Valued Laplace Transform and Cauchy Problems*. Springer, Basel AG, 2011.

[10] A. V. Balakrishnan. *Applied Functional Analysis*. Springer-Verlag, New York, 1981.

[11] T. Björk. *Arbitrage Theory in Continuous Time*. Oxford University Press, 2004.

[12] H. Bremermann. *Distributions, Complex Variables and Fourier Ttransforms*. Addison-Wesley, 1965.

275

276   *Bibliography*

[13] R. A. Carmona and M. R. Tehranchi. *Interest Rate Models: An Infinite Dimensional Stochastic Analysis Perspective.* Springer, Berlin, 2006.

[14] J. Chazarain. Problemes de Cauchy abstraits et applications a quelques problemes mixtes. *J. Funct. Anal.*, 7(3):386–446, 1971.

[15] I. Cioranescu. Local convoluted semigroups. In *Evolution Equations (Baton Rouge, LA, 1992)*, pages 107–122. Marcel Dekker, New York, 1995.

[16] I. Cioranescu and G. Lumer. Regularization of evolution equations via kernels $K(t)$, K-evolution operators and convoluted semigroups, generation theorems. In *Seminar Notes in Func. Anal. and PDEs, 1993–1994*, pages 45–52. Louisiana State Univ., Baton Rouge, 1994.

[17] Ph. Clément, H. J. A. M. Heijmans, S. Angenent, C. J. van Duijn, and B. de Pagter. *One-Parameter Semigroups.* North-Holland, 1987.

[18] J. F. Colombeau. *New Generalized Functions and Multiplication of Distributions.* Noth-Holland Math. Studies **84**, 1984.

[19] J. F. Colombeau. *Elementary Introduction to New Generalized Functions.* North-Holland Math. Studies **113**, 1985.

[20] G. Da Prato. Semigruppi regolarizzabili. *Ricerche Mat*, 15:223–248, 1966.

[21] G. Da Prato. *Kolmogorov Equations for Stochastic PDEs.* Springer Basel AG, 2004.

[22] G. Da Prato and L. Tubaro (eds.). *Stochastic Partial Differential Equations and Applications.* Lecture Notes in Pure and Applied Mathematics, 245, Marcel Dekker Inc, 2006.

[23] G. Da Prato and J. Zabczyk. *Stochastic Equations in Infinite Dimensions.* Cambridge Univ. Press, 2014.

[24] E. B. Davies. *One-Parameter Semigroups.* Academic Press, London, 1980.

[25] E. B. Davis and M. M. Pang. The Cauchy problem and a generalization of the Hille-Yosida theorem. *Proc. London Math. Soc.*, 55:181–208, 1987.

[26] Th. Deck, J. Potthoff, and G. Våge. A review of white noise analysis from a probabilistic standpoint. *Acta Appl. Math.*, 48(1):91–112, 1997.

[27] R. deLaubenfels. *Existence Families, Fuctional Calculi and Evolution Equations.* Springer-Verlag, Berlin, 1994.

[28] N. Dunford and J. T. Schwartz. *Linear Operators. Part I–III.* John Wiley & Sons Inc., New York, 1988.

Bibliography

[29] K.-J. Engel and R. Nagel. *One-Parameter Semigroups for Linear Evolution Equations.* Graduate Texts in Math. **194**, Springer-Verlag, 1999.

[30] H. O. Fattorini. *The Cauchy Problem.* Encycl. Math. and Its Appl. **18**, Cambridge University Press, 1983.

[31] A. Filinkov and J. Sorenson. Differential equations in spaces of abstract stochastic distributions. A review of white noise analysis from a probabilistic standpoint. *Stoch. Stoch. Rep.*, 72(3-4):129–173, 2002.

[32] D. Filipović. *Consistency Problems for Heath–Jarrow–Morton Interest Rate Models.* Lecture Notes in Math. **1760**, Springer-Verlag, Berlin, 2001.

[33] D. Filipović. *Term-Structure Models. A Graduate Course.* Springer-Verlag, Berlin, 2009.

[34] L. Gawarecki and V. Mandrekar. *Stochastic Differential Equations in Infinite Dimentions.* Springer-Verlag, Berlin, 2011.

[35] I. M. Gelfand and G. E. Shilov. *Generalized Functions. Volume 1. Properties and Operations.* Academic Press Inc., 1964.

[36] I. M. Gelfand and G. E. Shilov. *Generalized Functions. Volume 3. Theory of Differential Equations.* Academic Press Inc., 1967.

[37] I. M. Gelfand and G. E. Shilov. *Generalized Functions. Volume 2. Spaces of Fundamental and Generalized Functions.* Academic Press Inc., 1968.

[38] I. M. Gelfand and N. Ya. Vilenkin. *Generalized Functions. Volume 4. Applications of Harmonic Analysis.* Academic Press Inc., 1964.

[39] J. A. Goldstein. *Semigroups of Linear Operators and Applications.* Oxford University Press, 1985.

[40] T. Hida. *Brownian motion.* Applications of Math. **11**, Springer-Verlag, 1980.

[41] T. Hida, H.-H. Kuo, J. Potthoff, and L. Streit. *White Noise. An Infinite-Dimensional Calculus. Mathematics and Its Applications.* Springer Science+Business Media, Dordrecht, 1993.

[42] T. Hida and S. Si. *Lecture on White Noise Functionals.* World Scientific, Hackensack, 1993.

[43] E. Hille and R. S. Phillips. *Functional Analysis and Semi-groups.* Colloquium Publ. **31**, American Mathematical Society, 1957.

[44] H. Holden, B. Oksendal, J. Uboe, and T. Zhang. *Stochastic Partial Differential Equations. A Modelling, White Noise Functional Approach.* Springer, New York, 2010.

278 Bibliography

[45] Z. Huang and J. Yan. *Introduction to Infinite Dimensional Stochastic Analysis*. Math. and Its Appl. **502**, Springer, Netherlands, 2000.

[46] V. K. Ivanov and I. V. Melnikova. New generalized functions and weak well-posedness of operator problems. *Soviet Math. Dokl.*, 43(2):315–319, 1991.

[47] V. K. Ivanov, I. V. Melnikova, and A. I. Filinkov. *Differential-Operator Equations and Ill-Posed Problems*. Nauka, Moscow (in Russian), 1993.

[48] V. K. Ivanov, V. V. Vasin, and V. P. Tanana. *Theory of Linear Ill-Posed Problems and Its Applications*. Inverse and Ill-Posed Problems Series **36**, Walter de Gruyter & Co, 2002.

[49] A. Kaminski, D. Kovačević, and S. Pilipović. The equivalence of various definitions of the convolution of ultradistributions. *Proceedings of the Steklov Inst. of Math.*, 203:307–322, 1994.

[50] D. Kannan and V. Lakshmikantham (eds.). *Handbook of Stochastic Analysis and Applications*. Marcel Dekker, New York, 2002.

[51] T. Kato. *Perturbation Theory for Linear Operators*. Springer-Verlag, Berlin, 1995.

[52] H. Kellermann and M. Hieber. Integrated semigroups. *Journal of Functional Analysis*, 84:160–180, 1989.

[53] A. N. Kolmogorov and S. V. Fomin. *Elements of Theory of Functions and Functional Analysis*. Martino Fine Books, 2012.

[54] H. Komatsu. Ultradistributions. I. Structure theorems and characterizatio. *J. Fac. Sci. Univ. Tokyo*, 20(1):25–106, 1973.

[55] Yu. G. Kondratiev and L. Streit. Spaces of white noise distribution: Constructions, descriptions, applications. I. *Reports on Math. Phys.*, 33:341–366, 1993.

[56] S. G. Krein. *Linear differential equations in a Banach space*. Am. Math. Soc., 1972.

[57] E. Kreyszig. *Introductory Functional Analysis with Applications*. Wiley & Sons, 1978.

[58] N. V. Krylov, M. Rockner, and J. Zabczyk. *Stochastic PDE's and Kolmogorov Equations in Infinite Dimensions*. Lecture Notes in Mathematics 1715, Springer, 1999.

[59] H.-H. Kuo. *White Noise Distribution Theory*. CRC Press, 1996.

[60] Hui-Hsiung Kuo. *Introduction to Stochastic Integration*. Springer, 2000.

Bibliography

[61] J. W. Lamperti. *Probability. A Survey of the Mathematical Theory.* Wiley & Sons, 1996.

[62] R. Lattès and J.-L. Lions. *The Method of Quasi-Reversibility. Applications to Partial Differential Equations.* American Elsevier Publishing Company, 1969.

[63] Yu. I. Lyubich. The classical and local Laplace transformation in an abstract Cauchy problem. *Russian Mathematical Surveys*, 21(3):3–51, 1966.

[64] P. Malliavin and A. Thalmaier. *Stochastic Calculus of Variations in Mathematical Finance.* Springer-Verlag, Berlin, 2006.

[65] S. Mandelbrojt. *Fonctions entières et transformées de Fourier applications.* Publ. of Math. Soc. Japan **10**, Mathematical Society of Japan, 1967.

[66] I. V. Melnikova. General theory of ill-posed Cauchy problem. *J. Inv. Ill-Posed Problems*, 3(2):149–171, 1995.

[67] I. V. Melnikova. Regularization of stochastic problems with respect to variables of different kinds. *Doklady Mathematics*, 79(3):408–411, 2009.

[68] I. V. Melnikova. Generalized solutions of differential-operator equations with singular white noise. *Differential Equations*, 49(4):375–486, 2013.

[69] I. V. Melnikova and U. A. Alekseeva. Solution of an abstract Cauchy problem with nonlinear and random perturbations in the Colombeau algebra. *Doklady Mathematics*, 87(2):193–197, 2013.

[70] I. V. Melnikova and U. A. Alekseeva. Weak regularized solutions to stochastic Cauchy problems. *Chaotic Modeling and Simulation*, (1):49–56, 2014.

[71] I. V. Melnikova, U. A. Alekseeva, and V. A. Bovkun. Solutions of stochastic systems generalized over temporal and spacial variables. *New Prospects in Direct, Inverse and Control Problems for Evolution Equations. Springer International Publishing, Switzerland*, 2014.

[72] I. V. Melnikova and M. A. Alshanskiy. Stochastic problems in spaces of abstract distributions. *Commun. Appl. Anal.*, 14(3-4):435–442, 2010.

[73] I. V. Melnikova and M. A. Alshanskiy. S-transform and Hermit transform of Hilbert space-valued stochastc distributions with applications to stochastc equations. *Integral Transforms Spec. Funct.*, 22(4-5):293–301, 2011.

[74] I. V. Melnikova and M. A. Alshanskiy. Generalized solution to equations with multiplicative noise in Hilbert spaces. *Rend. Sem. Mat. Univ. Politec. Torino*, 71(2):239–249, 2013.

Bibliography

[75] I. V. Melnikova and M. A. Alshanskiy. White noise calculus in applications to stochastic equations in Hilbert spaces. *Journal of Mathematical Sciences*, 2015.

[76] I. V. Melnikova and M. A. Al'shansky. Well-posedness of the Cauchy problem in a Banach space: Regular and degenerate cases. *Journal of Mathematical Sciences*, 87(4):3732–3780, 1997.

[77] I. V. Melnikova and U. A. Anufrieva. Pecularities and regularization of illposed Cauchy problems with differential operators. *Journal of Mathematical Sciences*, 148(4):481–632, 2008.

[78] I. V. Melnikova and A. I. Filinkov. Integrated semigroups and C-semigroups. *Uspechi Mat. Nauk*, 49(6):111–150, 1994.

[79] I. V. Melnikova and A. I. Filinkov. *The Cauchy Problem: Three Approaches*. Monographs and Surveys in Pure and Applied Mathematics **120**, Chapman & Hall/CRC, London, 2001.

[80] I. V. Melnikova, A. I. Filinkov, and M. A. Alshansky. Abstract stochastic equations. II. Solutions in spaces of abstract stochastic distributions. *Journal of Mathematical Sciences*, 116(5):3620–3656, 2003.

[81] I. V. Melnikova, A. I. Filinkov, and U. A. Anufrieva. Abstract stochastic equations. I. Classical and distributional solutions. *Journal of Mathematical Sciences*, 111(2):3430–3475, 2002.

[82] I. V. Melnikova and V. S. Parfenenkova. Relations between stochastic and partial differential equations in Hilbert spaces. *Int. Journal of Stochastic Analysis*, 2012:9, 2012.

[83] I. V. Melnikova and V. S. Parfenenkova. Feynman-Kac theorem in Hilbert spaces. *Electronic Journal of Differential Equations*, 208:1–10, 2014.

[84] I. V. Melnikova and O. S. Starkova. Infinite dimensional stochastic cauchy problems in ito and differential forms: Comparison of solutions. In *Proceedings of the 9th ISAAC congress*. Springer, Birkhauser Series Trends in Mathematics/Reseach Prospectivs, 2014.

[85] I. V. Melnikova, Q. Zheng, and J. Zhang. Regularization of weakly ill-posed Cauchy problems. *J. of Inverse and Ill-Posed Problems*, 10(5):503–511, 2002.

[86] G. N. Milstein and M. V. Tretyakov. *Stochastic Numerics for Mathematical Physics*. Scientific Computation Series, Springer, 2004.

[87] I. Miyadera. *Nonlinear Semigroups*. Translations of Mathematical Monographs **109**, American Mathematical Society, 1992.

Bibliography 281

[88] I. Miyadera and N. Tanaka. Exponentially bounded C-semigroups and generation of semigroups. *Journal of Math. Analysis and Appl.*, 143(2):358–378, 1989.

[89] S. Mizohata. *Theory of Partial Differential Equations*. Cambridge Univ. Press, Cambridge, 1973.

[90] N. Obata. *White Noise Calculus and Fock Space*. Lecture Notes in Mathematics, **1577**, Springer-Verlag, Berlin, 1994.

[91] M. Oberguggenberger. *Multiplication of Distributions and Applications to Partial Differential Equations*. Pitman Research Notes Math. **259**, Longman Scientific & Technical, Essex, Harlow, 1992.

[92] M. Oberguggenberger. Generalized functions in nonlinear models – A survey. *Nonlinear Analysis*, 47(8):5029–5040, 2001.

[93] M. Oberguggenberger. Regularity theory in Colombo algebras. *Bull. T. CXXXIII Acad. Serbe, Sci Arts, Sci. Math.*, 31:147–162, 2006.

[94] M. Oberguggenberger and F. Russo. Nonlinear SPDEs: Colombeau solutions and pathwise limits. In *Stochastic Analysis and Related Topics, VI, Prog. Probab.*, volume 42, pages 319–332. Birkhauser-Verlag, Boston, 1998.

[95] B. Oksendal. *Stochastic Differential Equations. An Introduction with Applications*. Springer, 2003.

[96] A. Pazy. *Semigroups of Linear Operators and Applications to Partial Differential Equations*. Springer-Verlag, New York Inc., 1983.

[97] R. S. Phillips. Inversion formula for Laplace transforms and semi-groups of linear operators. *Annals of Math.*, 59:325–356, 1954.

[98] S. Pilipović. Generalization of Zemanian spaces of generalized functions which have orthonormal series expansions. *SIAM, J. Math. Anal.*, 17:477–484, 1986.

[99] S. Pilipović and D. Seleši. Expansion theorems for generalized random processes. Wick product and applications to stochastic differential equations. *Infin. Dimens. Anal. Quantum Probab. Relat. Top.*, 10(1):79–110, 2007.

[100] S. Pilipović and D. Seleši. Structure theorem for generalized random processes. *Acta Math., Hungar. DOI: 10.1007/s10474-007-6099-1*, 117(3):251–274, 2007.

[101] S. Pilipović and D. Seleši. On the generalized stochastic Dirichlet problem. Part II: Solvability, stability and the Colombeau case. *Potential Anal.*, 33(3):263–289, 2010.

Bibliography

[102] M. Reed and B. Simon. *Methods of Modern Mathematical Physics. II. Fourier Analysis, Self-Adjontness*, volume 2. Academic Press, New York, 1978.

[103] R. D. Richtmyer. *Principles of Advanced Mathematical Physics*, volume 1. Springer-Verlag, New York, 1975.

[104] R. D. Richtmyer. *Principles of Advanced Mathematical Physics*, volume 2. Springer-Verlag, New York, 1981.

[105] A. P. Robertson and W. Robertson. *Topological Vector Spaces*. Cambridge Tracts in Math. and Math. Phys. **53**, Cambridge University Press, 1964.

[106] W. Rudin. *Real and Complex Analysis*. McGraw-Hill, Inc., 1987.

[107] D. Seleši. Hilbert space valued generalized random processes. Part I. *Novi Sad Journal of Math.*, 37(1):129–154, 2007.

[108] S. E. Shreve. *Stochastic Calculus for Finance II. Continuous-Time Models*. Springer Science+Business Media, Inc., 2004.

[109] D. W. Stroock and S. R. S. Varadhan. *Multidimensional Diffusion Processes*. Springer Verlag, 1979.

[110] N. Tanaka and N. Okazawa. Local C-semigroups and local integrated semigroups. *Proc. London Math. Soc.*, 61(3):63–90, 1990.

[111] P. P. Teodorescu, W. W. Kecs, and A. Toma. *Distribution Theory with Applications in Engineering and Physics*. Wiley-VCH Verlag GmbH & Co. KGaA, 2013.

[112] A. N. Tikhonov and V. Ya. Arsenin. *Methods for Solving Ill-posed Problems*. Wiley & Sons New York, 1977.

[113] S. Zaidman. *Functional Analysis and Differential Equations in Abstract Spaces*. Monographs and Surveys in Pure and Applied Mathematics **100**, Chapman & Hall/CRC, 1999.

[114] E. Zeidler. *Applied Functional Analysis. Applications to Mathematical Physics*. Applied Mathematical Sciences **108**, Springer-Verlag, 1995.

[115] A. G. Zemanian. *Generalized Integral Transformations*. Interscience Publishers, New York, 1969.

Index

Q-white noise, 198, 215, 222, 250

abstract Cauchy problem, 3, 13, 30, 43
  R-well-posedness, 34
  n-well-posedness, 22–24
  generalized, 46
    solution, 47
    solution operator, 47, 48, 53, 55, 57
    well-posedness, 47, 57
  generalized well-posedness, 47, 48, 53
  solution operator, 72
  uniform (n, ω)-well-posedness, 18, 20
  uniform well-posedness, 3, 5, 86, 90

abstract distribution
  convolution, 44, 45, 200
  space, 44
  support, 44

backward Cauchy problem, 182, 189
backward Kolmogorov equation, 189
Bochner integral, 115
Borel σ-algebra, 114, 230
Brownian motion, 123, 126–128, 180, 231, 239

Cauchy problem for differential systems, 43, 59, 72, 209
  characteristic roots, 62
  Fourier transformed, 211
  generalized, 61
  solution, 61
  solution operator, 62

generalized solution, 61, 66
solution, 60, 71
characteristic function, 115
Colombeau algebra, 218
condition
  C-summability, 10
  $\mathcal{A}$-summability, 11, 12
  (M), 220
  (N), 221
  (R1), 4, 15–17, 20, 48
  (R2), 4, 21, 22, 40, 53
  (R3), 4, 26, 27, 29, 54, 55
  (R4), 4
  MFPHY, xii, 3, 9, 15, 168
convolutor, 45, 67–70
correlation operator, 116
covariance operator, 116

differential system, 59, 90, 91
  conditionally correct, 63, 68, 70, 71, 165, 211
  hyperbolic, 63, 86
  incorrect, 63, 69–71, 165, 211
  parabolic, 63, 84
  Petrovsky correct, 63, 66, 70, 71, 84, 86, 165, 211
diffusion, 184
dissipative non-linearity, 169, 175
distribution law, 115, 120

equation of age structured population, 259
expectation, 116
  conditional, 120

filtration, 132
Fourier transform, 72

283

classical, 54, 93, 104, 106
generalized, 61, 85, 88, 105
generalized inverse, 93, 105
Frechet derivative, 137, 183, 194
function
$\Lambda(\cdot)$, 62, 63
$e^{t\mathbf{A}(\cdot)}$
estimation, 62, 64, 66–71
associated, 54, 100
functional
linear continuous on Φ, 60
multiplication by a scalar, 60

Gaussian measure, 120, 127, 230
Gelfand triple, 233
Gelfand–Shilov classification, 63, 211
Green function, 44, 61, 85, 89, 91, 93, 161

Hermite functions, 232
Hitsuda–Skorohod integral, 247, 252, 253, 255

Itô formula, 136, 138
Itô integral, 253, 255
Itô isometry, 134, 136

Kolmogorov equation, 180

Laplace transform
classical, 52, 55, 72, 106
generalized, 47, 51, 53, 55, 72, 107
of semi-group, 3, 8, 9, 13, 14, 32
Lipschitz condition, 167, 169, 176

Markov property, 182, 185, 186
matrix function $\mathbf{A}(s)$, 61
characteristic roots, 62
its conjugate, 61
moderate elements, 220
multiplication operator, 62, 64, 66–70, 73
multiplicative stochastic
perturbation, 265
multiplier, 62

noise
additive, 254
multiplicative, 259
null subset, 221
number p_0, 62, 64

operator
compact, 117
Hilbert–Schmidt, 117, 170, 236
nuclear, 118
trace class, 118, 128
operator $\mathbf{A}(i\partial/\partial x)$, 60

process
Q-Wiener, 123, 127, 128, 137, 187, 222
adapted, 132
cylindrical Wiener, 123, 129, 130, 137, 187
Gaussian, 127, 128, 130
predictable, 132, 143
stochastic, 123
property
exponential boundedness, 3, 7, 15, 17

random variable
H-valued, 114
Gaussian, 120
real-valued, 115
regularizing operator, 37, 38, 40, 41
regularizing parameter, 37
resolvent identity, 3, 9, 16

S-transform, 247, 249
semi-group
K-convoluted, 26–28, 36, 199, 202, 203, 207
generator, 26
R-, 31, 33–36, 38, 75, 82, 83, 91, 158, 161, 175, 207, 215
generator, 32, 33, 35, 82, 83
properties, 32
n-times integrated, 27, 36, 80, 89, 91, 199, 200, 206

Index

exponentially bounded, 15–18, 20, 75
  generator, 16, 20, 21, 35
  local, 21–24, 34, 81
  properties, 17
of class $(1, \mathcal{A})_4$, 170, 171
of class $\mathcal{A}$, 11, 14, 147, 150
  properties, 11–13
of class C_0, 3, 6, 9, 38, 41, 75, 86, 91, 144, 175, 187, 200, 204
  generator, 6, 9
  properties, 6–10
of class C_1, 11, 14
  properties, 11–13
of growth order α, 14, 35, 75, 86, 91
regularized, 36, 154, 156, 157, 159
  generator, 36
relation, 3, 6, 9, 11, 192
strongly continuous, 11
  generator, 11
  infinitesimal operator, 11
semi-linear stochastic Cauchy problem, 166
  mild solution, 166, 170, 171, 177
  weak solution, 166, 177
semigroup
  K-convoluted, 72
  R-, 72
  n-times integrated, 51, 72
space
  (L^2), 231
  W^{Ω}, 64, 102
  $W^{\Omega, b}$, 103
  $W^{\Omega, b}$, 106
  W_M, 64, 102
  W_M^{Ω}, 103, 106
  $W_{M,a}$, 106
  $W_{M, a}$, 102
  $W_{M, a}^{\Omega, b}$, 64, 65, 104
  Z, 97, 103, 105
  Z^b, 97
  Φ, 60

Φ', 44, 60, 66
$\Phi'(\mathcal{X})$, 44
$\left(\mathcal{D}_a^{\{M_q\}, B} \right)'(\mathcal{X})$, 54, 57
$\left(\mathcal{D}^{\{M_q\}} \right)'(\mathcal{X})$, 108
$\mathcal{D}$, 94, 95, 103, 105
$\mathcal{D}'$, 107, 108
$\mathcal{D}'(\Psi')$, 212
$\mathcal{D}'(\mathcal{X})$, 53, 108
$\mathcal{D}_a^{\{M_q\}, B}$, 54, 55
$\mathcal{D}_A$, 94
$\mathcal{S}$, 99, 104
$\mathcal{S}'$, 108
$\mathcal{S}'(\mathcal{X})$, 108
$\mathcal{S}^{\beta}$, 98, 105
$\mathcal{S}^{\beta, B}$, 98
$\mathcal{S}_{\alpha}$, 95, 105
$\mathcal{S}_{\alpha}^{\beta}$, 99, 104, 105
$\mathcal{S}_{\omega}'$, 108
$\mathcal{S}_{\omega}'(\mathcal{X})$, 48, 108
$\mathcal{S}_{\alpha, A}$, 96, 105
$\mathcal{S}_{\alpha, A}^{\beta, B}$, 99, 104, 105
$\widetilde{\Phi'}$, 61
$\widetilde{\Phi}$, 61
$\widetilde{(\Phi')}$, 66, 105
$\mathcal{D}'(\mathcal{X})$, 53
Beurling, 101, 202
countably Hilbert, 230
measurable, 114
of abstract stochastic distributions, 230, 235, 253
probability, 114
Roumieu, 101, 202
stochastic Cauchy problem, 139, 198
  generalized solution, 200, 203, 204, 206, 207, 212, 215, 253, 267
  quasi-linear, 218
  strong solution, 139, 141
  weak regularized solution, 153, 156, 157
  weak solution, 140, 141, 144, 147, 150, 267
stochastic convolution, 132, 142, 143

286   *Index*

stochastic heat equation, 254
stochastic Hermite polynomials, 232
stochastic integral, 131, 133

theorem
    Bochner–Minlos–Sazonov, 230
    Feynman–Kac, 179, 180, 193
    MFPHY, xii, 3, 10, 20
    Tonelli–Fubini, 136
topology
    of countably normed space, 94
    of inductive limit, 94
    of projective limit, 94
trace, 118, 130

white noise, 198, 218, 232
    probability space, 231
    singular, 241, 250
Wick product, 247, 251
Wiener–Itô chaos expansion, 234

Yosida approximation, 169, 176